FORENSIC SCIENCE

FORENSIC SCIENCE

An Encyclopedia of
History, Methods, and Techniques

William J. Tilstone, Kathleen A. Savage, and Leigh A. Clark

A B C  C L I O

Santa Barbara, California • Denver, Colorado • Oxford, England

Copyright © 2006 by ABC-CLIO

Library of Congress Cataloging-in-Publication Data

Tilstone, William J.
    Forensic science : an encyclopedia of history, methods, and techniques / William J. Tilstone,
    Kathleen A. Savage, and Leigh A. Clark.
       p. cm.
Includes bibliographical references and index.
ISBN 1-57607-194-4 (hardcover : alk. paper) — ISBN 1-57607-592-3 (ebook)
1. Forensic sciences—Encyclopedias. 2. Forensic sciences—History.
3. Criminal investigation—Encyclopedias. 4. Criminal investigation—History. I. Savage, Kathleen A.
II. Clark, Leigh A. III. Title.

HV8073.T55 2006
363.2503—dc22   2006001140

10 09 08 07 06 10 9 8 7 6 5 4 3 2 1

This book is also available on the World Wide Web as an e-book. Visit abc-clio.com for details.

ABC-CLIO, Inc.
130 Cremona Drive, P.O. Box 1911
Santa Barbara, California 93116–1911

This book is printed on acid-free paper.
Manufactured in the United States of America.

CONTENTS

Introduction, 1

INTRODUCTION

This is the story of forensic science. The book provides a historical and technical frame of reference for the interested lay reader looking for increased general knowledge or more detail on specific applications or famous cases.

Forensic science is difficult to define precisely. Broadly speaking, it is the application of scientific techniques and principles to provide evidence to legal or related investigations and determinations. Some things are obvious, such as DNA typing or the identification of drugs. These involve the use of specialized scientific equipment, and the testing is conducted by personnel with science degrees. Others are less obvious. For example, the detection of fingerprints at a crime scene is usually conducted by police officers who do not have science degrees, and recent court cases have questioned whether there is a true scientific principle underlying the discipline.

Although it may seem odd to associate fiction directly with science, particularly forensic science, the influence of Sir Arthur Conan Doyle, via his fictional character Sherlock Holmes, in popularizing the use of science for the solution of crimes cannot be underestimated. He introduced his many readers to the usefulness of fingerprinting and firearms examination and questioned document examination and serology sometimes even before they were familiar to real-life detectives. For example, in his first novel, *A Study in Scarlet* published in 1887, Holmes tells Dr. Watson

"I have found a reagent which is precipitated by haemoglobin and by nothing else." In fact, reliable identification of a stain as blood and the invention and application of precipitin tests in forensic science are separate discoveries that were not made until the turn of the century.

Applications are wide-ranging, too. Environmental and wildlife enforcement, control of dog and horse racing, immigration documentation, and parentage testing are all generally accepted as employing forensic science but clearly do not fit the image of the scientific sleuth.

The late Stuart Kind (1925–2003), who was one of the key figures in British forensic science in the second half of the twentieth century, wrote eloquently about the "scientific investigation of crime." However, he was considering the principles that would result in successful police investigations, not the work of the forensic science laboratory. Kind published his thoughts in *The Scientific Investigation of Crime* (Forensic Science Services) in 1987. The book is no longer available for purchase, but some libraries may have copies.

Despite the uncertainty about what it is, forensic science is catching the public imagination as never before. There are several reasons for this. Without a doubt, one is the considerable popularity of the "CSI" television shows. Although they have some technical flaws from a scientific perspective, they capture the excitement of the scientific sleuth without being bogged down with obscure and hard-to-grasp technical issues. Moving from mass-media entertainment to social significance, DNA testing has been so successful that its applications are familiar to most laypeople. Its power to exonerate the wrongly convicted and to identify literally thousands of offenders who would otherwise have gone free is far beyond what could have been imagined as recently as the mid-1980s. (See **DNA in Forensic Science**.) Mainstream press articles about cases solved by DNA testing have played their part in the popularization of forensic science.

Unfortunately, another emerging theme in media coverage of forensic science that has made the subject front-page worthy is not so positive. Flawed procedures and less-than-acceptable testing methods by laboratory personnel have been the subject of many reports. Cases worked in laboratories in many states and in federal agencies—including the FBI—have been critically questioned, and the previously solid foundation of forensic science has been severely shaken.

The common thread that runs through all of this—the imprecision in definition, the stunning successes, the popularity of entertainment shows based on such an arcane subject, the investigative journalism—is the growth of science and its role in society. There can be no forensic science without there first being science. There can be no developments in forensic science until the fundamental research has produced novel techniques and applications. There is nothing to catch the public attention, unless the writers, readers, and viewers have at least some degree of comfortable understanding with the subject matter. It is the aim of this work to provide that understanding.

The story is told chronologically, starting with the few known examples from early times. However, significant progress was only possible because of the wide range of discoveries in many and varied fields of science during the Industrial Revolution. Forensic science is

essentially parasitic. As we shall see, other than some aspects of fingerprinting, ballistics, and questioned document examination, it has depended on discoveries in other fields. The importance of the Industrial Revolution to forensic science is not that it was a time of absolute advances in science, but that it was a time characterized by the application of science to real-life problems. It was a period of ingenuity and inventiveness, and many of the highly specialized instruments used in today's forensic science laboratories contain components that are by-products of that era.

The history therefore begins in the late nineteenth and early twentieth centuries, with an account of a wide range of general and specific milestones, mainly from Europe. The period between approximately 1930 and 1980 saw the United States become a significant center of development, as it still is today at the start of the twenty-first century when forensic scientists are using methods and examining materials that were not even thought of a generation ago.

"The world's history is the world's judgment."
—Friedrich von Schiller, *Lectures,* May 1789

Many essays on the history of forensic science (such as the review by Douglas M. Lucas published in the newsletter of the American Society of Crime Laboratory Directors and available through their website at http://www.ascld.org/pdf/lucas.pdf) refer to ancient Roman and Chinese examples of practices that might be viewed as forensic science. These are anecdotal, and there was no such thing as "science," far less a forensic variant, in those days. Indeed, the earliest history of forensic science is really the story of forensic medicine. Medicine and law have been closely related since the earliest days of recorded history. Even with the most primitive of aboriginal tribes, these functions were often integrated with religion in the person of the priest who was physician, judge, and spiritual leader combined. The earliest documented code of law, promulgated by Hammurabi, king of Babylon around 2200 BC, contained laws governing the practice of medicine, while in Greece, Hippocrates (460–355 BC) showed how medicine could influence legal matters by discussing the lethality of wounds. In ancient Rome, after Julius Caesar was murdered in the Forum in 44 BC his body was examined by a physician who subsequently opined that only one of the twenty-three wounds was fatal. In the sixth century AD, the emperor Justinian recognized the special position of the expert witness (which persists to this day) when he declared that physicians were not ordinary witnesses but rather persons who gave judgments rather than testimony. In Italy in 1209 Pope Innocent III appointed doctors to the courts to perform autopsies for the purpose of establishing the nature of various types of wounds, and Italy, notably at the University of Bologna, became the first country in which legal medicine was recognized as a specialty.

Around 1250 AD, a classic Chinese work, the *Hsi Yuan Lu* (translated approximately as "Instructions to Coroners"), was first published, while in the Western world in 1507, the Bamberg Code in Germany required that medical evidence be presented in all cases of violent death. These provided a legal basis for the practice of forensic medicine and were the precursors of legislation regulating modern coroners and medical examiners. Ambrose Paré (1510–1590) in France was probably the first true practitioner of forensic medicine, following in the footsteps of the great anatomist Andrea Vesalius (1514–1564). The first systematic course of lectures on legal medicine was instituted at the University of Leipzig in 1642 and the first medico-legal journal was published in Berlin in 1782.

The Industrial Revolution of the nineteenth century gave birth to many scientific discoveries and inventions, driven by a thirst for knowledge but more importantly by a market economy in which scientific discoveries

were converted into useful products. The birth of applied science to service the needs of the Industrial Revolution also provided the base from which forensic science grew to eventually become identifiable as a separate discipline of science and medicine.

"The brightest heaven of invention . . ."
—Shakespeare, *King Henry V,* Chorus, 1

Europe during the Industrial Revolution was indeed "the brightest heaven of invention" and provided the tools needed for science to solve crime as well as responding to the needs of economic development. Most of the inventions and discoveries linked to the Industrial Revolution were in the fields of physics and chemistry. The ultraviolet spectrophotometer, the workhorse of the forensic chemistry laboratory in the early and mid-twentieth century, depended on Johann Ritter's discovery of ultraviolet radiation in 1804. The UV spectrophotometer is used to identify organic chemicals, and the first working model was the UV spectrograph invented by Walter Noel Hartley in London in 1877. Hartley collaborated with Adam Hilger to produce a commercial instrument in 1907. However, the instrument was cumbersome to use and required manual comparison of spectra recorded on photographic plates. The first working spectrophotometer was invented by Arthur Hardy working at MIT in the early 1920s. Hardy used cesium photocells to detect the spectra. Although he commercialized his invention together with the General Electric Company in 1930, the first "workhorse" UV spectrophotometer was the Beckman DB, invented by Arnold Beckman in 1941.

Joseph von Fraunhofer was interested in the dark lines he observed in light from the sun. His research led to the invention of the spectrograph, around 1814. Various forms of inorganic spectroscopy have been used by forensic chemists throughout the twentieth century for the identification of metals in glass. The chemist Robert Bunsen (1811–1899) and

Gustav Kirchhoff (1824–1887), a physicist trained at Königsberg, met and became friends in 1851 when Bunsen spent a year at the University of Breslau, where Kirchhoff was also teaching. Bunsen was called to the University of Heidelberg in 1852 and he soon arranged for Kirchhoff to teach at Heidelberg as well. Bunsen's most important work was in developing several techniques used in separating, identifying, and measuring various chemical substances. He also made a number of improvements in chemical batteries for use in isolating quantities of pure metals—including one known as the Bunsen battery. He created the Bunsen burner for use in flame tests of various metals and salts: Its nonluminous flame did not interfere with the colored flame given off by the test material. This line of work led to the spectroscope. When an element, for example sodium, is introduced into a Bunsen burner flame, radiation at one or more wavelengths that are characteristic to the material burned is emitted. This is called the emission spectrum of the element. In the case of sodium, it results in the flame turning a bright orange color. It was Kirchhoff who suggested that similarly colored flames could possibly be differentiated by looking at their emission spectra through a prism, thereby identifying the specific wavelengths emitted. When he shone bright light through such flames, he found that the light was absorbed at wavelengths (the absorption spectrum of the element) that corresponded to the wavelengths of the bright, sharp lines characteristic of the emission spectra of the same test materials. Various derivatives of the spectroscope were used throughout the twentieth century to characterize materials such as soils, paint, and glass in crime laboratories by identifying the metallic elements that they contained.

In 1825 William Sturgeon invented the electromagnet and so laid the foundation for electronic communication using microphones and loudspeakers. Although it was more than a hundred years later before their invention, the mass spectrometer and nuclear magnetic resonance spectrometer (see **Mass Spectrometry**) also depend on the electromagnet, as do other modern scientific instruments.

The polarizing microscope is widely used in forensic science laboratories in the characterization of soils, drugs, and fibers (see **Drugs; Fibers; Soil**). The technique was made possible by the invention of the Nicol prism, by William Nicol in Edinburgh in 1828. Nicol (frequently incorrectly spelled Nichol) found that a prism created from two pieces of Iceland spar resulted in the polarization of the beam of light that passes through the crystal. The impact of the invention was considerable—for example, even today many laboratories use microcrystal tests (see **Microcrystal Tests**) to deal with the great volume of drug cases submitted by law enforcement agencies.

Elsewhere, the science of optics was being advanced by the work of Giovanni Battista Amici in Italy around 1827. Amici's interest in optics ranged from the celestial (a minor planet and a crater on the dark side of the moon are named in his honor) to the microscopic. Probably the most significant of his inventions as far as forensic science is concerned is the oil-immersion lens. Invented by him in 1840, the technique revolutionized high-power optical microscopy by minimizing the optical aberrations that severely limited the magnifying power. Every forensic biologist today who identifies human sperm in samples from a rape victim is using a technique traceable to Amici's work.

Nineteenth-century inventions include several that are the objects of forensic science investigation, rather than instruments to assist investigations. Samuel Colt invented the revolver in 1836, Richard Gattling patented the machine gun in 1862, Alfred Nobel invented dynamite in 1866, and Sir James Dewar and Sir Frederick Abel invented cordite in 1889. This is the same Dewar who in 1892 invented the dewar flask. More familiar by its generic name of vacuum flask, dewars are used in several applications where liquid nitrogen is required, such as the scanning electron microscope (SEM).

"By a set of curious chances . . ."
—W. S. Gilbert, *The Mikado*

The same principles of applied science that underpinned the discoveries and inventions made by nineteenth-century researchers laid the foundations for the scientific characterization of evidence right through to the present day. Careful observation, rational thinking, and awareness of surrounding circumstances have consistently proved to be at least the equal of technological advances in their value to forensic science. This is certainly true of firearms and fingerprint examinations. The story surrounding the establishment of their foundational principles involves more than one "set of curious chances" or coincidences.

The first recorded case of a successful investigation of a murder, applying the very scientific processes of careful observation, rational thinking, and awareness of surrounding circumstances, was in 1835. Henry Goddard, a Bow Street runner (an early London police officer) solved a murder by identification of the source of the fatal projectile. There are several accounts of the case; none of them name the victim or the accused, and one reports the event as a staged burglary. However, the account by Hamby and Thorpe is from a credible source (Hamby and Thorpe 1999). The most likely reconstruction is that a servant shot and killed his employer. Goddard was able to identify the source of the projectile (a ball shot) by matching an imperfection on it to one on the mold that the servant used to make his shot. Goddard also was able to trace paper wadding involved in the shooting to paper in the possession of the servant. As an aside, the servant seems to have been employed as a butler, a position frequently featured in detective fiction from the United Kingdom and one of the suspect characters in the popular board game Clue

or Cluedo. If so, this may be the first instance of a real murder where the butler actually did it!

Henry Goddard depended on good sense, a good eye, and the good fortune that the mold defect was large enough to see without aids. It was nearly 100 years later that the fundamental tool of the firearms examiner, the comparison microscope (see **Comparison Microscope**), was invented. The inventor was one Calvin Goddard (1891–1955), a medical doctor from the Johns Hopkins medical school. (The two Goddards are not related in any way.)

Fingerprints were the result of even greater coincidence. An Englishman working in India, a Scot working in Japan, a Croatian working in Argentina, and a knight of the realm and relative of Charles Darwin working in England were independently laying the basis for the most significant of all techniques used by forensic scientists to establish identity. (See **Fingerprints; Latent Prints**.) The physiologist Johan Purkinje, who gave his name to the specialized muscle fibers (Purkinje fibers) essential for the functioning of the heart, described the basic ridge patterns of fingerprints in 1823. However, he did not show any awareness of their unique nature. That was left to two Britons working abroad. In 1858 William Herschel (1738–1822), a senior civil servant working in Bengal, India, required people to "sign" for their pensions by leaving their fingerprint. In 1880 Henry Faulds (1843–1930), a Scottish physician working in Japan, published an article in the scientific journal *Nature,* which is the first recorded suggestion that fingerprints could be used for identification. Perhaps both were aware of the practice of their countryman, the gifted naturalist and engraver Thomas Bewick (1753–1828). Bewick used an engraving of his fingerprint, which he called "his mark," to sign his wood carvings.

It was just ten years later that Sir Francis Galton (1822–1911), working in England, proposed a scientific classification system of fingerprints, still recognized today. At the same time, on the other side of the world in Argentina, the determination of the Croatian immigrant Juan Vucetich (1858–1925) resulted in the first major success of fingerprinting. Vucetich developed a more extensive classification based on Galton's system. Introduced to the Buenos Aires police force in September 1891, his system resulted in the identification of twenty-three felons. He was able to identify a woman by the name of Rojas who had murdered her two sons and cut her own throat in an attempt to place blame on another.

It is amazing that Herschel, Faulds, Galton, and Vucetich, working in relative isolation and in locations that span the globe, produced such a compelling body of evidence that established the discipline of fingerprinting. Their work spans barely forty years in a time of no faxes, no air transportation, not even worldwide radio. (It was not until 1910 that the radio telegraph was used to prevent the murderer Dr. Hawley Crippen from fleeing English justice on the SS *Montrose*, bound for Quebec.)

The way that events fell into place to create fingerprinting as the premier forensic science tool for individualization makes it easy to forget the controversy that surrounded the subject in the late 1800s. It was by no means straightforward, as there was a significant movement for forensic identification to be based on anthropometry. This was quite understandable, as anthropometry—the measurement of physical characteristics—is an objective and quantifiable translation of how we recognize one another. The leading proponent of the anthropometric school was Alphonse Bertillon (1853–1914). Bertillon's system, which used physical characteristics such as ear shape, was highly regarded and promoted in the 1880s and 1890s. It had several successes, and he deserves recognition as the father of forensic criminal identification. The problem with anthropometry, of course, is that it leaves nothing at the scene of the crime and is only useful if there are reliable witnesses capable of an effective

Alphonse Bertillon, a French police employee in the 1800s, who identified the first recidivist based on his invention of anthropometry. (Courtesy of the Universite de Lausanne Institut de police scientifique et de criminologie)

reconstruction. This is not so with fingerprints, and inevitably anthropometry lost out. The principles were sound, however, and anthropometry is enjoying a resurgence today as security systems are being developed to screen transportation passengers against databases of known terrorists.

"Turning to Poison . . ."
—Keats, "Ode on Melancholy"

The problems presented by poisons and poisonings have plagued medicine and the law for centuries. The use of drugs has been a source of comfort to the sick but also a source of despair to victims of their misuse. Opium has been used for therapeutic purposes since at least the eighth century BC and knowledge of it passed directly from one ancient civilization to another, primarily in Egypt, Byzantium (Syria), and Persia (Iran).

The early Greeks were familiar with poisons and there are references to Ulysses and Hercules anointing their arrows with serpents' venom. In 339 BC Socrates was executed by poisoning with hemlock. As with other forms of art, the art of poisoning blossomed during the Renaissance. Poisons initially were of vegetable origin (hemlock, aconite, and opium) but arsenic, corrosive sublimate (mercuric chloride), lead, and chloroform gradually appeared.

Judging from the ancient Romans, the Borgias and Medicis in Renaissance Italy, and the French School it is clear that eliminating rivals by poisoning had a long and successful history in love and in politics prior to the scientific advances of the nineteenth century. Indeed, the French School of poisoners, which originated with the marriage of Catherine de Médicis to the future King Henry II of France, had become so prolific by the end of the sixteenth century that it was normal for the death of anyone of standing in Paris at that time to be considered a poisoning. Two things characterized the heyday of the poisoner. Living conditions were generally poor, which meant that early and unpleasant death from disease was hard to differentiate from deliberate poisoning, and poisons were impossible to detect in the body of the victim. Evidence of poisoning was purely clinical and circumstantial until 1781 when Dr. Joseph Jacob Plenck (1739–1807), a professor of surgery in Vienna, published the statement, "The only certain sign of poisoning is the chemical identification of the poison in the organs of the body" (Elementa medicinae et chirurgiae forensic, 1781 cited by Jaroslav Nemec, "Highlights in Medicolegal Relations," National Library of Medicine, http://www.nlm.nih.gov/hmd/pdf/highlights.pdf), even though at that time there were very few chemical tests for poisons available.

By the nineteenth century, chemistry was beginning to produce analytical methods of sufficient sensitivity and specificity to detect poisons in the blood. Medical schools were recognizing forensic medicine as a special discipline. And although there was a long way to go, improved living conditions meant that unexplained deaths were more obvious. The history of arsenic poisoning illustrates all three factors.

Arsenic (see **Poisons**) itself is a gray metal and not particularly toxic. In contrast, its oxide, known as white arsenic, is very poisonous. It works by disrupting the digestive system, and so its symptoms are very similar to those of many pre-twentieth-century common diseases of the intestine, such as cholera. Add to this its ready availability as a medicine (treating syphilis), a rat poison (and there was ample need for that), an ingredient of tonics, a component of green dyes in wallpapers, and a component (with vinegar and chalk) of a mixture that women took to whiten their complexion, then there is little wonder that it became the top choice for the poisoner, and earned the nickname "inheritance powder."

Until the work of the Scottish chemist James Marsh (1794–1846), it was not possible to confirm the traces of arsenic that remained in the body of the deceased. Marsh's first encounter with arsenic was in 1832 when he was a witness at the trial of James Bodle, who had been charged with the murder of his grandfather by administration of arsenic in his coffee. Marsh conducted what at that time was the standard test for arsenic. Hydrogen sulfide was bubbled through a solution and if arsenic was present it would react to form a yellow precipitate of arsenic sulfide. The test proved positive when applied to samples from the body of the deceased, but by the time of the trial, the precipitate had decomposed and there was no physical evidence to present to the jury, who acquitted Mr. Bodle. The result motivated Marsh to produce a better forensic test for arsenic, which he did in 1836. The test starts by treating samples, for example of blood or decomposed tissue from a corpse, with zinc and sulfuric acid, which results in conversion of any arsenic in the sample into the gas arsine. The Marsh apparatus then directs the gas through a heated tube, which decomposes it into arsenic and hydrogen, and then onto a cold glass surface where the metallic arsenic condenses and forms a black mirror. The mirror is stable and the glass plates can be preserved as evidence.

Inevitably, the invention of the Marsh test was soon followed by challenges. In 1840, Marie Lafarge was tried for the murder of her husband by poisoning him with arsenic. The circumstantial evidence was compelling—it was not a happy marriage, Marie was known to have purchased white arsenic ostensibly as a rat poison, and the family maid testified that she had seen Marie sprinkle white powder on her husband's food. The Marsh test gave negative results on body samples. Mateu Orfila (1787–1853) testified at the trial that the Marsh test was unreliable if used by unskilled personnel, and further, that when he had applied the test correctly, it detected traces of arsenic in body samples. Marie was found guilty and sentenced to life imprisonment.

The growing availability of analytical methods for detecting metallic poisons around the middle of the nineteenth century made them less popular as poisons, and practitioners of that art turned their attention to the alkaloids that were being isolated from plants (morphine from opium, strychnine from nux vomica, quinine from cinchona, nicotine from tobacco, and atropine from belladonna). These were considered to be undetectable in human organs. In 1850 this began to change with the development by Jean Servais Stas (1813–1891), professor of chemistry at L'École Royale Militaire in Brussels and a former student of Orfila, of a method for the extraction of nicotine from the organs of a murder victim. Stas was somewhat fortunate in that the murderer had poured vinegar down the victim's throat after death and the acidity of the vinegar made the nicotine soluble in water. Stas's procedure (as modified by Friedrich Julius Otto, professor of chemistry at Brunswick in 1856) is based on the differing solubilities of many drugs in acids and bases and, with further modifications, is still in use today.

In the second half of the nineteenth century, forensic toxicology slowly began to develop in North America. Initially, analyses for poisons were conducted on an intermittent basis by individual professors of chemistry or pharmacology at the request of investigating officials on a case-by-case basis. An early example of this was in Canada in 1859 when Professor Henry Holmes Croft (1820–1883), professor of chemistry and experimental philosophy in the University of Toronto, testified at the trial of a physician for the murder of his wife that he found eleven grains of arsenic in her stomach. This early example of forensic toxicology emphasized the fact that, although tests for inorganic poisons were available, tests for vegetable poisons were also needed. The doctor, as he was about to be executed, confessed to the murder but said he had used morphine for the deed.

The trend toward specialization in toxicology began near the end of the nineteenth century with the emergence of Sir William Willcox (1870–1941) at Guy's Hospital in London and the most prominent of the early American toxicologists, Dr. Rudolph August Witthaus (1846–1915). Witthaus had studied chemistry at Columbia University in New York and at the Sorbonne in Paris. He became professor of chemistry and physiology at New York University in 1876 and published one of the earliest American books on chemistry and toxicology in 1879.

As a result of the limitations of the investigative systems at that time, requests for analyses for poisons were often an afterthought and frequently the toxicologists had the additional challenge of having to work with exhumed organs. It was not until Dr. Charles Norris (1868–1935) established the medical examiner investigative system in New York City that improvement in toxicological service began to occur. Dr. Norris recruited Dr. Alexander O. Gettler (1883–1983), a pathological chemist at the Bellevue Hospital, to establish a toxicology laboratory for the chief medical examiner's office in 1918. Born in Austria, Dr. Gettler was raised on the lower east side of Manhattan Island and educated in the New York City public school system. He received a BS degree from the College of the City of New York in 1904 and his PhD from Columbia in 1912. There was no model for Gettler to follow when he set up the toxicology laboratory for the chief medical examiner, so much of his early work was of a groundbreaking nature.

Virtually all of the early analytical methodology was based on wet chemistry and microscopic procedures that required large samples—500 grams of tissue was the typical starting point as contrasted with the 1 or 2 microliters of blood that is the norm today using sophisticated instrumental techniques. In addition to his innovative analytical procedures, Gettler's principal contribution to modern forensic toxicology was his superb teaching. In 1935 he started a graduate course in toxicology at New York University that he continued until his retirement in 1959. His graduates went on to become an outstanding "second generation" of American forensic toxicologists who started laboratories across the country, thus spreading Gettler's influence.

"Experto credite."
—Virgil, *Aeneid*, xi, 283

The nineteenth century gave birth to the "expert." For the first time, individuals were making a name for themselves as sources of expertise in forensic matters. The Spanish toxicologist Matthieu (or Mateu) Orfila was perhaps the earliest and greatest of them all. Published in 1814, his "*Traite des poisons tires des regnes mineral, vegetal et animal, ou toxicologie general*" (Thesis on poisons of animal, vegetable and mineral origin, or general toxicology) did more than establish forensic toxicology as a legitimate scientific discipline, it was the first time in the history of science that someone earned recognition for being a forensic expert. Orfila was followed by the Austrian Hans Gross (1847–1915), Alphonse Bertillon (1853–1914), and Francis

Galton (1822–1911), each of whom became famous for contributions to the new discipline of forensic science.

Orfila and Gross stand out from the others because of their writings. Each authored an authoritative text that became the standard reference work in the field, Orfila's for toxicology, Gross for what we would now call criminalistics (see **Criminalistics**). Interestingly, Gross was a lawyer, and his book *Handbuch fur Untersuchungsrichter als System der Kriminalistic* (Handbook on Criminalistics for the Examining Magistrate, first published in 1893) was written to describe what the investigator could expect from forensic science.

Early warnings of the problems that the cult of the expert were to bring to forensic science were already evident in the respect accorded to Orfila in the Lafarge case. The problem of the expert personality being more visible than expert science would be seen again and again right up to the present day.

"For the red blood reigns . . ."
—Shakespeare, *The Winter's Tale* iv, ii

The nineteenth century was the time for forensic chemistry and toxicology, but the dawn of the twentieth century saw the foundational discoveries for the identification of blood. Blood, the bright red mark of violent crime, had bedeviled forensic science for centuries: shed in copious amounts by victims, obvious to the naked eye, but completely silent as to the origins of stains on the clothing or body of suspects and witnesses. Any forensic scientist schooled in the practice of forensic serology today will give the same response to the question, "How do you go about identifying blood?" Testing is conducted in the sequence presumptive or screening test, then confirmatory test for blood, followed by species testing, and finally by grouping. However, the science did not develop in the same measured way (and DNA testing has short-circuited the traditional sequence). A flurry of activity in the late nineteenth and early twentieth centuries simultaneously established the basis of most of the scientific tests that were the mainstay of forensic serology in the second half of the twentieth century.

Most of the classic work on blood identification depends on some form of immunological reaction. Immunology can be traced back to the late nineteenth century and the work of Paul Ehrlich (1854–1915). His work on immunology resulted in Ehrlich sharing the 1908 Nobel Prize for physiology or medicine with Ilya Mechnikov (1845–1916). A series of publications between 1885 and 1891 on what he called the "side chain reaction" established the chemical nature of the reaction between antigens and antibodies. A genius and prolific researcher, Ehrlich has had his work remembered in the many reagents and reactions named after him. Two of them, Ehrlich's reagent and Ehrlich's reaction, are still used today in forensic biology to screen evidence samples for feces. Ehrlich was a prolific researcher in many fields and is credited with coining the expression "magic bullet" to describe drugs that can target and kill specific disease-causing organisms.

Ehrlich was working at a time that saw major advances in immunology arising from research in several European centers. The most important for forensic science was the work of Karl Landsteiner (1868–1943) on blood groups. Landsteiner completed his illustrious career in the United States, but it was while he was an assistant at the University Department of Pathological Anatomy in Vienna that he published (1901) his work on ABO blood groups. He was awarded the 1930 Nobel Prize in physiology or medicine in recognition of this outstanding research. Landsteiner's research made safe blood transfusions possible; it was the start of work that ended with successful tissue matching and thus organ transplants. And it allowed forensic scientists to conclude whether or not a bloodstain found on an accused could have come from the victim.

Landsteiner's research was all devoted to clinical applications, and it was left to others, including Leone Lattes, to develop the forensic applications. Landsteiner had separated blood samples into red cells and serum and shown that some of the sera could cause cells to clump together, or agglutinate. He used the reaction to classify blood as type A, B, AB, or O. The cells from type A blood agglutinated with the serum from type B or type O blood; cells from type B agglutinated with serum from type A or O; cells from type AB agglutinated with serum from type A or B; and cells of type O did not agglutinate with serum from either type A or type B persons.

The problem is that Landsteiner's classification depends on the reaction of cells with the corresponding antiserum (see **Antiserum**). Red blood cells are destroyed in stains, and so there is nothing to see when an extract of the stain is mixed with antiserum. The problem was solved in 1915, by Leone Lattes, working in the Institute of Forensic Medicine in Turin, Italy. Lattes realized that although the cells in the stain were destroyed, the antibodies would still be active and he introduced the idea of reverse grouping whereby bloodstains were characterized by identifying the antibodies present. Although a great advance in its time, the Lattes test has drawbacks. It is not particularly sensitive and weak stains may give false results due to undetectable amounts of antibody in the extracts. This compounds the problem that it is not scientifically valid to call an AB type because the characterization depends on the absence of reaction (there being no antibody to agglutinate test cells). It would take more than half a century before a sufficiently reliable technique for typing bloodstains was developed.

It might have taken forensic science a long time to translate Landsteiner's research into a routine and reliable test, but other discoveries in immunology being made at the same time had a more rapid impact on forensic biology. In 1901—the same year that Landsteiner published his research on ABO blood groups—the research of Paul Uhlenhuth (1870–1957) laid the foundation for the identification of the species of origin of bloodstains. Uhlenhuth's work can be traced back to the research of the Belgian Jules Bordet (1870–1961), winner of the 1921 Nobel Prize in physiology or medicine for his discoveries in immunity. While working at the Pasteur Institute in Paris between 1894 and 1901, Bordet showed that vaccination produced a specific antibody and demonstrated a range of responses when an antibody was mixed with protein antigens. One of these was the precipitation of a colloidal complex—the precipitin reaction. The value of the finding is that it offered a tool that gives a visual indication of an antigen-antibody reaction. Thus, if an extract of a bloodstain is mixed with a solution containing an antibody for a protein that is present in the extract, it will result in a visible reaction, namely formation of an insoluble complex that precipitates out of solution. Uhlenhuth used this research as the springboard to investigate what happened when proteins from one species are injected into another. Specifically, he injected egg white into rabbits, and found that serum taken from the rabbits would form a precipitate when mixed with extracts of chicken tissues.

Forensic scientists were quick to see the potential of the research, and it was used in a case the very same year of Uhlenhuth's discovery (1901). There are several accounts of the Tessnow case; one of the more readable is found in Colin Wilson's book *Written in Blood, Book 3, The Trail and the Hunt* (New York: Warner Books, 1989). Two small girls failed to return home from school in the village of Lechtingen, near Osnabruck in Germany. A carpenter, Ludwig Tessnow, had been seen in the woods where their bodies were later found and he was questioned by police. His clothes bore dark brown stains that could have been blood but that he claimed resulted from spills of wood stain. They did indeed look exactly like the marks produced by the stain that he used. He was released. All this took place in 1898, and things settled down until three years later, when two young school boys went missing on the island of Rugen. Their mutilated bodies were found in the woods. A man seen near where the bodies were found turned out to be Tessnow. He was questioned and asked about brown stains on his clothing. Once again he explained the stains as being from wood dye. However, a magistrate recalled his name as being associated with the Lechtingen murders and discussed the case with a friend, Ernst Hubschmann. Hubschmann had heard of Uhlenhuth's work and arranged for Tessnow's clothing to be sent to him for testing. The tests identified seventeen stains of human blood, totally disproving Tessnow's story. Tessnow was duly tried and convicted.

The Tessnow case began with the assumption that the brown stains on Tessnow's clothing could be blood—but the first and second steps of the sequence from screening to typing were missing. At least one specific test for blood was available in 1898. Ludwig Teichmann had shown in 1853 that hemoglobin could be converted to hemin, which forms characteristic crystals in the presence of halides. Teichmann's reagent used a solution of potassium chloride, iodide, and bromide in glacial acetic acid. It was the combination of identifying hemoglobin or a chemical derivative together with the specificity of the crystals formed that made the Teichmann test a benchmark in forensic serology for many years. The same basic principles were used by Masaeo Takayama when he introduced a somewhat more reliable test in 1912. Takayama showed that mixing an extract of blood with pyridine results in a chemical reduction of the hemoglobin and the formation of feathery pink crystals.

The history of the first step—screening—is less certain. Somewhere in the period between 1818 (when hydrogen peroxide was discovered by Louis-Jacques Thenard) and the mid-nineteenth century, people became aware of the peroxidase-like activity of hemoglobin. In 1863 Christian Schonbein advocated using the observation of frothing when hydrogen peroxide is applied to a possible bloodstain as a presumptive (or screening) test. This was not at all a reliable process. The first reported screening test for blood was that of Oskar and Rudolf Adler, who found that the new chemical benzidine, synthesized by the Merck Company in 1904, was converted into its colored oxidized form when exposed to hydrogen peroxide and even minute traces of blood. Unfortunately, benzidine is highly carcinogenic. A safer and reliable screening test, and one that is still used, is the Kastle-Meyer test, which replaces benzidine with phenolphthalein. Hemoglobin exhibits peroxidase activity, that is, it converts hydrogen peroxide into water. It will therefore oxidize the colorless reduced form of phenolphthalein in the reagent into its pink oxidized form.

There are many variants of these screening tests, but almost all depend on the conversion of a colorless compound to its colored, oxidized form by the peroxidase activity of hemoglobin and the presence of hydrogen peroxide. All of them are sensitive, and none are entirely specific. For example, the original work in the "Dingo Baby" case in 1982 in Australia, which claimed to have found blood on the car of the Chamberlains, was found to be flawed. Exhaustive testing conducted some ten years later during a Royal Commission hearing found no blood and indicated that the earlier tests might have been false positives due to copper in the soil around the crime scene.

"The True University of These Days Is a Collection of Books"
—Thomas Carlyle, "Heroes and Hero Worship Lecture V. The Hero as a Man of Letters"

The history of a body of knowledge can usually be constructed around its development in universities and the publication of research in the forms of books and journals. This is not the case for forensic science because it hardly features in any academic environment right up to the end of the twentieth century, and the literature is scant. There is no coherent body of knowledge accompanied by a research agenda that identifies it as a valid discipline within science. For example, the writer was informed by a senior officer in the National Science Foundation (NSF) in 1996 that the NSF had no funds to support research in forensic science "because it was applied and not truly a branch of science."

There is, however, a valid trail of fundamental research in other disciplines that lays the foundation for the testing that makes forensic science today such a powerful tool in the administration of civil and criminal justice. The discoveries of Ritter, Sturgeon, Nicol, Kirchoff, Bunsen, Marsh, and other chemists have already been mentioned, but it was biology and medicine that were creating what academic presence there was for matters forensic.

Leone Lattes deserves credit for his research, but it is worth noting that the work was conducted in continental Europe, where medical specialists were trendsetters in recognizing "forensic" as a viable discipline within a larger body of knowledge. The first ever academic course in forensics was in the University of Leipzig, where forensic medicine was taught as long ago as 1642. There was a department of forensic medicine in the University of Paris in 1794. What may have been the world's first forensic science laboratory, the Institut de medicine legale de Paris, was established there in 1868. The University of Lyon was one of the first academic institutions that recognized forensic medicine as a subject. Andre Lacassagne (1823–1924) was appointed to the post of professor of legal medicine in 1880. Lacassagne had a varied career including military service. He is recognized as being one of the first to show the correlation between markings on a bullet and the rifling inside the barrel of the weapon from which it was fired.

This landmark observation illustrates the eclectic atmosphere of the early practitioners in forensic medicine. A few years later, another French doctor, Victor Balthazard (1872–1950), followed suit. Balthazard was professor of forensic medicine at the Sorbonne, in Paris. He conducted research on a surprisingly wide range of topics, including a complex statistical evaluation of the effects of a second impairment on a patient already suffering from a separate impairment. However, it seems that his main interest was in physical evidence and the evaluation of how distinct features have to be to allow conclusions of identity to be drawn. He published a paper on the statistical basis of fingerprint identification in 1911. The previous year he had published the first expert treatise on human hair identification. Balthazard was busy in another field of physical evidence and conducted several studies on bullets and firearms between 1909 and 1913. He was probably the first to show that each gun imparts a unique pattern on a projectile fired from it due to manufacturing and wear impressions.

Meanwhile, Scotland was setting the pace in the English-speaking world. Andrew Duncan (1744–1828) lectured on forensic medicine at the University of Edinburgh in 1789, and his son became the first holder of a professorship that included the word *forensic* when he was appointed Regius Professor in 1807.

Dr. William A. Guy (1810–1855) was appointed to professor of medical jurisprudence at Kings College, London, in 1838. Guy had a distinguished career that included publication of his text *Principles of Forensic Medicine* in 1844. The University of Glasgow followed with the appointment of Dr. R. Cowan (1796–1841) as its first professor of forensic medicine in 1839 and the creation of its own Regius Chair in forensic medicine, with the appointment of Dr. John Glaister (1856–1932) in 1889. Glaister senior was succeeded by his son, who held the chair from 1931 until 1962.

The most widely known of the Edinburgh school is probably Sir Sydney Smith (1883–1969). Born in New Zealand to English parents, Smith received his MD from Edinburgh in 1914 where he studied under Professor Henry Duncan Littlejohn (1828–1914).

In 1917 he took up the position of principal medico-legal expert in Egypt where, in addition to his expertise in forensic medicine, he also became an expert in the emerging field of forensic ballistics. In 1925 he published the first edition of his *Textbook of Forensic Medicine* based largely on his experience in Egypt. He returned to Edinburgh in 1927 to succeed Littlejohn and held that chair until his retirement in 1953. In the midthirties, Smith was among the first of the medical specialists to realize the importance of other disciplines in forensic science. His autobiography, *Mostly Murder*, was published in 1959.

In England, developments in forensic science came a bit later. The most prominent early name was that of Alfred Swaine Taylor (1806–1880), who became professor of medical jurisprudence at Guy's Hospital Medical School in 1834. Taylor was a prodigious author and his *Principles and Practice of Medical Jurisprudence,* first published in 1865 and continuously revised and updated by other authors, is still recognized as a standard work.

The next giant in forensic medicine in England was Sir Bernard Spilsbury (1877–1947). Dr. Spilsbury received his medical training at St. Mary's Hospital Medical School in London and from 1902 until his death in his laboratory at University College in 1947, devoted his life to pathology and histology. He was not considered to be a great innovator but rather was a practical man with keen powers of observation. After becoming a Home Office pathologist in 1908, he dominated forensic medicine in England for almost forty years and convinced Scotland Yard detectives of the importance of having the specialist at murder scenes. In the courtroom he seemed able to exert a spell over judges and juries. His reputation was not based on his writings, because he wrote virtually nothing, but rather on his vast experience of over 25,000 cases during his career.

Following Spilsbury in London, a small group of forensic pathologists dominated the scene in England. They were led by Professor

Keith Simpson (1907–1985) at Guy's Hospital and Professor Francis Camps (1905–1972) at the London University Hospital.

In North America, the earliest indication of forensic medicine is in records of autopsies carried out by one of the explorer Samuel de Champlain's surgeons early in the seventeenth century at a settlement in what is now the Canadian province of New Brunswick. In the United States, the early English settlers brought English law and the coroner system with them, and inquests were held as early as 1635 in New England.

Edinburgh also influenced academic forensic medicine in the United States. Born in New York City, James S. Stringham (1775–1817) studied medicine at the University of Edinburgh from which he graduated in 1799. On his return to the United States he held various positions before being appointed professor of medical jurisprudence at the Columbia College of Physicians and Surgeons in 1813. Another Edinburgh graduate, Benjamin Rush (1745–1813), lectured on medical jurisprudence in Philadelphia and was one of the signers of the Declaration of Independence in 1777. The first North American textbook was *Elements of Medical Jurisprudence,* a 900-page, two-volume work written by Theodoric R. Beck (1791–1855) and published in 1823 at Albany, New York.

More than 100 years after Stringham's appointment, Dr. Charles Norris (1868–1935) was appointed the first chief medical examiner to the City of New York in 1918. Norris was at the time a physician at the Belleview Hospital Center, which had been affiliated with the Columbia College of Physicians and Surgeons since 1860. It is tempting to postulate a linkage through the college's history to Stringham, but perhaps this is just another of forensic science's coincidences. Although not strictly a university, the Office of the Chief Medical Examiner (OCME), which Norris led, made significant contributions to forensic medicine research and service development under several leaders and was a key factor in the growth of organized,

high-quality forensic science in the United States.

Dr. Milton Helpern (1902–1977), a native-born New Yorker, took the foundation established by Norris and built the structure for a world-renowned center for service, research, and training. A graduate of Cornell Medical School, Helpern studied pathology at Bellevue Hospital under Norris and joined the Medical Examiner's Office in 1931 where he served as its chief until his death. He was responsible for training many young pathologists who went on to establish medical examiner's offices across the United States. An indication of his stature is the fact that he was referred to by everyone as "The Chief," even by those who never worked for him.

Back in Europe, the nonmedical sciences were starting to develop some form around the same period. And, just as with Norris and the OCME, a key figure in the history of forensic science was responsible for establishing one of the earliest centers of excellence, but not in a university. The figure was Edmond Locard and the institution was the Laboratoire de Police Scientifique, or Police Scientific Laboratory.

Locard was born in France in 1877 and died in Lyon in 1966. He obtained his doctorate in 1902 and went on to work with Lacassagne at the University of Lyon. Locard broadened his forensic knowledge by spending some time with Bertillon and became interested in identification. He moved from forensic medicine and the university to science and the police force in Lyon in 1910, setting up what became the official police laboratory in 1912. Locard realized that dactyloscopy, or fingerprinting, was the best approach to identification. He made several significant contributions to the new science of fingerprinting and first used it successfully in a case in 1910.

Locard can justifiably claim to be the first true forensic scientist. He was concerned about the scientific principles behind his work. Thus he was the first to set down rules for the application of Galton points, or points of identification such as bifurcations (see **Fingerprints**) for definition and the observation of detailed minutiae in a fingerprint necessary to establish identity. He is credited with establishing the concepts of poreoscopy (the pattern of pores in a fingerprint) and edgeoscopy (the shape or contour of the edges of the ridges in a fingerprint), which are the basis of what today's fingerprint examiners term ridgeology or the study of the uniqueness of the ridges in a fingerprint, and are the basis of modern approaches to fingerprinting. His tripartite rule set three levels of interpretation: more than twelve Galton points in a sharp and clear print is enough for finding identity. The conclusion that could be drawn from a print with eight to twelve points depended on other factors, and no reliable conclusion could be drawn from fewer than eight points.

Locard is best known for the principle identified by his name: the Locard exchange principle. In summary, this principle holds that every contact leaves a trace. Locard's principle is behind all trace evidence, which depends on comparing traces of materials found on a suspect with bulk material from the scene of the crime. The reasoning goes that the various contacts between the perpetrator and objects at the scene will result in a transfer of materials between the two on each contact.

There are several things wrong with Locard's exchange principle. First, and most practical, is that it is not always possible to establish a reliable association between the object making the contact and the material transferred to it. Traces may be lost with time and movement of the perpetrator away from the scene. The materials exchanged may not be sufficiently unique, either alone or in combination, to provide reliable evidence of contact, and the techniques used to make the comparisons might not be capable of sufficient discrimination.

The second problem is that Locard never said that every contact leaves a trace. What he did say, in several ways in his publications, was

that it is impossible for a criminal to perform an act of violent crime without leaving some trace of his or her presence. He also repeatedly asserted that the microscope was a powerful weapon to characterize the debris deposited on the clothing of people as they move through different environments. It is thus reasonable that the concept is named after him.

Locard's reputation is also based on something he did write, his extensive treatise on forensic science *Traité de criminalistique,* published in parts between 1931 and 1935. Locard was fifty-eight when the last part of the treatise was published. It was his last personal contribution to forensic science.

Bragging rights for the first university department of forensic science probably belong to the University of Lausanne in Switzerland. The forensic program in the School of Forensic Science and Criminology can trace its history to 1902 and Professor Rudolphe Archibald Reiss. Reiss, who was born in Lausanne in 1876 and died in Belgrade in 1929, was yet another of the pioneers of forensic science who had worked with Bertillon. He established a forensic photography course at the university around the turn of the century and developed it into a full forensic science program in 1909. His forensic science work includes a review of the Paris police (1914) and several manuals of forensic science techniques and photography published in Lausanne and Paris between 1905 and 1914. Reiss's contributions to forensic science ended in 1914 because of his devotion to another cause, investigation of war crimes committed against the Serbs. He reported on "The atrocities committed by the Austro-Hungarian army during the first invasion of Serbia" in 1915. He is honored by a statue in Belgrade. War crimes, genocide, and similar atrocities have existed as long as war and conflict and are still with us. Although his report did not deal with any highly technical issues, it is the first time that a forensic scientist was involved. Today, the special skills of forensic medical and scientific teams are regularly incorporated into such investigations.

R.A. Reiss, a professor at the University of Lausanne in the early 1900s and a pupil of Alphonse Bertillon, set up one of the first academic curricula in forensic science. His forensic photography department grew into Lausanne Institute of Police Science. (Courtesy of the Universite de Lausanne Institut de police scientifique et de criminologie)

Brave New World
—Aldous Huxley, 1932

The foundations of forensic science were laid in Europe in the late nineteenth and early twentieth centuries, but the United States was where most advances took place in the following fifty years.

After the pioneering work of Lacassagne and Balthazard, the field of ballistics, or firearms examination, is dominated by American contributions (see **Firearms**). Various reports, such as an article, "The Missile and the Weapon" by Dr. Albert Llewellyn Hall published in the *Buffalo Medical Journal* in 1900, and the official report "Study of the Fired Bullets and Shells in Brownsville, Texas, Riot" included in the annual report of the chief of ordnance, U.S. Army in 1907, show that

there was a growing general awareness that ammunition could be linked to the gun from which it was fired by means of markings on shell cases and projectiles. It was not yet a sound science, however, and there was a near miscarriage of justice in New York State in the Stielow case in 1915. Stielow was convicted of the murder of his employer and his employer's housekeeper. Evidence at trial included testimony that the fatal bullets had been fired from a gun found in Stielow's house. There were several doubts surrounding the whole investigation and the attorney general conducted an investigation while Stielow awaited execution. One of his special investigators, Charles E. Waite, reexamined the firearms evidence together with a microscopy expert, Dr. Max Poser. They proved that the bullets could not have been fired from Stielow's gun, and Stielow was pardoned and released.

The first significant institution devoted to firearms examination was the Bureau of Forensic Ballistics in New York. The bureau was formed by Waite, together with Calvin Goddard, Philip Gravelle, and John Fisher in 1925. Fisher had invented the helixometer for examining the interior of gun barrels in 1920, and the group perfected the firearms comparison microscope. However, Goddard was the doyen of the group.

Goddard's history encompasses notorious cases and major forensic science institutions beyond the Bureau of Forensic Ballistics. The first of the notorious cases was that of Nicola Sacco and Bartolomeo Vanzetti. Sacco and Vanzetti were tried and convicted of robbery and murder in 1921. The evidence included testimony on examination of weapons and ammunition that may have been involved in the crime. However, the Sacco-Vanzetti case had become a cause célèbre largely due to social and political circumstances—media attention consistently referred to the Italian background and anarchist political affiliations of the accused. Both sides vigorously pursued their cases, including reexamination of the firearms evidence. One of the original defense experts was Albert H. Hamilton—the examiner whose prosecution evidence had been discredited in the Stielow case. There is some evidence to assert that Hamilton was the first forensic "hired gun"—that is, he was ready to manipulate his story to fit the requirements of the side that employed him. However, the trial judge found him to be not reliable, and his testimony probably damaged the Sacco-Vanzetti defense.

A committee appointed to review the case in 1927 engaged Goddard. His tests concluded that one of three shell cases recovered from the scene had been fired from a gun owned by Sacco. The defense experts (not Hamilton) present during Goddard's examinations agreed, as did further retesting conducted in 1961 and 1983. The guilty verdict was confirmed, and Sacco and Vanzetti were executed on August 23, 1927.

Goddard's second notorious case was the St. Valentine's Day Massacre. The case is part of American folk history as an illustration of the gangster era. On February 14, 1928, a group of hit men from Al Capone's criminal gang gunned down seven members of a rival gang in Chicago. The gang turned up in a fake police car, with three of them wearing police uniforms. Even in those violent days, the killing of seven people in a public place by gunning them down with over 70 rounds from machine guns was a headline-grabbing case. The magnitude and flagrant nature of the crime, together with allegations of police involvement, led to a rapid and thorough response by the authorities. Goddard was engaged to conduct an examination of the firearms evidence. He showed that the killers had used a 12-gauge shotgun and two Thompson submachine guns. He further tested all the Thompson submachine guns used by the police and showed that none of them was involved in the killings. In contrast, he found positive matches to guns found in the home of one of the suspects.

The physical tests used by Goddard had been developed to the point that a confident conclusion associating a bullet or shell case with the weapon from which it had been fired

could be made. Things were different when it came to associating an individual with a fired gun, other than through the weapon being in the possession of the individual. Gunshot residue tests (see **Firearms**) are used for that purpose. When a gun is fired, the various components of the primer and propellant in the ammunition produce a gaseous cloud of residues, including nitrates, that escapes from the gun and can condense on the hands and clothing of the person firing the weapon. One of the earliest tests for these residues on skin was the paraffin test. This test recovers surface contaminants from the hands by applying a film of warm paraffin wax and peeling it off. The wax is then treated with diphenylamine, which reacts with nitrates to give a blue color. The test was introduced to the United States in 1933 by Teodoro Gonzales of Mexico City. Although not very reliable—it can give false positive and false negative results—the paraffin test was to be used for some fifty years.

Part of the reason for the longevity of the paraffin test is that there was nothing better to replace it. It was not until 1958 that H. C. Harrison and R. Gilroy introduced a colorimetric test for antimony, barium, and lead on hand swabs. These metals are typically found in ammunition primers. If anything, the Harrison-Gilroy test was even less reliable than the paraffin test, but it did produce a focus on alternate procedures. Instrumental methods for detection of residues of the three metals were a considerable advance. Today, it is widely accepted that the combination of scanning electron microscopy (to identify the characteristic particles) and energy dispersive x-ray analysis (to identify the elements present in the particles) is the method of choice. This method was introduced into routine use by the Metropolitan Police Forensic Science Laboratory in London, England, in 1968.

Although he already had an assured reputation from his technical expertise and his position in the Bureau of Forensic Ballistics, Goddard had more to contribute to the development of forensic science in America.

Goddard's work in the St. Valentine's Day Massacre led to the next phase of his association with forensic science institutions. Through the foreman of the investigating grand jury, he was asked to establish a crime laboratory to serve the citizens of Chicago. The foreman, Burt A. Massee, guaranteed financial support. The result was the creation of the Scientific Detection Crime Laboratory at Northwestern University near Chicago in 1930. Goddard served as its director until 1934. Goddard also advised the FBI when it set up its laboratory in 1932 and the army when it established its Tokyo crime laboratory in 1948.

Goddard's Scientific Detection Crime Laboratory in Chicago was not the first crime laboratory in America. That honor goes to California and August Vollmer (1876–1955). Vollmer was appointed chief of police in Berkeley in 1909. He was an energetic innovator who had the vision to see how advances in science and technology could be applied to public safety. His leadership included pioneering systems of fingerprint and handwriting evidence, use of the polygraph, and the application of forensic science to investigations. He is variously reported as having established a crime laboratory in Berkeley around 1910, but it is most likely that there was no physical entity but rather a range of activities carried out from around 1907, probably in conjunction with scientists at the University of California at Berkeley. Vollmer understood the value of training and created a police academy in his department, with lectures from visiting scientists included in the curriculum. Vollmer was appointed chief of police in Los Angeles in 1923, and the history of the Los Angeles Police Department Scientific Investigation Department records that he established a crime laboratory soon after his appointment there. On his retirement, he created a program in police sciences at the University of California at Berkeley.

The university at Berkeley merits recognition in any historical account of forensic science. Vollmer's interests alone would assure that. One of those—the polygraph (see **Lie Detector Test**)—was invented at Berkeley and is almost uniquely associated with forensic science in the United States. The polygraph makes simultaneous recordings of the pulse rate, blood pressure, breathing rate, and perspiration (skin resistance) of the test subject. The core of the instrument is the measurement of pulse rate. In fact, the name *polygraph* was first given to an invention of the Scottish physician James Mackenzie (1853–1925) to simultaneously measure arterial and venous pulses. Mackenzie published his work in 1902. A Berkeley medical student, John Larson, adapted Mackenzie's polygraph to make the first "lie detector" in 1921. The basis of Larson's work was that lying induces some degree of stress, even at a subconscious level, and that the stress will result in involuntary changes in one or more of the physiological parameters measured. The apparatus was further developed over the next year or two by Larson's associate, Leonarde Keeler. However, the *Frye* decision of 1923 (see **Frye Rule**) made it clear that the courts were less than convinced of the validity of the research. Notwithstanding the *Frye* case and other controversies as to its reliability, the polygraph is a widely established tool used in civil and criminal investigations in the United States.

Berkeley itself is most widely recognized within the forensic science community for its illustrious history as a center of academic excellence. Vollmer may have planted the seed, but it was Paul Kirk (1902–1970) who was responsible for forensic science flourishing there. Kirk was a gifted and industrious scientist who was associated with Berkeley from 1927. He is credited with using his skills and knowledge to convert the "art" of microchemistry into the science of trace chemistry. He created a new branch of science, criminalistics (from Gross's *kriminalistik*), and established it as a credible academic discipline. Kirk believed that the fundamental distinguishing characteristic of criminalistics is individualization or identity, a concept of great merit and typical of his high scientific standards.

It is sad that this account of the contributions to forensic science in the United States and throughout the world that came from the San Francisco Bay area has to close by recording that the forensic science program at Berkeley has not been able to compete with the pure sciences for available funding. What could and should have been a model program in an ideal location has closed, despite the leadership of Dr. George Sensabaugh.

"The American system of rugged individualism . . ."
—Herbert Clark Hoover, campaign speech, New York, 1928

Paul Kirk was certainly in the right place to assert that identification is the purpose of forensic science. The history of the International Association for Identification (IAI), records that

> On August 4, 1915, Inspector Harry H. Caldwell of the Oakland (California) Police Department's Bureau of Identification wrote numerous letters addressed to "Criminal Identification Operators" asking them to meet in Oakland for the purpose of forming an organization to further the aims of the identification profession. A group of twenty-two men met and, as a result, the International Association for Criminal Identification was founded in October 1915, with Inspector Caldwell as the presiding officer." (Cited from http://www.theiai.org/history/)

Today, the IAI is a prestigious professional association with more than 5,000 members in the United States and worldwide. Its membership and professional development activities encompass a wide range of applications, but fingerprinting is one of the core subjects. The story of fingerprinting in the United States is one with curious beginnings that echo Doyle's Sherlock Holmes, because one of the first references to it is in a work of fiction.

Mark Twain's *Life on the Mississippi,* written in 1883, identifies a murderer from his fingerprints, and a later book (*Pudd'n Head Wilson*) contains a courtroom scene in which fingerprinting is debated. This was before Galton's classification and not all that long after Herschel and Faulds.

In real life, some U.S. institutions were among the first users of fingerprinting. In quick succession from 1902 to 1905, the New York Civil Service Commission, the New York State Prison service, the Leavenworth State Penitentiary, and the U.S. Army all introduced fingerprinting as a means of identification. In 1905 the Department of Justice formed the Bureau of Criminal Identification in Washington, D.C., to provide a central collection of fingerprint cards. Two years later, the bureau was transferred to Leavenworth State Penitentiary. In 1924 an act of Congress amalgamated the bureau and the Leavenworth collections into the FBI's Identification Division.

Meanwhile, the case history of fingerprinting was being written with some memorable events. The story of Will and William West perhaps goes a little way to explain why Leavenworth Penitentiary features so strongly in the history of fingerprinting in the United States. Will and William West were unrelated, but were both incarcerated in Leavenworth at the same time—roughly between 1903 and 1909. They were both of similar build and the Bertillon records used by the prison to record the identity of inmates on the basis of physical characteristics such as ear shape were sufficiently close that they could have been mistaken one for the other. Leavenworth was one of the first institutions to use Bertillonage for this purpose and one of the first to convert to fingerprinting. It appears that Will was indeed mistaken for William, either at the time of admission or in a subsequent but uncertain criminal investigation, and that he was only cleared of involvement in the crime because of the fingerprint records.

However, the story of the two Wests is as close to Mark Twain's fiction as it is to reality. In fact, there is no account of mistaken identity or other supporting evidence anywhere in the Leavenworth records. The truth is that the staff at Leavenworth included some extremely farsighted and able officers, particularly Major Robert. W. McClaughry (1839–1920), who were pioneers in creating reliable identification records systems and ready to implement and evaluate new techniques as they became available. McClaughry was the first person to introduce Bertillonage into the United States, in 1887. At the time he was warden of the Illinois State Penitentiary and he persuaded the Warden's Association of the United States and Canada to adopt the system in the same year.

His son, M. W. McClaughry, was his records clerk at Leavenworth at the time of the West fable. The younger McClaughry had attended the World's Fair in St Louis in 1904 and there met Sergeant John K. Ferrier of Scotland Yard, London, England. Ferrier had given a course of instruction on fingerprinting at the fair and McClaughry junior was one of his students. As a result, and still in 1904, Major McClaughry was granted permission to introduce fingerprinting for identification of inmates and had Ferrier conduct a course of instruction at Leavenworth that year. The West fable is a wonderful account of what fingerprinting can do, but the credit must go where it belongs, to Ferrier and the McClaughrys.

Indisputably true and significant is the case of *People v. Jennings,* which trial took place in 1911. One night in September 1910, a Mr. Clarence Hiller was awakened by his wife, who believed that there was an intruder in the house. He went to investigate and disturbed an armed burglar, who shot and killed him after a struggle. Later that night, police questioned Thomas Jennings and found that he was injured and also was carrying a loaded revolver. His fingerprints were on record. Comparisons showed they were a match for four prints recovered from the scene of Hiller's

murder. Jennings was tried and convicted. Jennings appealed his conviction to the Illinois Supreme Court on the basis of the admissibility of the fingerprint evidence. The court found that "there is a scientific basis for the system of fingerprint identification, and that the courts are justified in admitting this class of evidence; that this method of identification is in such general and common use that the courts cannot refuse to take judicial cognizance of it" (*People v. Jennings,* 96 N.E. 1077 [1911]). This case established the admissibility of fingerprint evidence for the first time in U.S. courts. Other states and the federal government quickly followed suit in admitting fingerprint evidence in court, and it remained unshaken and unchallenged as the gold standard of identification for 100 years.

The responses of so many parts of the justice system to the advances in chemistry, biology, and physics that were taking place support Kirk's perception of forensic science being about identification. "What is it?" and "Who is it?" readily merge into "Who did it?" and "How did they do it?" Questioned document examination (see **Document Examination**) is often regarded as more of a subjective art than objective science. Its practitioners will soundly defend its objective basis and certainly no one would question the scientific validity of the chemical and physical tests used to characterize the materials used in creating documents: paper, ink, printers, copiers, and so on.

The issue of authenticity of writing is as old as writing itself. Thus, the Justinian Code of Roman law in the year 539 recognized that there were "experts" in handwriting analysis and required that they be sworn, presumably in an attempt to ensure some reliability in the procedure. But the story of QDE as an organized discipline within forensic science is essentially a story of the twentieth century, and one with a considerable American content.

It can be said that modern QDE began in 1910 when Albert S. Osborne (1858–1946) published *Questioned Documents,* a book that is still recognized as authoritative on the examination of handwriting and traditional documents. Osborne probably did more than any other one person to advance the subject, including founding the American Society of Questioned Document Examiners in 1942.

Despite the work of Osborne and others, there remains some suspicion about the reliability of QDE. This is in part due to confusion with what is known as graphology, which is the process of drawing conclusions about the character of a writer based on her or his handwriting. Apart from its confusion with graphology—which has no scientific foundation whatsoever—QDE has some problems associated with the fact that it absolutely obeys the falsification principle (see **Daubert Ruling**). Falsification was proposed by the eminent scientific philosopher Karl Popper (1902–1993) as a fundamental principle of science. In summary, no amount of testing can ever absolutely prove a scientific law to be true, because there may always be alternative explanations for a phenomenon that would lead to the same experimental results or observations. However, one well-designed experiment that produces a result contrary to a law establishes it as false, or "falsifies" it. To be accepted as "scientific," therefore, a theory must be capable of testing and falsification.

In a way, the principle of falsification is the basic principle of questioned document examination. Returning to the Justinian Code of Roman law, we find it expresses "hatred for the crime of forgery," and the question behind all QDE cases is "is this document authentic or has it been forged (falsified)?" We only need to look at a selection of our own signatures to see the problem that arises—each is unquestionably authentic, but none is identical to any of the others. The examiner has to make a judgment call as to whether differences are due to natural variation or to falsification. Osborne's book addresses this well, but it remains a difficulty. False claims made by unqualified examiners and exacerbated by circumstances are another matter.

Questioned document examination remained an experience-based craft with little new by way of technology or scientific advances until the 1980s, when techniques for examination of writing materials, indented impressions, and laser- and bubble-jet copy and print media were developed. Until then the discipline advanced case by case, and some of those raised more questions than they answered.

We will begin with the kidnap and murder of Bobby Franks in the affluent suburb of Kenwood, near Chicago. Bobby was picked up in a car on his way home from high school in 1924. When he had not returned home by dinnertime, his parents started a search, thinking he may have been with a friend. But then they received a phone call that Bobby had been kidnapped. They received a typewritten ransom demand the next morning. They notified the police, but at the same time as they were considering their response to the kidnappers' demands, a dead body found a little distance away was identified as Bobby's. Eventually suspicion centered on two young men, Richard Loeb and Nathan Leopold. Both were well educated and from well-off families. They had alibis that appeared strong. Leopold owned a typewriter, but it was excluded as the source of the ransom note. By chance, police found that he had used another, portable, machine to type notes from a law study group to which he belonged. Comparison of pages of notes with the ransom demand showed them to be identical. Leopold and Loeb were tried and convicted. There were many reasons why this became a headline case: The families were affluent and well-connected; the crime was presented as a cold-blooded intellectual exercise pursued by two privileged youths; and the defense was represented by the legendary Clarence Darrow. Publicity dogged the case even after sentencing—Richard was murdered in prison and the accused assassin (against whom there was considerable evidence) was found not guilty.

The Bobby Franks case is significant to forensic science as it established the acceptability of typewriter examination as evidence. The story illustrates a different point also, namely that much of forensic science evidence depends on leads produced by careful investigation by police officers (and, as was the case here, investigative journalists).

Sometimes the two merge with the telling evidence reported by the laboratory being more a matter of everyday commonsense than specialized knowledge. Such was the situation in the 1928 contested will of James Biddle Duke (the same family whose endowments established Duke University) where no less than 107 people were fighting for shares in the multimillion-dollar estate. One family submitted a copy of what they claimed to be the family Bible with birth dates of the children inscribed in it in 1887 and 1889. The entries purported to have been made at the time of each birth, but examination of the printed book showed it to be copyrighted in 1890, and so the claim was invalidated.

The Franks and Duke cases are typical of those where the authenticity or writer of a document is a central issue—ransoms or wills. Document examination was one of three items of evidence central to a front-page case just four years after the Duke case, when Charles A. Lindbergh Jr. was kidnapped from the family home in New Jersey. A handwritten ransom note was found at the scene. The case soon became one of murder with the discovery of the child's body. Eventually suspicion fell on Richard Bruno Hauptmann, an illegal German immigrant. The handwriting on the ransom note was compared to samples from Hauptman by several document examiners, including Albert Osborn. They all concluded that the note had been written by Hauptman. The ransom note was one of three planks of the prosecution case; the others were tracing of special bills used to pay the ransom and the physical nature of a ladder found at the scene, part of which could be traced to timber at Hauptman's home. Like many cases involving celebrities, the trial and conviction of Hauptmann did not end the public interest. There were many subsequent essays written and investigations conducted, but the handwriting evidence was not seriously challenged in any of them.

Possibly the most valuable of the countless challenged wills that document examiners have testified on would be that of Howard Hughes. Innovator, billionaire, and ultimately recluse, Howard Hughes died, apparently without leaving a will, in 1976. Before long there were lots of documents purporting to be the true will of Howard Hughes, the most notorious being the "Mormon" will. The document purported to be a holographic will, that is, handwritten by Hughes. It was found on a desk at the headquarters of the Mormon Church, in Salt Lake City. This was not in itself a surprise, as many of Hughes's aides toward the end of his life were Mormons. What was a surprise was that the will left one-sixteenth of the estate to a Utah gas station operator, Melvin Dummar. Because the purported will was three pages long, there was an ample volume of material for document examiners to study. Their unanimous conclusion was that it was a forgery, a view upheld by the Las Vegas court. Yet again, other material buttressed the questioned document examiner's investigation and its conclusions. For example, investigators were able to develop a fingerprint on the envelope in which the will was found, and it was a match to Dummar.

The results of the investigation of Dummar leads us to another document case involving Howard Hughes. Mr. Dummar was a college student at the time. The college library contained a copy of the book *Hoax: The Inside Story of the Howard Hughes-Clifford Irving Affair* (Stephen Fay, Lewis Chester, and Magnus Linklater, Viking Press, NY, 1972). Mr. Dummar's fingerprints were found on the book. *Hoax* tells the story of Clifford Irving and how he almost succeeded in selling a fake biography of Hughes. As well as telling the story of the hoax, the book contains illustrations of the handwriting of Howard Hughes. It is clear how it suggests the idea of forging a will and provides writing examples to copy.

The hoax itself was well assembled. Irving claimed that Hughes, who at that time (1971) was a recluse who did not make public appearances, had engaged him to write the biography. He forged letters from Hughes to support the claim, and obtained a contract with the McGraw-Hill publishing company. The contract required Hughes's signature, which Irving duly forged, too. Irving and his friend and partner Dick Suskind claimed to be collecting taped interviews with Hughes, but in reality were inventing the biography from whatever material they could find. This included the manuscript of a biography of Hughes held by an old friend of Irving's, Stanley Meyer. The material in the manuscript allowed the conspirators to present enough material to McGraw-Hill for the publisher to announce the forthcoming book in December 1971.

Close associates of Howard Hughes responded quickly that he had not authorized the biography and had not provided any material for it. However, the journalist Frank McCulloch, a media authority on Hughes, vouched strongly for the validity of the manuscript. It was at this point that the document examiners became involved, with the Osborne name as prominent as ever. Paul and his brother Russell Osborne examined authentic writing of Howard Hughes together with the questioned writing. They reported that the fluidity and speed of writing throughout the considerable volume presented to them led them to be certain that the known and questioned material was written by the same hand. Things were looking good for the schemers. But Mr. Hughes was not amused, and on January 7, 1972, he broke fourteen years of silence to speak to journalists, denouncing Irving's claims. By the end of the month, Irving and his coconspirators confessed.

If Hughes had been dead or had chosen not to speak out, the hoax would have succeeded. Document examination was certainly at the core of this case and did not come out of it at all well. Searching for an explanation takes us just over ten years later, to the somewhat similar case of the Hitler Diaries. In April 1983 the German magazine *Der Stern* paid almost 10 million marks for a sixty-two-volume collection that purported to be the diaries of Adolph Hitler. Linguistic and historical experts vouched that the content was authentic, and document examiners vouched that the diaries matched an example of Hitler's handwriting that they were given. The historians were wrong. More extensive examination showed the material to contain errors, and indeed the errors pointed to the source of the fake content. The handwriting experts were right—the various samples had indeed been written by the same person, just not Adolph Hitler. In fact, the diaries were forgeries, and the so-called authentic sample of Hitler's writing was itself also a forgery. Further testing showed that the materials were anachronistic. The paper contained additives that were not introduced until 1954, and the ink was less than a year old and of a type not available at the supposed time the diaries were written.

The explanation is simple. Examination of writing materials can identify chemicals and processes that provide objective evidence that inks or papers do or do not share characteristic properties. Handwriting comparison is much more subjective, and detection of forgery often relies on the presence of features that result from the process of forgery. For example, a forger usually draws a signature and will pause to compare the drawing with the original. This gives rise to hesitancies and pen lifts in the forgery that are telltale signs of a forged signature. Freehand forgery possesses none of these and requires a sufficient volume of writing possessing sufficient characteristics to allow a determination of origin to be made.

The Hitler diaries were free-form, and during the first round of examination the examiners were fooled by the fluidity of the writing and the apparently authentic script for the time period. Irving's hoax almost succeeded also because of the fluidity of the forged writings, and because he had made a credible facsimile of the real thing.

This brings us full circle and back to Osborne's 1910 treatise. Authentication of handwriting is indeed possible. Brian Found's proficiency testing research in Australia shows that you do not even have to have many years of experience to get it right—but conversely, even with many years of experience you can get it wrong. All of the advances in the field have been in the technologies applied to writing materials, although reliable computer analysis of handwriting is close to being a realistic proposition at the beginning of the twenty-first century.

"We have drunken of things Lethean."
—Swinburne, Hymn to Proserpine, 1866

Greed may be the motive behind much of the work of the document examiner, but Douglas M. Lucas opined that the forensic scientist would have comparatively little to do were it not for alcohol and sex (American Society of Crime Laboratory Directors, October 2003, see http://www.ascld.org/pdf/lucas.pdf). The two often come together in the investigation of rape, when the victim and/or the perpetrator may have been under the influence of alcohol. Alcohol is covered in the headwords, as is rape, but the United States has a very special place in the history of a relatively modern offense involving forensic science and alcohol—that of drunk driving.

There were enough automobiles on the roads of America by the early 1930s to raise concerns about drunk driving. The country was approaching the end of the Prohibition era, and the Bureau of Prohibition in the Department of Justice took on the task of investigating the question of alcohol impairment of driving ability. At that time, the leading authority on the relationship between drinking, blood alcohol levels, and their effects was the Swedish scientist Dr. Erik Widmark (1889–1945). Widmark's name is still associated with the topic today through the Widmark factor (a mathematical factor used to correlate blood alcohol and amount of alcohol consumed). He was the first scientist to develop a reliable test sufficiently sensitive to permit the measurement of the amount of alcohol in blood using a sample as small as a drop.

Widmark's papers describe his contacts in 1931 and 1932 with E. P. Sanford, the head of the research division of the Department of Justice's Bureau of Prohibition (see the article by R. Andreasson and A. Jones in the *Journal of Analytical Toxicology*, Vol. 20, pp. 207–208, 1996). Sanford had first contacted Dr. Walter Miles of the Institute of Human Relations at Yale University. Miles was a prolific researcher and writer, with an interest in aging and color vision, as well as alcoholism. His works include reports for various military committees and a collection of sermons. Miles in turn recommended that Sanford write to Widmark.

According to Andreasson and Jones, Sanford wrote to Widmark on July 22, 1931, requesting assistance with the problem of testing for drunkenness: "Such tests in our various states vary from smelling the offender's breath to making him walk a chalk line, but no scientific test apparently is applied." The letter from the Department of Justice continued: "We are particularly anxious to know what the alcoholic content of the blood must be before a person can be described as being under the influence of alcohol."

That question could not be answered in 1931, because the required data did not exist. Sanford had further communications with Widmark on his new method. Andreasson and Jones record that Widmark sent him a manuscript to submit to the *Journal of the American Medical Association*, and that it was rejected. The Widmark method was adopted by many European countries, all of which went on to establish legal limits for blood alcohol and driving. The principle that the levels of a drug in blood reflect the level of its effect is well established. Widmark's work gave researchers and public safety officers a tool to answer Sanford's question, but research and development in the United States was to become focused on the indirect and less precise area of breath testing.

The first breath-testing instrument to be used by police was the Drunkometer, invented by Dr. Rolla N. Harger and introduced to New Year's revelers by the Indiana police on December 31, 1938. Yet again, a development in forensic science was associated with more than one person. At the same time that Harger was constructing the Drunkometer, Dr. Glenn C. Forrester was busy in St Louis, Missouri, developing his Intoximeter. Forrester incorporated a company under that name in 1945, and Michigan began using the Intoximeter in 1947.

Harger was a professor at Indiana University with an active research interest in the biochemistry of alcohol, and his Drunkometer included tables to convert breath readings to the equivalent in blood. The relationship between Harger, Indiana University, and the Indiana State Police flourished, and by 1948 the university had an active teaching program on breath-alcohol testing, sponsored by the National Safety Council. Faculty included Lt. Robert F. Borkenstein (1912–2002) of the Indiana State Police.

Borkenstein had no formal qualifications, but avidly embraced the science and technology associated with the measurement and public safety applications of breath-alcohol testing. So much so, that in 1954 he was responsible for the invention of the Breathalyzer. A logical development of the instruments of Harger and others, the Breathalyzer was the first truly portable and reliable breath-testing device. It was to become the "Hoover" of the field, widely adopted, and eventually the name became synonymous with any breath-testing device.

Borkenstein went on to become head of the Indiana State Police crime laboratory. He received an AB degree from Indiana University in 1958 and joined the faculty of the university's Department of Forensic Studies in the same year.

Harger, Borkenstein, and Indiana University remained at the center of developments in breath testing and alcohol-induced driving impairment for the next twenty years. Until very recently, the nationally accepted limit for drunk driving was 0.10 gram alcohol per 100 milliliters of blood (or the breath equivalent). The 0.10 limit originated at a 1958 conference at Indiana University, led by Harger and others from the department. The conference expressed the opinion that 0.05 percent was a reasonable threshold level and that most people would show impairment at 0.10 percent.

Practical confirmation of the opinion, and a reliable answer to Sanford's question, had to wait until 1964 and the Grand Rapids study, directed by Borkenstein. The team measured blood alcohol levels in drivers involved in road accidents and in a control group that matched in all regards other than the accident. They produced quantitative data relating likelihood of causing an accident and blood alcohol level. The odds do indeed begin to increase at 0.05 percent and are around five times higher at 0.1 percent.

Andreasson and Jones imply in their interesting historical note that the failure of the *Journal of the American Medical Association* to publish Widmark's paper resulted in the United States not adopting direct measurement of alcohol in blood and instead relying on the somewhat less accurate and more variable breath testing. However, the breathalyzer gave law enforcement a powerful tool that was economical and effective. The European countries that were leaders in blood testing now all use breath analysis as the primary enforcement tool.

Modern advances have been directed to making breath testing more reliable. Some jurisdictions have legislated direct breath levels of alcohol, which takes away any argument centered on the reliability of the conversion from breath to blood. Modern instruments are much more specific and are fast enough to permit duplicate testing. For example, the testing is not affected by ketones in the breath

Edmond Locard first suggested 12 matching points as a positive fingerprint identification. He also published L'enquete criminelle et les methodes scientifique, *in which appears a passage that may have given rise to the forensic precept that "every contact leaves a trace." (Courtesy of the Universite de Lausanne Institut de police scientifique et de criminologie)*

of diabetics, the instruments recognize the presence of alcohol from mouth washes or regurgitation of stomach alcohol, and they will not return a reading unless a reliable breath sample has been given.

"Our little systems have their day"
—Tennyson, In Memoriam, 1850

The half century between 1920 and 1970 can be regarded as the golden years of forensic science. Bertillon, Locard, Vollmer, Goddard, Kirk, Borkenstein, and others were creating a new branch of science and establishing its research, teaching, and service programs. They shared something else—they were forging partnerships between law enforcement investigators and university scientists.

It should be possible to trace a lineage from the pioneers to today's leaders, but it is not, despite the American Academy of Forensic Sciences being able to list eighty forensic science programs in U.S. universities and colleges. There are no significant partnership services. There are only three significant journals publishing refereed papers in forensic science, and only one of these is American. And every significant advance for the last fifty years prior to

2005 has come from someplace other than a forensic science laboratory or university program. The analysis of DNA came from medical research, as did advances in toxicology; automated firearms and latent print analysis came from private industry; the techniques used in trace and drug chemistry came from the world of pure and applied chemistry; and there is nothing new in questioned document examination. We are back where this history began, with developments in forensic science depending on discoveries in other fields of science.

Today's story of forensic science is the story of its case successes and failures, the story of its institutionalization, and the story of human frailty.

No case illustrates the success of modern forensic science better than that of Debbie Smith. Debbie has had the courage to tell her story and moved all those who have heard it. On May 3, 1989, she was alone in her house in Williamsburg, Virginia, attending to routine household chores. She had to go outside to check the laundry drier vent and when she came back inside, left the door unlocked as she knew she would be going right back outside with trash. In the fleeting moment available, a masked man entered the house, forced her outside into a wooded area, and repeatedly raped her at knife-point. Debbie would always remember his dreadful parting words: "Remember, I know where you live and I will come back if you tell anyone." Debbie Smith is married to a police officer, and her husband Rob was asleep in the house after a spell on nights. Debbie ran home and told him what had happened. Rob insisted that she report the rape and go to a hospital for examination. She did, and found the experience almost as violating as the crime itself.

Debbie's life was ruined. She could not escape the horror and relived the rape every day. She had no will to keep going and only the thought of the impact on her husband and two children kept her from suicide. The happy family home became a place of fear and guilt. Rob was devastated that he, a police officer, had not been able to protect his wife,

her daughter was afraid to go out, her son was beset with guilt that the rapist had used his baseball bat to intimidate his mother. Only the apprehension of the rapist could bring peace; and as long as he was free, his threat hung over Debbie. Months became years but nothing changed—time was not a healer. Then, on July 26, 1995, Rob came home with the news that a scientist in the Virginia State Laboratory had obtained a DNA match between Debbie's assailant and an entry in the state database. The rapist had been jailed just a few months after the attack, and Debbie and her family had endured six needless years of fear.

Debbie's story illustrates several points about forensic science. First, it is a reminder that forensic science is about people—principally victims and their families. It shows that while society and the justice system rightly have an expectation that scientific evidence is fault free, there is more to the "right" answer than one that is accurate. Timeliness matters too; Debbie and her family suffered for six years before obtaining closure. Finally, there are lessons to be learned from the factors that made resolution possible. Yet again, the science behind the case did not originate in the forensic field, but the pure knowledge would have been of no value without the organized response of the forensic science community in the United States to develop standardized systems that in turn permitted the implementation of databases (see **Combined DNA Index System (CODIS); DNA Databases; DNA in Forensic Science**).

That organization had to be created from the ground up. In a way, the story of the organization of forensic science paused after the time of Vollmer, Goddard, and Reiss in the United States and Europe. The value of the forensic laboratory was clear in both regions by the mid-1930s, and law enforcement agencies were enthusiastic in embracing the new tool. They did so by creating their own service, either as a dedicated laboratory within the law enforcement agency or by contracting with others, such as university departments. We saw an example of both when discussing Borkenstein, the development of the Breathalyzer, and the role of the Indiana State Police. In general, however, agencies in the United States went down the "own service" path, while European countries either developed national laboratories or services associated with universities.

The European approach is best illustrated by the story of the Scotland Yard crime laboratory and the Home Office Forensic Science Service. Scotland Yard, or more correctly the London Metropolitan Police, investigated how it could introduce training in forensic science as long ago as 1929. At that time, police forces in England used contractors to perform their scientific tests. The interest of Arthur Dixon (1881–1969), assistant secretary in the Home Office, resulted in the establishment of the Metropolitan Police Laboratory, within the Police College at Hendon, in 1935. Regional laboratories funded by the Home Office followed. The Metropolitan Police laboratory moved from Hendon to police headquarters in 1965 and again in 1974 when the host agency relocated to the Lambeth district in London. The Home Office laboratories came under a common administration following the Police Act of 1964 and the Home Office Central Research Establishment (CRE) was created as part of the service in 1967. The CRE was located at the United Kingdom's Atomic Weapons Research Establishment in Aldermaston, as part of the transition of the facility to peaceful purposes. It became a world leader in forensic science research and information dissemination. In 1991 the Home Office re-engineered the Forensic Science Service as an executive agency that was required to operate under strict economic guidelines, including recovery of costs through charging users for services. The CRE was an early casualty and was closed down. In 1996 events turned full circle and the Metropolitan Police Laboratory was integrated into the Forensic Science Service (FSS) agency. Although many lamented the loss of independence of the "metlab" as it was affectionately known, there is no doubt that the amalgamation produced a critical mass that enhanced service delivery in later years.

In particular, the FSS was extremely active and effective in implementing DNA testing. It launched the world's first national DNA database in 1995. The subsequent success of the FSS DNA database has far exceeded expectations, and it is only now that the U.S. CODIS database is approaching its effectiveness. (Note that the DNA match in the Debbie Smith case was in the local state database and there was not even a nationally agreed set of standard tests in the United States then.) The annual report issued in 2004 (available online at http://www.forensic.gov.uk/forensic_t/inside/about/docs/NDNAD_AR_3_4.pdf) shows that there is a better than 40 percent chance of an immediate match between a profile from a scene and information in the database. The success is not cheap—the annual cost of labor and testing is £182 million (over $300 million).

The ability to fund the database and the ability to convert research and development into a uniform set of scientific and operational conditions were critical success factors in implementing the U.K. database. There is no doubt that the speedy and effective implementation of the database is a testament to the value of having a single national forensic science service. (Note that strictly speaking the FSS covers only the countries of England and Wales; the countries of Scotland and Ireland have their own legal systems and their own forensic science laboratories. However, there is a high degree of coordination between them and the much larger FSS.)

"All men are liable to error: and most men are, in many points, by passion or interest, under temptation to it."
—John Locke, "Essay on the Human Understanding," 1689

Size and coordination do not guarantee success in every sphere, and the steady advance of forensic science in the United Kingdom was rocked in the 1980s by the discovery of flawed testing by one of the most senior analysts in the FSS, Dr. Alan Clift. General concerns about the quality of Clift's work had led to a detailed review of his files. One of the most important was the case of John Preece who had been convicted of rape and murder some ten years earlier, largely on the basis of the evidence produced by Clift. The case itself began with the discovery of the body of a young woman by the roadside in northern England. The clothing of the victim, along with samples collected postmortem, were sent to the FSS laboratory in Preston where they were examined by Clift. He confirmed that the victim had been raped and reported that the assailant was an A secretor (see **ABO Blood Groups**). Police investigations centered on John Preece, a truck driver, and samples of body fluids were taken from him for typing. He was an A secretor. Clift had also collected fibers from the clothing of the deceased and matched these to sources in Preece's truck. Clift prepared a report for the Lancashire police and Preece was charged with the rape and murder. However, further police inquiries showed that the crime had been committed in Scotland and the body dumped in England. Eventually, the trial was held in Edinburgh, Scotland, in 1973. Clift's report was submitted to the prosecution in Scotland, and pursuant to Scottish law, copies were made available to the defense. There was little questioning of the scientific evidence at trial and Preece was convicted.

The internal FSS review identified several areas of concern in Clift's work and testimony. The review findings resulted in the Scottish courts instigating an appeal of the 1973 verdict. Most of the appeal hearing centered on the blood group tests, on Clift's interpretation of the results, and on the responsibilities carried by the expert. Specifically, the original testing had shown that the deceased was also a group A secretor. Clift had not revealed this in his evidence nor had the defense explored the possibility. At the appeal hearing, Preece's attorney put it to Clift that the test results in the vaginal swab could have come from the body fluids of the deceased. This was precisely the view held by the FSS experts who had reviewed Clift's work and by international experts called on behalf of Preece at the appeal.

Clift's response was that he had not raised the matter in his testimony at the trial that the blood group results could have been from the deceased because no one had asked. He further claimed that in his experience the magnitude of the reaction he had seen in the test was only found with the blood group chemicals in semen and would have been much less if it had been due to the material in the vaginal secretions. The defense produced a series of witnesses, including colleagues who had worked alongside Clift in 1973, all of whom said that his theory was totally wrong, and that, if correct, it was of such significance that it should have been published for the benefit of the forensic serology community. The three appeals court judges not only agreed with that view, they found that Clift had erred to such an extent that they considered him discredited as a witness and reversed the conviction of Preece without examining any of the other evidence. They also made an explicit comment on the "no one asked me" excuse. The judges were of the view that the body of knowledge of fact and interpretation lies with the expert witness, that Clift—not the defense—therefore had the responsibility to inform the court of the possible limitations on his interpretation of the blood-grouping evidence.

The case illustrates clearly the dangers that come from reliance on a single expert, no matter what his or her history is. There is no doubt that the cult of the expert grew as forensic science grew. Mention was made earlier that Orfila was possibly the first example of a professional forensic scientist whose opinion was sought because of his reputation. Fortunately, Orfila's reputation as a scientist was justified, but that was not so with many others. Alan Clift was highly regarded by the investigators for whom he worked. He was also active in professional circles and well regarded by his peers. In essence the weakness in his work was that he became a one-man show, proud of his successes and ready to back his personal judgment without being checked by peers.

This is a pattern that, sadly, has been found in other places, including the United States and Australia. One of the most notorious forensic science debacles of modern times is the so-called dingo baby case from Australia. Michael and Lindy Chamberlain were camping with their ten-week-old daughter, Azaria, at Ayers Rock in the central Australian outback in August 1980. Early in the night of the seventeenth, Lindy cried out that a dingo—a wild dog—had seized Azaria and run off with her. A subsequent coronal inquest agreed.

If true, this would have been the first recorded instance of a dingo taking a human. Police doubted the mother's story and their skepticism was partly responsible for a second inquest into the disappearance of Azaria. There was no eyewitness evidence of any note, but a substantial amount of scientific evidence was being amassed. The body of Azaria was never found, but there were many physical objects subjected to testing to re-create the possible events. These included the tent, the bedding, Azaria's clothing, and the Chamberlains' car. Damage to the blanket from Azaria's bed and on her clothing was reported as more likely to have been caused by scissors than dog teeth. Scissors found in the Chamberlains' car were reported to be stained with fetal blood. (Fetal blood contains a somewhat different form of hemoglobin from that found in adult blood, and the fetal hemoglobin persists for some months after birth.) Investigations conducted at the time of the child's disappearance found blood that could have come from her in the tent. Examination of the family car a year later resulted in a report that there was fetal blood found in several places in the Chamberlains' car, including the wheel arch. The original inquest held in December 1980 found that Azaria had been taken by a dingo. Northern

Territory police continued their investigations, including the work on the car, and in November 1981 the supreme court of the Northern Territory squashed the findings and ordered a second inquest. In February 1982 the coroner reported that a prima facie case had been made out for the involvement of Lindy in the disappearance of Azaria, and she was eventually charged, tried in September/October 1982, and convicted. Appeals were heard in February 1983 and February 1984, and rejected on each occasion. In February 1986, a jacket that was said to have been worn by Azaria at the time of her disappearance but not found during the investigation was discovered by a tourist at Ayers Rock. This was enough to reopen the case, and a judicial enquiry (Royal Commission) opened in May 1986.

The case commanded public attention right from the start. Ayers Rock is a mystical site with sacred connotations for the Aboriginal populations of Australia. The Chamberlains were Seventh-Day Adventists, which added to the mystical overtones. Truth and fantasy became intertwined, such as the story that *Azaria* means "sacrifice in the wilderness," when in fact it means "spared by Jehovah" or "helped by God." As well as the public fascination with the case, there was a continuing undercurrent of concern in the scientific community about the quality of much of the laboratory work that was presented to the coroners' inquests and at trial. The public and scientific pressures were considerable and, together with the discovery of the jacket referred to above, led to a royal commission being set up under Justice Trevor Morling to review the circumstances that led to the convictions of Lindy and Michael Chamberlain. By then, just about everyone in forensic science in Australia had had some involvement in the case. The New South Wales Department of Health Forensic Biology Section, the Forensic Biology section of the Institute of Medical and Veterinary Sciences in South Australia, and the Northern Territory Forensic Science Laboratory had worked for the police. Academics at the University of New South Wales and the Australian National University in Canberra had been involved in the pressure for retrial. Morling secured the services of the Victoria Police Forensic Science Laboratory, the only major facility with no prior involvement, to review earlier reports and records and to conduct new examinations of the physical evidence.

The review was thorough and compelling. Practically all of the expert evidence presented previously and that went to paint the Chamberlains as guilty was flawed. Morling concluded that the strength of the evidence presented at trial had been so damaged that it was unsafe to let the conviction stand.

If we strip away the air of mystique that surrounded the Chamberlain case from the start, it becomes an excellent vehicle to show how forensic science can go horribly wrong. The most simple description of events, from a police point of view, is as follows:

A baby disappears from a family on a camping vacation. The disappearance clearly involves violence, as there is considerable blood spread around the inside of the tent. Tests that were never disputed show the blood could have been that of the missing baby. The most likely possible explanations are that a dingo seized the baby, as claimed by the mother, or that the mother killed the baby and, aided by her husband, disposed of the body. The police are skeptical about the dingo option as this has never happened before. On the other hand, mothers have killed their babies, and there is some anecdotal evidence that Lindy was finding baby Azaria

quite stressful, with many sleepless nights. The investigation odds are therefore on the version of events involving human intervention rather than canine. The inquiry therefore centers on using the silent witness of forensic science to test that hypothesis.

Some of the obvious options to be studied were:

The key evidence differentiating the two options is that relating to the car and to the clothing. The coroner and the trial jury heard from forensic biologist Joy Kuhl, from the New South Wales Department of Health, that she had found human fetal blood on a pair of scissors in the possession of the Chamberlains; on hinges, brackets, and bolt holes around the front seats and the carpets of the car; and in the form of a spray such as would be formed from arterial blood spurting from a wound, under the dash. Dr. Ken Brown, a dentist at the University of Adelaide who had an interest in forensic science, testified that the damage to the clothing could not have been caused by dog bites. This was supported by a British university pathologist, Dr. James Cameron, who went further and showed the jury marks on the clothing that he said were blood imprints from a small—human—hand. There is hardly any wonder that the trial verdict was guilty.

However, the work conducted at the Victoria Forensic Science Laboratory during the royal commission painted an entirely different picture. The records of the tests conducted at the time of the investigation did not support the strong conclusions drawn. Retesting did not confirm that any of the material in the car was blood, and certainly

Table 1 Evidence

Evidence	Lindy murdered Azaria	Dingo carried her away
Car	Baby blood stains, as the car would almost certainly have been used to take the body away for disposal. Requires confirmation of blood and of presence of fetal hemoglobin.	No blood
Bedding and cloting	Damage faked to look like tooth marks.	Damage caused by dog incisor and canine teeth
Inside tent	Extensive blood	Extensive blood

not fetal. The spray pattern was caused by paint treatment during the manufacture of the car. The positive screening test was most likely due to chemicals in the soil and dust that permeated the vehicle. Sufficient doubt existed about the conclusions of Brown and Cameron that the damage could not have been caused by a dog to result in that evidence being discounted. Most damaging of all, the alleged bloody handprint—which had always been less than compelling as a mark made by a hand—turned out not to be bloody either.

What went wrong? Joy Kuhl was a very experienced biologist. Ken Brown was a respected teacher and researcher. James Cameron's resume was so lengthy that the coroner's hearing set aside the normal procedure of reading it out during qualification of the witness. Looking back, we can see two fatal flaws. Firstly, the "dingo did it" hypothesis was always going to present a challenge. The reason the police and prosecution authorities were so skeptical about it was because it had never been known to have happened before. But because it had never happened before, there was no recorded history of what to expect, and the scientists had to break new ground all the way as they tried to figure out what might have gone on and subject their ideas to testing.

Second, Joy Kuhl was not in any way breaking new ground. However, she was working in tantalizing circumstances. Many of her tests gave results that today would be regarded as inconclusive and therefore not reportable. Unfortunately she chose not to take that conservative approach and proceeded to report findings from results that did not justify the strength of her reported conclusions. If you are brought up in a working environment that places a premium of squeezing every possible piece of information from the material submitted, and not one that is driven by quality assurance procedures, then the Chamberlain case was not a happy place to be—not for Lindy and not for Joy Kuhl either.

The final comment on Chamberlain is on the evidence of Cameron. It can be argued that he really should have known better. The greater concern is that right from the start there was skepticism about his "bloody handprint." Even when he tried to trace the outline for the jury, they had problems seeing it. However, they had no problems being impressed by his impressive resume and were ready to accept that the doctor knew best.

The case is a wonderful illustration of a quite obscure philosophical question about forensic science: Is it an inductive process or a deductive one? If it is inductive, the scientist conducts tests, makes observations, and draws conclusions. If it is deductive, the scientist knows what to expect in the specified circumstances and conducts experiments to see if the predicted events or properties can be demonstrated or discounted. The essential difference between the two approaches is that the deductive approach is buttressed by a solid framework that governs the expectations. It may be fundamental chemical properties of a substance as with identification of drugs or a database of thousands of observations as with firearms or fingerprints, but the framework is sufficiently reliable to make the selection and application of tests and the conclusions drawn reliable. No such framework exists in the inductive approach. Indeed, the results are what go to eventually permit a reversal of principle and the work to become deductive.

Many of today's critics of standards and practices in forensic science (e.g., Saks 2002) argue that knowledge of events will lead to bias in the scientist, albeit unconscious bias. They would like to see all tests conducted without the scientist having any information about the circumstances of the case. In other words, the work should be done from an inductive perspective. The Chamberlain case provides strong support for that view.

"[T]hey are good servants but bad masters."
—L'Estrange, *Aesop's Fables*, no. 38

Some critics take the "unconscious bias" argument to the limit and claim that anyone working within a police department and whose regular activity is conducting tests requested by the investigators and prosecutors must become tainted. The story of Fred Zain fits that mold. It is a story that shares something with *Chamberlain,* namely reports that promoted the police case but that independent review showed were flawed. It is very different from *Chamberlain* in that the flawed evidence all came from just one person but covered many cases.

The essentials of what was wrong with Fred Zain's work are contained in an opinion of the Supreme Court of Appeals of West Virginia, filed in November 1993 (No. 21973 "In the matter of an investigation of the West Virginia State Police Crime Laboratory, Serology Division"). The court accepted a report of an investigation into Zain's work conducted by a panel of peers. Their report found many instances that they described as acts of misconduct by Zain, including: (1) overstating the strength of results; (2) overstating the frequency of genetic matches on individual pieces of evidence; (3) misreporting the frequency of genetic matches on multiple pieces of evidence; (4) reporting that multiple items had been tested, when only a single item had been tested; (5) reporting inconclusive results as conclusive; (6) repeatedly altering laboratory records; (7) grouping results to create the erroneous impression that genetic markers had been obtained from all samples tested; (8) failing to report conflicting results; (9) failing to conduct or to report conducting additional testing to resolve conflicting results; (10) implying a match with a suspect when testing supported only a match with the victim; and (11) reporting scientifically impossible or improbable results.

The opinion raises concerns, not only because it identifies wrongdoing by the individual scientist that had major consequences for many people—the wrongly convicted and the victims who now have to face the fact that their assailants are still at liberty—but because of the question, "How could the system have allowed this to happen?" Judge Holliday addressed this to a degree, noting that many

of Zain's former supervisors and subordinates regarded him as "proprosecution." The report further states: "It appears that Zain was quite skillful in using his experience and position of authority to deflect criticism of his work by subordinates." Holliday commented that Zain's supervisors may have ignored or concealed complaints of his misconduct, and that the laboratory operating procedures undoubtedly contributed to an environment within which Zain's misconduct escaped detection.

The report was the end of the line for Fred Zain. The investigation into his work arose because of a separate West Virginia review of one of his cases. Glen Woodall had been convicted of two instances of sexual assault in 1987. The evidence linking him to the victims came from Zain's report that the blood types found in the semen were identical to those of Woodall. A postconviction action on behalf of Woodall resulted in DNA testing that showed that the semen could not have been his, and he was released in 1992.

Reinforcement, if any were needed, came from a second case covering the same time period. Jack Davis had been found guilty of the murder of a Texas woman, Kathie Balonis. Zain testified that blood under the body of the victim was that of Davis. There were no witnesses, and the prosecution case centered on the blood grouping. Davis was convicted and escaped receiving the death penalty by just one juror vote. The case was reopened in 1992 to hear a claim of prosecutorial misconduct, and it then came out that Zain had mixed up the samples and the blood was not that of the accused. The conviction was overturned.

At the time of the West Virginia Supreme Court of Appeals hearing, Fred Zain was the chief of the Serology Section of the West Virginia State Police Crime Laboratory. He had risen to that position on the basis of thirteen years of highly regarded work. He was regarded as an excellent witness and as the serologist who could be depended on to find the evidence. He was much sought-after and in 1989 he moved to Bexar County, Texas, as chief of physical evidence in the medical examiner's laboratory. Zain was dismissed in 1993, because of a question about his role in a case in which evidence had disappeared. A subsequent internal investigation of at least 180 cases in which Zain had been involved found reports from tests that were never done, negative results that would have cleared a suspect reported as positive, and inconclusive results described as conclusive.

The West Virginia investigation identified 134 specific cases for further review, giving a total of over 300 cases containing questionable reports from Zain's work. These numbers are staggering, and raise the question not just "How?" but "How many?"—namely "How many other Fred Zains have there been?" No one knows. The leaders in the forensic science community are as outraged as anyone (but Alan Clift was a leader in the community) and rightly argue that modern quality assurance systems have created an environment that is as effective as possible in protecting against human errors in testing and reporting. What is certain is that forensic science is in the crosshairs as it has never been in its history, and nothing else of the dimension of the Zain situation has been found. The closest is the work of Oklahoma City Police Department crime laboratory worker Joyce Gilchrist.

Gilchrist worked for twenty-one years in the laboratory, with generally good appraisals for her work. Not everyone agreed, and certainly Judge Ralph Thompson criticized her for untrue testimony and the blatant withholding of unquestionably exculpatory evidence in the 1999 hearing of the case of Alfred Mitchell. Other cases where her work was overturned included those of Jeffrey Pierce, convicted of rape with evidence from Gilchrist that placed him at the scene but released when DNA tests showed that the semen was not his; and Robert Miller, released from death row when DNA tests showed that hairs from the scene that Gilchrist had said were his were actually from someone else. Not so fortunate was Loyd LaFevers, executed for first-degree murder,

before DNA tests showed that blood on his pants identified by Gilchrist as being from the victim was from someone else.

Many of the questioned findings of Gilchrist related to her work on hair identification. Hair comparisons (see **Hair**) are traditionally conducted by microscopy and can involve a quite high degree of judgment on the part of the examiner. However, an FBI review of Gilchrist's work on eight of her hair cases concluded that in five she had either made errors or had overstepped the limits of what peer standards would find acceptable. She was fired in September 2001.

"Times change and we change
with them, too"
—John Owen, Epigrams 58, Vol. I, 1603

This history began by paying homage to the inventors of the Industrial Revolution whose work provided the basis for so many advances in science and technology. We then visited more specific highlights in the development of forensic science. By the time that we get to the current era, roughly 1985 onwards, we find that times have not so much changed as they are being revisited. Analogous to technological advances in many fields, forensic science has witnessed seemingly exponential growth in the integration of sophisticated equipment and techniques since the 1980s.

During the mid-1980s, trace evidence and associated techniques, along with fingerprints, composed the bulk of forensic investigations. Photography was achieved using film and chemical processing. Drug chemistry and toxicology depended upon tried-and-true methods, and biology was limited to conventional serology and protein comparisons. The microscope was still the tool du jour, and interagency cooperation was somewhat rare.

The integration of computers and the acceptance of techniques used in molecular biology brought forensic science and scientists hurtling into the information age, sometimes more rapidly than the court systems were able to accommodate. At present, a mix of both excitement and scrutiny surrounds many of the techniques common to forensic science. In the realm of digital evidence and computer forensics, a great deal of concern envelops the capabilities of the technology, or lack thereof. Conversely, with respect to natural and chemical sciences, such as forensic biology and DNA analysis, less controversy arises from the technology and science than from questions of its appropriate use and human error.

"It is a capital mistake to theorize before one has data."
—Arthur Conan Doyle, *The Adventures of Sherlock Holmes,* "Scandal in Bohemia" (1891)

Technological advances have impacted forensic science beyond the direct ability to identify and characterize evidence by chemical and physical examination. The computer age has provided a more effective tool to realize the potential of databases, particularly for identification of individuals from material left at the crime scene. We shall consider fingerprints and biometrics first.

At the time of this writing in 2005, the integrated automated fingerprint identification system (IAFIS) contains information for more than 47 million individuals in its criminal master file. Law enforcement agencies voluntarily submit fingerprints and criminal history information to this file. How did this national fingerprint and criminal history system, maintained by the Federal Bureau of Investigation Criminal Justice Information Services Division, come to be the world's largest biometric database?

Automated fingerprint identification systems (AFIS) are not new to the millennium. AFIS technologies have been around since the late 1970s, but with high costs and few vendors, their potential was largely untapped. The first automatic fingerprint recognition system was installed at Scotland Yard in 1984. Having first heard about AFIS in 1980, detective and former science teacher Ken Moses convinced the mayor of San Francisco and the San Francisco Police Department to put out the first request for proposals for an AFIS in 1984. After installation, the department literally solved thousands of crimes, and an AFIS became the "must have" item for larger jurisdictions all over the United States.

After the International Association for Identification (IAI) released a report in 1988 discussing the American National Standards Institute's (ANSI) recommendations for files, the early 1990s saw a great deal of work on how to compress and decompress electronic fingerprint images to allow for sharing among agencies. Clearly, any degradation of Galton details would render a print less useful. It was July 2000 before ANSI approved and issued the National Institute of Standards (NIST) Special Publication 500–245 addressing standards for a data format for interchange of fingerprint and similar files.

Since the first judicial ruling accepting fingerprint evidence in U.S. courts, *People v. Jennings,* all states have come to have their own AFIS databases. These and the AFIS databases of most larger cities may contain fingerprint records not stored anywhere else, so standardization of database interfaces is imperative to record sharing and searching.

As the result of approximately ten years of work by the FBI and other law enforcement agencies, FBI Director Louis Freeh inaugurated IAFIS for full operation on August 10, 1999. A dozen or so states and several federal agencies were prepared to submit electronic fingerprint images at that time. To accommodate states without electronic capabilities, the FBI contracted Lockheed Martin Information Services for card scanning. Even with this detour, agencies saw significantly shorter turnaround times for fingerprint comparisons than previously experienced. Other industry players in the building of IAFIS again included Lockheed Martin (developer of the AFIS), Science Applications International (developer of the interstate identification system), and Litton PRC (developer of identification tasking and networking). These three segments came together to form the IAFIS at a cost of approximately $640 million (U.S. Department of Justice). The system is maintained and operated by a division in Clarksburg, West Virginia.

The greatest technological leap in AFIS has been in reducing the amount of time required for a fingerprint check. For example, with IAFIS, agencies can expect an electronic response to criminal ten-print (that is, a fingerprint record card with impressions of all 10 fingers) submissions within two hours. Civil fingerprint submissions, such as those required by law for job applicants, receive a response within twenty-four hours.

Civil files, including enlisted military service member print cards since 1990, are computerized as well but are only searchable internally by the FBI. With an elevated threat of terrorism and the ever-changing role of federal agencies such as the Department of Homeland Security and the Department of Defense, the FBI may ultimately make civil file AFIS searches available to other entities. In 2004 the Department of Homeland Security launched its US-VISIT program requiring index finger scans to aid in verifying the identity of visitors with visas.

Identification services offered by the Royal Canadian Mounted Police (RCMP) date to approximately 1910. The first AFIS overhaul took place in 1987, but even with the cooperation of the Identification Services Committee, Canadian Association of Chiefs of Police, and the then newly formed Canadian AFIS Users Group, police forces throughout Canada largely were not utilizing the RCMP services as of 1990. In the early 2000s the Canadian AFIS, composed predominantly of workstations manufactured by Printrak, operated through the RCMP Ottawa site and three other central databases in Vancouver, Toronto, and Montreal. Born of this system, a regional automated fingerprint identification access system from Motorola, Inc., was accepted for implementation in 2002. Based on the Printrak system, this technology was part of Canada's National

Security Enhancement Initiative and offered agencies web-enabled capture, submission, and searching of latent prints and ten prints against the central RCMP AFIS database.

Australia's national automated fingerprint identification system (NAFIS) was born in 1986. When this system was taxed to capacity around 2001, a new NAFIS was developed under CrimTrac. This system complies with the ANSI/NIST standards, allowing Australian agencies to exchange print records with the FBI and Interpol.

The United Kingdom being slightly ahead of the curve, the Association of Chief Police Officers (APCO) mandated a requirement for automated fingerprints in 1987. Subsequently, the Police Information Technology Organisation, part of the Biometrics Working Group, rolled out a system nationally in 2001, providing service to all of England and Wales. In a reminder of the fragile nature of these database technologies, a computer virus brought down this system in 2004, leaving England and Wales unable to run national checks for more than a week (see **Digital Evidence**). In late 2004, the organization signed a hefty contract with Northrop Grumman Corporation to bring in IDENT1 with the retirement of the national AFIS. IDENT1 will integrate automated fingerprint recognition services used by Scotland, England, and Wales since 1991 to permit identification of suspects throughout the United Kingdom.

Although fingerprints have been considered an acceptable means of identification for centuries, today's legal climate continues to scrutinize fingerprint evidence. In 1993 the case of *Daubert v. Merrill Dow Pharmaceuticals* set the precedent for standards of admissibility that came to replace the *Frye* standard in federal courts and those states that abide by the federal rules of evidence or model their own rules thereafter. In *United States v. Byron Mitchell* (1999), FBI fingerprint examiners identified Mitchell as the contributor of latent fingerprints developed inside a stolen vehicle used as an armed robbery getaway car. This marked the first case in which the defense cited *Daubert* to challenge fingerprint evidence. A *Daubert* hearing was held, and fingerprint evidence was accepted by the court as a reliable science. A subsequent appeal in 2004 accusing the judge of error by admitting testimony on the fingerprint evidence again resulted in acceptance and affirmation of the judgment.

The Scientific Working Group on Friction Ridge Analysis, Study, and Technology (SWGFAST) was spawned from the Technical Working Group on Friction Ridge Analysis, Study, and Technology (TWGFAST), which first met in 1995. It was the *Mitchell* case that prompted the SWGFAST to characterize a proper comparison as being carried out via the ACE-V method, meaning analysis, comparison, evaluation, and verification. This moved the discipline from Galton details and "points" toward ridgeology, which even sounds more scientific. Championed by Roy Huber and David Ashbaugh, both of the RCMP, literature and training guidelines now cite ACE-V as fitting the steps of the scientific method.

Dissension amongst experts is not uncommon in the world of fingerprints with respect to training. For example, harsh statements were made by Allan Bayle in 2002 regarding the FBI's fingerprint examiner proficiency tests. Bayle worked as a fingerprint expert with the Metropolitan Police in London from 1975 to 1996 before leaving and setting up business as an independent consultant. Given Bayle's expertise and fame in presenting the supporting evidence that aided in the conviction of one of the Libyan suspects in the 1988 Pan Am Flight 103 bombing, these criticisms opened the door for defense attorneys to discredit fingerprint examiners, if not the methodology itself. Bayle's statements were part of his expert defense testimony in a landmark case wherein U.S. District Judge Louis H. Pollak barred experts from testifying that crime scene latent prints matched those of an individual defendant under the premise that fingerprint evidence had not undergone scientific testing, had no calculated error rate, and lacked standards to determine what constitutes

a "match." This indeed paved the way for other courts to reconsider the admission of latent fingerprint match testimony. Often referred to as *Plaza I,* in *United States v. Llera Plaza,* 181 F. Supp. 2d 414 (E.D. Pa. 2002), Judge Pollak ruled that the fingerprint expert witnesses could not indicate that a particular latent print could be unequivocally identified as having been made by a particular individual. However, approximately two months later in *Plaza II,* Pollak reversed himself. Witnesses did indeed testify in the second trial, and a conviction was secured. Both *Plaza* opinions are notable as a demonstration of the judge's alternative applications of *Daubert.* In contrast to *Plaza I,* the court had held that fingerprint evidence satisfies *Daubert* with respect to reliability in *United States v. Havvard,* 117 F. Supp. 2d 848 (D. Ind. 2000).

Another point of contingency demonstrated by these cases surrounds the issue of peer reviews. *Havvard* advised that a second fingerprint examiner could compare the prints as well, but this is not in keeping with the intended use in *Daubert,* in which peer review meant "refereed to scientific journals." Debate ensues still as to whether fingerprint examiners can be called a scientific community as per *Daubert.* Changes in underlying academics and training of new examiners may ultimately alleviate such discrepancies.

In *Regina v. Buckely,* 143 SJ LB 158 (1999), the British Court of Appeal indicated that at least eight similar ridge characteristics should be available for the prosecution to successfully seek to proffer such evidence. At last issuance in 2004, the SWGFAST Standards for Conclusions 1.2.1 held that "[t]here is no scientific basis for requiring that a predetermined number of corresponding friction ridge details be present in two impressions in order to effect individualization."

A modern example of the fallibility of fingerprint identification is observed in the commuter train bombing case in Madrid, wherein the ridgeology analysis conclusion of identification was flawed. The FBI built its case against an Oregon lawyer, Brandon

Mayfield, on the basis of fingerprint evidence, but ultimately apologized for the errant identification, blaming it partly on the image's lack of quality. Although Mayfield was released and an Algerian arrested, the debacle widely manifested as a red flag for latent print comparisons. Both scientific and popular literature alike have since been riddled with expositions questioning the practice of fingerprint matching. Whereas *Regina v. Buckely* called for eight characteristics, this is agency specific and may be eight or ten or twelve, etc. Because the application of statistical probability is not possible, one agency may call two prints a match, although a second agency may not.

"I have been looking for a person . . . all my life."
—Sydney Smith, *Memoir*, Chapter 9, 1855

Some consider fingerprint identification part of the broader field of biometrics, which also addresses automatic examination and identification based on geometric physical features such as the hand, face, iris, retina, and vasculature. Biometrics as a whole includes both physical and behavioral features that can be identified via pattern recognition, but physical characteristics such as a fingerprint are most applicable to forensics. With respect to forensics, biometrics is simply the newest form of identification.

Fingertip scanners are increasingly more common in security, verification, and criminal justice arenas. People are more comfortable submitting to fingertip scans than, for example, retinal scans. Although the scanners are less expensive and more portable than other technologies, scanners may differ with respect to their matching approaches, making some incompatible with AFIS. Many larger or technologically advanced agencies have switched from rolled ink prints to fingerprints captured via a scanning device. In fact, such scanners can often immediately detect discrepancies between a scanned print for a particular finger and the full hand slap of the hand supposedly bearing that finger. This type

of detection could prevent mix-ups in the collection of offender data at an intake facility or jail, for example.

Because personal identification numbers, passwords, cards, or badges have historically been forged, forgotten, or falsified, biometrics is being incorporated into the realm of security and forensics as both verification and identification tools. Although more prevalent in criminal identification and prison security at present, the integration of engineering and computer technology into identifying people for legal purposes is forwarding biometrics as a new genre of forensics. Because biometrics technologies are being integrated into the security infrastructure of businesses, health care facilities, and banks, criminals will of course follow suit in adopting techniques to compromise such systems or re-create tools to permit access to secure information or areas. Thus, a new role for the forensic examiner is born, requiring in-depth knowledge of biometrics and technology and skill in detecting breaches and authenticity.

Although hand geometry was often used to open doors on the popular science-fiction series *Star Trek,* the measurement of physical characteristics of the hand and fingers for identification and verification purposes is not entirely science fiction. Hand geometry is well-established technology but less widely deployed for use in computer security or forensics applications, in part due to the need for a larger scanning surface and camera. However, research and development continues in the area of hand geometry in hopes that it may play a role in crime solving when gloves are worn, as well as for security access systems.

Not entirely new in theory, a North Atlantic Treaty Organization (NATO) Advanced Study Institute on Face Recognition was held in 1997. The meeting brought together research and development experts from universities, industry, and governments. Not unlike other areas of forensics, facial recognition technology may be borrowed from areas such as computer science, neuroscience, and engineering and applied to matters of the law. For example, in late 2004 the United Kingdom began testing three-dimensional biometric facial recognition software. Two-dimensional image technology marking characteristics such as interpupillary distance is already in place, but cost will probably prove the prohibitive factor in launching the software widely.

The Massachusetts Institute of Technology has patented a technology that uses two-dimensional global grayscale images. Known as *Eigenface* ("one's own face" in a blending of German and English words), the images represent distinctive facial image characteristics. Approximately 100 to 125 Eigenfaces can be used to reconstruct enough features to constitute a face. Less sophisticated than Eigenfaces, Automated Face Processing uses distance ratios between basic features such as eyes, nose, and mouth. Neural Networks technology has also been employed to compare faces, referred to as enrollment and verification faces. An algorithm is applied to determine if a live face's features match that of a reference face. Last, the most commonly used recognition technology is Feature Analysis. Related to Eigenface, Local Feature Analysis can accommodate changes in expression and uses multiple features from regions throughout the face, recognizing their relative relationships.

The iris serves to control the amount of light that enters the eye through the pupil. An internal organ, the iris is also visible from the outside. Iris patterns demonstrate a high degree of randomness, and iris patterns are not known to change over life, although some pigment change may occur. Also, the detailed iris texture is not genetically determined.

Degrees of randomness can be determined and mathematically calculated based on the number of degrees of freedom or values that are free to vary in a template. It is commonly accepted that fingerprints have about 35 degrees of freedom. This is less for faces, but approximately 250 degrees of freedom can be identified in iris patterns. Thus, due to random variation among people, iris patterns can serve as a reliable biometric identifier at chosen measurements. Laboratories, such as the United Kingdom's National Physical Laboratory, conducting comparisons using this technology in IrisCode® software have yet to report a false match. Because iris patterns are often resolvable from distances as great as a meter with good video cameras, this technology is applicable to crime prevention and security, as well as digital evidence and forensics.

Retina scanning is more advanced, yet also more intrusive, than iris scanning. Unique patterns of the retina, the layer of blood vessels at the back of the eye, may be scanned with a low-intensity light source and optical coupler. Although highly accurate, the process does require that an individual place his eyes very close to the device, something a criminal is not likely to do during the commission of his crime. One of the oldest forms of biometric identification, dating back as early as the 1930s, retina scanning is probably only applicable in very high-end security situations.

In an ever-evolving attempt to thwart criminals—believe it or not—some entities such as banks are opting for vein recognition systems. The term *vein* is a misnomer, because the technology scans for all vasculature. Most often, the scan is taken of the palm, which is placed immediately above a scanner. Vasculature patterns in the hand are developed *in utero* and remain constant throughout life except for the increase in size associated with normal growth. The image may be taken using a charged coupled device camera under infrared lighting. Again, this technology is less than useful in identifying a subject from crime-scene evidence but may prove valuable in verification and security processes.

Voice recognition deserves mention with other biometric technologies, even though it is addressed separately in the section on digital evidence. Speech is natural and often a tool in the commission of a crime. The verification component of a voice recognition system may authenticate a voiceprint as a particular individual's, while a recognition component checks for phrases. Audio is often incorporated into security systems with recording devices and alarms. The technology is inexpensive and widely available, but the voice is highly difficult to measure with enough consistency to call two recordings a match. Environmental and physiological characteristics can thus make identification difficult using voice alone.

Advantages noted in the use of biometrics, specification and compliance, are inherent to its developers and proponents. Because these technologies are largely developed in high-tech industrial and research environments, conformity to certain standards is expected. Several self-regulating bodies have arisen in the area of biometrics, including the BioAPI Consortium, the International Biometrics Industry Association, and the Biometrics Consortium. With technology advancing daily in industrial and research entities and thus ever changing criminal instruments and opportunities, the criminal justice system and forensics must strive to keep up the pace.

"That's something I could not allow to happen."
—HAL in Arthur C. Clarke's *2001 A Space Odyssey*, 1968

Many of the biometric techniques listed above are employed in e-commerce and Internet applications in attempts to thwart cybercrime. For example, banks are readily embracing smart-card and print-scanning technology on site. Universities and industry can already select from various means of securing websites, including requiring a biometric check such as hand geometry on a peripheral device. Similar peripheral-based devices may soon be required for employees in secure environments to access files and folders within company Windows-based or terminal services systems. Voice-checking systems are even available for integration with modern telephony.

Although crime prevention and security are booming with technological advances, the application of forensic techniques to solving cybercrime is riddled with problems. For example, the nature of cybercrime leaves jurisdiction highly questioned. Also, cybercrime is inextricably linked to other types of crime, such as sex crimes, financial crimes, and other white-collar offenses, none of which are particularly new to the information age. In fact, the concept of the Internet is much older than one might think.

The communications concept of "packet switching" can be traced to the 1960s. The Department of Defense Advanced Research Projects Agency (ARPA) used the technology as the basis for its network ARPANET. Not to be left behind, the academic world followed quickly and UCLA, Stanford, UC-Santa Barbara, and the University of Utah joined ARPANET. These early links between government scientific entities and universities laid the groundwork for developments such as electronic mail, which dates to the early 1970s. As will be discussed in the section on **digital evidence**, e-mail has been both the avenue and the bane of many criminals. The United Kingdom followed suit in the 1970s with a network known as the Joint Academic Network (JANET), and the U.S. MILNET split the defense components from ARPANET in 1983. Due to similarities in construction, all of the mentioned networks and more of the time were able to talk to or interconnect with one another, hence, the Internet.

The 1990s saw huge increases in awareness and use of the Internet, with the number of Internet hosts increasing nearly sixfold between 1991 and 2000, according to the Internet Software Consortium (see http://www.isc.org/). With the advent of cybercrime, forensics has been forced to keep up by introducing a new type of examiner, versed in computer science, legal matters, and communications. Internet access is increasingly available to the public at large, including the criminal element, and even modern refrigerators come equipped with connections. It is society's reliance on information technology that makes cybercrime particularly heinous.

Fortunately, computer forensics has advanced at a rate analogous to that of cyber-crime. Law enforcement officials have prosecuted many cases under the federal computer crime statute 18 U.S.C. §1030. Consider "Mafiaboy," a fifteen-year-old Canadian student who launched an assault blocking victim websites (including Yahoo, Amazon, and eBay) with so much data that customers could not access them for e-commerce transactions and news in February 2000. The brief "Distributed Denial of Service" attacks reportedly cost companies millions of dollars, and the case was investigated through a joint effort of the RCMP's Computer Investigation and Support Unit and the U.S. FBI and Justice Department. "Mafiaboy" was caught and charged after the FBI was able to obtain chat-room logs demonstrating his plans and log files of a computer at UC-Santa Barbara that was hacked and used to attack CNN.com. "Mafiaboy" has since been sentenced to eight months in juvenile detention, but the ease with which he committed the crime and the authorities identified him are both enlightening.

Another interesting case that affected several aspects of the criminal justice and public safety systems involved David Jeansonne, a Louisiana man who sent e-mail to WebTV service users in July 2002 that, when opened, programmed their computers to dial 911 instead of an Internet access number. The executable attachment containing the virus also e-mailed hardware serial numbers to a free webmail account, which allowed federal investigators to track Jeansonne, who was arrested on charges of cyberterrorism under the USA PATRIOT Act. He pled guilty in February 2005 to violating 18 U.S.C. §1030(a)(5)(A)(i) for intentionally damaging protected computers, causing a threat to public safety, and losses of over $5,000. The case was overseen by the Computer Hacking and Intellectual Property Unit of the U.S. Attorney's Office.

Forensics in the world of computers is not strictly limited to offenses associated with the Internet and the World Wide Web. The Computer Crime and Intellectual Property Section (CCIPS) was ultimately incorporated into the U.S. Department of Justice Criminal Division resultant to the National Information Infrastructure Protection Act of 1996. Arising from a Computer Crime Unit established in 1991, CCIPS focuses exclusively on computer and intellectual property crime.

The National Hi-Tech Crime Unit in the United Kingdom, set up pursuant to recommendations by the Association of Chief Police Officers (ACPO) and upon approval by Parliament in 2000, classifies cybercrime in a very simple fashion—new crimes, new tools, and old crimes, new tools.

Because modern computers run various operating systems and platforms, such as Windows, Macintosh, and UNIX, the forensic computer expert must either specialize or receive extensive training across the board. Unlike the trace evidence examiner or biologist who has been forced to embrace computers over time, the forensic computer expert is inherently high tech. Computers and information systems have presented criminals with a means of conducting their transgressions, but these same advances have allowed for more information sharing among criminal justice agencies. Similarly, techniques and discussions can be shared between forensic scientists and examiners via platforms formerly reserved for annual meetings and the occasional hard-copy newsletter.

Not unlike those in other disciplines, the forensic computer examiner traditionally received mentor-based training, with most examiners being members of law enforcement agencies. Like most areas of forensic science, computer examination is probably far more tedious and less instantaneously gratifying than depicted by modern books and television programs. Individuals who commit computer crime may be self-taught, trained through technical programs, or have academic credentials. Essentially, the criminal and the examiner may have taken courses together, which is less likely in other fields. In recent years, the forensic computer examiner has increasingly received instruction and education prior to beginning mentorship and employment. Colleges and universities have integrated forensic courses or emphases into their computer science tracks. Students may enroll in a variety of course work, ranging from online seminars to formal classes designed to culminate in a certificate or associate's, bachelor's, or master's degree.

Today's forensic computer examiner must be able to find evidence on a computer and subsequently articulate how it was found and identified. As with other areas of forensic science, documentation is key. A great deal of knowledge and experience goes into simply understanding how to power down systems, such that evidence can be protected and preserved. The actual analysis is conducted on a physical level and then a logical level, meaning that initial searches look to clusters and sectors for possible evidence, followed by a search of what the user would see when operating the computer normally. Examiners may also employ commercially available forensic tools for cracking passwords, imaging, and retrieval.

File retrieval is a form of forensic analysis with which most people are familiar. As will be noted in the **digital evidence** section, e-mails and graphics are rarely truly deleted from a computer system. The field has become so linked to traditional crime scenes that the U.S. National Institute of Justice has issued literature such as the "Electronic Crime Scene Investigation: A Guide for First Responders" (2001) and "Forensic Examination of Digital Evidence: A Guide for Law Enforcement" (2004). Both are available via the National Criminal Justice Reference Service at http://www.ncjrs.org/.

Whereas computer forensics encompasses the use of analytical techniques to identify, collect, and examine evidence and information stored in computer formats, traditionally magnetically stored or encoded, the discipline can be further broadened by the preservation and investigative techniques associated with retrieved digital evidence. Other types of recordings, such as photographic, video, and audio also fall into this category.

Digital evidence, such as recovered files, is inextricably linked to computer forensics. In keeping with the TWG/SWG convention, the Technical Support Working Group (TSWG) was originally developed in 1986 in the United States. More recently, the group has added its Investigative Support and Forensics subgroup and has added electronic evidence as a focus area. The Scientific Working Group on Digital Evidence was formed in 1998 in keeping with the trend that traditional audio and video was moving toward digital media and computer forensics. Child pornography, fraud, and piracy were once the concentrations of digital evidence; however, digital evidence now weighs heavily in all types of crimes. Examiners may follow an electronic trail or identify and extract specific information, such as a sound byte, still image, or e-mail.

On February 1, 2004, a video surveillance camera at a Florida carwash captured the abduction of eleven-year-old Carlie Brucia. The tape was considered crucial evidence by law enforcement and prosecutors. Several agencies were called upon to assist in making the surveillance images more usable. Ultimately, examiners were able to discern clothing and tattoo characteristics from the still frames. Repeatedly, worldwide, criminals are being identified based on enhanced photos shown to the public, who recognize the image and call in tips. Not only do such images aid in identification of possible perpetrators, video often catches the entire commission of a crime, which often facilitates a plea or conviction.

Arguments surround the fine line between discovering or enhancing evidence and creating it. However, the courts simply dictate that admission of a photograph depends upon the requirements of relevance and authentication. In past admissibility hearings, experts have indicated that digital photographs used for comparison purposes are, in fact, accurate representations of the actual captured images. Still, those with conservative views toward evidence integrity differentiate between images captured on film, video images, and digital images.

Controversy related to fingerprints has been discussed; however, digital enhancement of latent print images adds another layer of debate. *Commonwealth v. Knight* (1991) was the first case to establish precedence for acceptance of digitally enhanced evidence in the United States. The experts used a frequency filter, a Fast Fourier Transform, to enhance a fingerprint in blood from a pillow case. At the time, it may have been possible to type the DNA present in the blood, but doing so would have ruined the print and may have proved to be the victim's blood anyway. The technology allowed experts to remove the background "noise," the pattern of the fabric, allowing for a better image of the print itself. Upon identification of the print as Knight's the defense attorney moved for a *Kelly-Frye* hearing to determine if the evidence met the standard of general acceptance in the field. The court determined that the techniques were acceptable as photographic processes, and a conviction was achieved. Fast Fourier Transform is one of the principal algorithms or encoded finite set of instructions of the popular Adobe Photoshop software.

A second case of this nature, *State v. Hayden* (1998), affirmed the Virginia ruling. Eric Hayden was charged with a 1995 murder of a woman who was found with a bloody sheet around her neck and head. Because latent fingerprints on the sheet could not be identified by conventional methods, an expert used digital imaging enhancement techniques including Fast Fourier Transform to filter out the sheet's background color and texture. A *Kelly-Frye* hearing was held in 1995 regarding these techniques and the argument was raised that the chain of custody was broken through use of the software and that the fingerprint image had been manipulated and altered to match that of the defendant. Both arguments were unsubstantiated, and the print evidence was again allowed.

Yet another, more modern case dealing with these same issues is *State v. Reyes* (2003).

In 1996 the body of Henry Guzman—a drug addict and drug dealer—was found lying on the side of a road in Pompano Beach, Florida. The victim had been killed execution-style with a gunshot to the head. His body had been concealed inside a blanket and his head was inside a plastic bag. The bundle was wrapped with duct tape. At that time, latent prints on duct tape were deemed essentially useless. However, in 1999 the latent images were reanalyzed using technology commonly referred to as "dodge and burn," which can lighten and darken images in an attempt to bring out contrast and detail. This allowed for an identification of Reyes. Although Reyes was ultimately acquitted due to insufficient evidence that he committed the murder, the admissibility of the digitally enhanced print was significant.

Surprisingly little legal drama surfaced after certain jurisdictions in the United States and abroad installed cameras at specific junctions such as intersections and toll booths to capture images of vehicles and their plates. It is unlikely the traffic law courts would entertain a plea that someone had digitally altered a still or video image of a car's license plate numbers or other distinguishing characteristics. In the United Kingdom automated number-plate recognition is increasingly prevalent, with some agencies employing mobile units for collecting intelligence on passing automobiles. These specially equipped vans are able to read passing plates and detect cars on which no tax has been paid. An officer can then pull over the vehicle without the lengthy process of involving a human to consult a computerized database. Such equipment is valuable for citing speeders, but it is easy to imagine how a fixed camera might detect criminal activity in the absence of an officer.

Even more sophisticated technology, global positioning systems (GPS) and mapping techniques are used by crime prevention experts and in policing, but GPS has also found its way into the courtroom as forensic digital evidence. Applicable for locating fleet vehicles and documenting crime-scene locations, GPS can also be used to track offenders such as parolees. More relevant to this discussion, investigators may use GPS devices to track suspects. For example, GPS devices were placed on Scott Peterson's vehicles during the well-known 2002 to 2004 investigation. In his ruling on admissibility of the GPS data, Judge Al Delucchi stated, "The generic methodology is generally accepted and fundamentally valid" (CNN February 17, 2004). This set a California precedent. Other states using the GPS technology to track offenders have since deemed the data admissible in court.

In January 2005 Attorney General Thomas Reilly of Massachusetts announced arrests in an international marijuana trade ring, advising that authorities had tracked the accused using GPS technology and by monitoring their cell-phone text messages. Time will tell if this data is admitted into evidence during the trials of the various accused drug dealers.

Yet another term linked to digital evidence is *steganography*. Dating to the ancient Greeks, steganography refers to encryption or hiding of words. Once likely to include tactics such as tattooing a message on the head of a courier who let his hair grow back during the journey to be shaved upon arrival at the recipient's location, today steganography refers to both encryption and forms of digital watermarking. Forensic examiners may study securities and documents to detect such watermarks. Examples of this type of crime and detection in modern times may involve credit-card fraud and identity theft. Also, it has been reported that some terrorist groups may use high-frequency encrypted voice/data links in order to communicate undetected. When Japanese authorities seized the Aum Shinrikyo cult's computers after suspected involvement in the Tokyo subway gassing incident of 1995, authorities decrypted the electronic records and found evidence crucial to the investigation. This may have taken significantly longer or been impossible had the key to the encryption not been found on one of many confiscated floppy disks.

Other notable cases in computer forensics, cybercrime, and digital evidence include that of Zacarias Moussaoui, the alleged twentieth hijacker in the 9/11 terrorist attacks in the United States. In the U.S. Criminal Case 01–455-A, much of the incriminating evidence was digital in nature. For example, Moussaoui's laptop, at least four other computers, and several e-mail accounts were accessed and searched. The FBI later announced that Kinko's computers were used throughout the states for Internet access, whereby the nineteen hijackers conducted planning for the attacks.

In the heartland of America, an Internet Protocol (IP) address led authorities to Lisa Montgomery, charged with killing Bobbie Jo Stinnett and kidnapping her fetus. The case, expected to go to trial in 2006, will likely see the admission of e-evidence obtained by investigators who examined the Stinnett computer.

With the number of cases involving digital evidence increasing daily, proponents and examiners alike are striving toward a consensus on standards. Issued in the FBI's April 2002 edition of *Forensic Science Communications*, SWGDE and the International Organization on Computer Evidence issued a document entitled "Digital Evidence: Standards and Principles." However, this did not completely address competency, standard operating procedures for examinations, or information sharing, all of which are still on the table for discussion. Because any crime that leaves an electronic trail produces some form of digital evidence, computer crime is no longer limited to fraud or child pornography.

"This entrusted with arms . . . should be persons of some substance."
—William Windham, Speech to the British Parliament, July 1807

The Association of Firearm and Tool Mark Examiners (AFTE) was born of members of the American Academy of Forensic Science (AAFS) and the Chicago Police Department in the late 1960s, while published papers regarding firearms identifications date to at least the 1950s. Initial meetings of the AFTE group included members of the Chicago Police as well as law enforcement investigators and firearms specialists from public crime laboratories nationwide, with the Midwest being heavily represented. The U.S. Bureau of Alcohol, Tobacco, and Firearms (ATF) is another player in the firearms world. Although historically linked to organizations dating to the eighteenth century, the ATF officially took over powers related to those three items from the Internal Revenue Service (IRS) in 1972. Responsibility for the investigation of commercial arsons nationwide was received by the ATF in 1982. In 2003 the law enforcement functions of the ATF were transferred to the Department of Justice (DOJ), while the tax and trade aspects of the bureau remained under Treasury Department control. Although it may seem strange to link alcohol, tobacco, and firearms, consider the early 1900s era involving Prohibition (repealed in 1933), and it is easy to relate these taxable items and their association to crime.

It is posited that individual firearms have characteristics with uniqueness analogous to human fingerprints. The characteristics are transferred to projectiles and casings when fired, setting the stage for forensic comparisons. In 1992 the ATF developed an enforcement program known as CEASEFIRE, which laid the groundwork for collecting information obtained from seized firearms in a database. Forensic Technology Incorporated (FTI) presented the ATF with information on a proprietary system called Bulletproof in 1993. The ATF leased a machine from FTI and begin investigating possible uses of the technology. Soon thereafter, the ATF began planning deployment of the technology to certain sites based on specific criteria, such as incidence of firearms-related crime. On only a slightly earlier time frame, beginning around 1989, the FBI had developed DRUGFIRE, a database for linking serial shootings and the identification of weapons used in drug-related

and gang crime. Although both ballistic imaging systems used database technology to search for matches between crime-scene evidence and known information, the platforms were not exactly compatible. FTI then developed Brasscatcher, a platform for the evaluation of both projectiles and casings. The combination of Bulletproof and Brasscatcher became known as the Integrated Ballistic Identification System (IBIS) in 1996. Because it was further determined that the new IBIS and the FBI's DRUGFIRE needed to be interoperable, 1996 saw modifications to the systems. IBIS units were soon thereafter upgraded to Windows platforms, and standard operating procedures for use were developed.

The National Integrated Ballistic Information Network (NIBIN) arose in 1997, and the ATF and FBI abandoned their program names CEASEFIRE and DRUGFIRE respectively. A NIBIN board was formed to bring together the federal efforts and to assist state and local police. After two years of changes to the memorandum of understanding between the agencies, the NIBIN board, with recommendations from the National Institute of Standards and Technology, decided to pursue deployment of a single system dependant upon interagency cooperation. FTI remained on board to assist in developing servers to meet the needs of the data storage and communications.

Initially, NIBIN units were set up with regional hubs with data-acquisition stations on a local-area network. With nationwide roll-out, the regional servers were afforded more advanced storage and management capabilities. An examiner completes entries for a casing or projectile and captures related images. The station remote sends this data to the regional server for comparison. If the system is able to produce a list of candidates for the match, a firearms examiner then examines the individual specimens. Obviously, with rapid and reliable searching, a match through IBIS can provide information to link or solve more than one case. Participation in the network is highly regulated, however, and a specific memorandum of understanding must be accepted and executed before the ATF deploys IBIS equipment to a state or local agency or laboratory.

As of January 2005 the NIBIN program reported 232 sites with IBIS equipment and 182 agencies participating. Also, the life of the technology has seen some 853,000 pieces of crime-scene evidence entered, resulting in over 10,799 hits, that is, matches between evidence items and information in the database (see https://www.nibin.gov/nb_success.htm for regular updates on the success of the NIBIN program). This is due to the technology of the program that eliminates many nonmatches, freeing time for the examiner to concentrate on possible match confirmation and entry of even more evidence.

Agencies often praise NIBIN in tracking firearms across multiple crimes, some of which are not specifically gun related. For example, in a jurisdiction rife with crime, such as Los Angeles, California, confiscation of a firearm during a property crime such as vandalism could lead investigators to an attempted murderer, as was the case in early 2005. The shooting victim could not readily identify his assailant, but the gun being carried by a man arrested for vandalism proved to be the same gun that fired the bullet into the victim a few hours earlier. The New Orleans Police Department (Louisiana) obtained an IBIS in 1996 and within a month received a hit involving the drive-by shooting of a child. Over the years, IBIS technology in New Orleans and a partnership between the New Orleans Police Department and the ATF under the NIBIN program have been associated with a reduction in gang-related gun violence. One set of linked crimes put eleven gang members behind bars in 2002. Various agencies report that literally hundreds of violent crimes have been solved in New Orleans through IBIS (see https://www.nibin.gov/nb_success.htm).

The Hamilton County Crime Coroner's Office (Ohio) got its first NIBIN hit in 2003 in a homicide case. The Nebraska State Patrol got its first hit in 2004, some three years after it first obtained a hit in the combined DNA index system (CODIS). The Detroit Police Department (Michigan) received a firearms database hit in 1996. In 2004 the New York City Police Department recorded its 1,400th hit. Clearly the efficacy of NIBIN is yet to be fully appreciated. One roadblock, though, stems from the limited number of qualified firearms examiners available to make the microscopic comparisons required to confirm a match. The ATF has offered a National Firearms Examiner Academy, but the number of qualified examiners is still presumed to be dangerously low. Trainees are typically subjected to mentor-based training, which can be extremely lengthy and produce examiners who are only as good as their mentors. Other agencies and organizations are formulating plans to help combat the shortage. In Atlanta, the ATF's local NIBIN coordinator actually picks up bullets on a weekly basis from a drop box installed at a local hospital operating room. The bullets are deposited there upon extraction from gunshot victims. The ATF is thus able to obtain projectile data without impinging upon the hospital staff in instances when the bullets are not specifically requested by a responding law enforcement agency.

Many recall the three weeks of terror in 2002 in the Washington, D.C., area caused by the sniping spree of Lee Boyd Malvo and John Allen Muhammad that left ten dead in and around the District of Columbia and one in Alabama. It was an ATF representative who made the announcement that the rifle in custody was forensically matched to bullets in eleven of the fourteen shootings. Although the weapon used in the Alabama shooting was not the rifle in question, Malvo's fingerprint was lifted from that scene. However, at the time of the incidents, NIBIN was not operational to the extent that bullets from other potential scenes as far away as Georgia and Louisiana were not entered or identified as possibly being related to the Malvo and Muhammad shootings until after their arrest. This is a testament to the fact that forensic techniques are only as effective as an agency's ability to employ them. While techniques, technology, and databasing continue to progress, the criminal justice system is tied to resources such as personnel and budget.

This ballistics information database technology, attributed largely to FTI, is not sequestered in the United States. In fact, Spanish authorities using an IBIS were able to link an attempted murder, shots fired into a home, and a third shooting although eleven years had passed between the first and last incidents. Suspects apprehended at the third shooting were ultimately linked to both previous offenses. Also, the ATF itself at its Rockville, Maryland, laboratory processed a large bulk of firearms evidence retrieved from the Ovcara mass grave in Bosnia on behalf of the International Criminal Tribunal for the former Yugoslavia. The comparisons resulted in a conviction for war crimes.

"... summon up the blood."
—Shakespeare, *King Henry V*, III, I, 1

In the early to mid-1980s, forensic biologists borrowed techniques from serology in order to differentiate between people or to exclude a subject from having contributed to the blood or semen at a crime scene. Characterization of stains as being composed of blood or semen or other bodily excretions dates back over a century.

Serology is the science concerned with antigens and antibodies in the sera or fluid components of bodily fluids. For example, most are familiar with blood typing or at least that people have different blood types known as A, B, AB, or O. Karl Landsteiner first identified blood types in the late nineteenth century. This difference in phenotypes, outward physical expressions of traits under genetic control, could be employed by investigators to categorize people. It was 1925 when scientists discovered that approximately 80 percent of the population were "secretors" or those whose characteristic

blood proteins could also be found in other bodily fluids and tissues. For the investigator, this meant that a semen or saliva sample could perhaps be as revealing as blood left at a crime scene.

Special serological tests are still occasionally used to determine the species of origin of a sample or at least to determine if a sample is human or not. The most prevalent of these techniques, the Ouchterlony procedure, has been in use since its first description in 1948. Carried out on a plastic or glass dish or plate of a gel medium, a visible precipitin line forms where the antigens and antibodies of different species meet.

Presumptive tests for the possible presence of a biological substance or fluid may date back more than a century, but it is modern forensic DNA analysis that allows scientists to further examine those substances and fluids to identify who may have possibly contributed them. Thus, the likely chain of events in the modern laboratory involves screening evidence for possible biological stains of bodily fluids or tissues, followed by the process of forensic DNA analysis. The once impressive serology techniques are now largely historical.

When DNA typing was first introduced in forensics in the mid-1980s, its effect was widespread, and the ability to individualize or attribute the DNA found in a bodily fluid or crime-scene stain overwhelmed the criminal justice system and its participants. Forensic DNA analysis was born of techniques used in the biomedical world. Whereas previous tests enabled determination of phenotypes based on examination of proteins, DNA tests delved deeper, revealing genotypes, the actual inherited designations imparted by a person's DNA. Such tests were based on nucleic acid blueprints rather than on proteins found in an individual's bodily fluids. A brief overview of the progression of techniques used in forensic DNA analysis is found below, followed by legal considerations, and case examples.

All people inherit two copies of DNA, one from each biological parent to make up his or her genome. Genotypes are descriptors of genes or locations within the genome. If a particular genetic variant is observable in at least 1 percent of the population, it is considered polymorphic. Thus, detection of the type of polymorphism an individual bears may be used to differentiate between that person and others. Under this convention, ABO blood types are considered protein polymorphisms. Because proteins are products of genes, versions of proteins reflect differences in the DNA or gene. These differences or variant forms of genes or DNA sequences at particular locations on chromosomes are known as alleles. To reiterate, one allele is inherited from each parent.

Ray White, an American geneticist at the University of Utah, identified regions of DNA that did not code for proteins but were highly variable between individuals. Restriction enzymes are proteins that cut strands of DNA at specific locations to produce DNA fragments of defined lengths. Using these conventions and a technique known as gel electrophoresis, White separated the fragments based on size, calling the variations restriction fragment length polymorphisms (RFLP). In 1980 White described the first polymorphic RFLP marker. Later that same year, researchers David Botstein, Ronald Davis, Mark Skolnick, and Ray White proposed methods for mapping the human genome based on this RFLP technology, lending fuel to the Human Genome Project, which was ultimately launched in 1990.

Back on the forensics front, in 1984 Alec Jeffreys discovered methods for identifying individuals based on RFLP and dubbed it DNA fingerprinting. His work at the time focused on paternity testing, but in 1985 the West Midlands police approached Jeffreys to try his hand at testing samples from a rape case. Ultimately, using RFLP technology, Richard Buckland was exonerated of committing two rape-murders and Colin Pitchfork was convicted in 1988, in a veritable double-whammy of forensic DNA analysis. Yet another British case of this time frame involved Robert Melias, who in 1987 became

the first person in England convicted of a crime based on DNA evidence. Soon thereafter, Tommie Lee Andrews was convicted in Florida and Timothy Wilson Spencer in Virginia based heavily on DNA evidence. Author Joseph Wambaugh wrote of the Colin Pitchfork case and the first time DNA testing was used to catch a criminal in his famous book, *The Blooding* (Wambaugh, J. *The Blooding*. New York: William Morrow & Company, 1989).

It was Jeffrey's work with probes that made identification of RFLP variants feasible. Probes, pieces of DNA complementary to the fragment of interest, were used with RFLP to examine variable number of tandem repeats (VNTR) loci. The southern blot method was originally developed in 1975 and adapted for use in RFLP analysis. In short, the technique calls for extracting DNA, cutting it into fragments using restriction enzymes, separating them based on size using gel electrophoresis, and finally transferring the fragments to a nylon membrane. The membrane is then subjected to radioactive probes, which bind specifically to applicable VNTR fragments. X-ray film is exposed to the membrane and the profile visualized by autoradiography. Jeffrey developed a multilocus probe, which allowed visualization of more than one variable region at once. Also in the late 1980s, scientists developed and employed probes based on chemiluminescence instead of radioactivity.

Although highly discriminatory, RFLP analysis of VNTRs had several drawbacks. First, the process was extremely laborious and time-consuming. Radioactive probes posed health and disposal risks, and a relatively large amount of sample was required to perform the tests.

RFLP proved to be the forensic DNA testing mainstay for many years, but as with all technology, advancement was inevitable. In 1983 scientist Kary Mullis (b. 1944) developed the technique known as the polymerase chain reaction (PCR) that ultimately revolutionized molecular biology, including forensic

DNA analysis. Mullis and members of the Cetus Corporation first described the process in 1985, for which he was awarded the Nobel Prize in Chemistry in 1993. PCR allowed scientists to make millions of copies of specific DNA sequences of interest in a relatively short time. As previously mentioned, RFLP analysis required a relatively large starting sample. Because crime-scene samples are often minute, degraded, or otherwise compromised, PCR offered the forensic DNA analysis world an opportunity to create copies of isolated DNA. Through PCR, forensic DNA analysis essentially became more rapid and sensitive.

With PCR available, scientists sought other markers by which to differentiate and identify individuals. The human leukocyte antigen involved in the immune response was known to be a polymorphic protein. Thus, forensic scientists looked to the DNA coding for this protein, and in 1991 developed the DQ-Alpha test, named after the variable region of DNA at this location. The DQ-Alpha test examined a poly-allelic locus, meaning a location on the genome with variable alleles seen in different people. Although this afforded a fair amount of discrimination, forensic scientists soon incorporated a number of other loci to be typed concomitantly with DQ-Alpha. In 1993 the Polymarker system was developed using additional bi- and tri-allelic loci in addition to DQ-Alpha. The FBI began casework with DQ-Alpha in 1992 and added Polymarker in 1994. Due to sensitivity imparted by PCR, forensic DNA analysts were able to obtain a set of types of profile for these loci from a very small starting amount of sample. However, the entire process still proved labor intensive due to required probing and detection processes, and the discriminatory power of the test was less than optimal for forensic analysis.

Scientists returned to a VNTR in typing the D1S80 locus, and the FBI incorporated it into casework in 1995. The naming convention simply follows that "D" stands for DNA, "1" for chromosome 1 on which the locus is found, "S" for single copy sequence, and "80"

for a sequential number imparting uniqueness. Although sensitive, the discrimination potential of D1S80 and other amplified fragment length polymorphisms (AMPFLP) was still less than that of RFLP, and no great strides in detection had been incorporated.

Later in the 1990s, short tandem repeats or STR testing appeared in forensic DNA analysis. In keeping with the name, STRs are short sequences, ranging from approximately two to six base pairs (bp), which repeat over and over a given number of times at specific locations in the genome. People differ with respect to alleles at these loci, based on the number of times the sequence is repeated. Although individual loci were not as discriminatory as RFLP markers, the short size and number of available STRs allowed scientists to multiplex the STRs, or amplify and analyze three or more simultaneously. Multiplex kits became available as early as 1996. Much earlier scientists had developed means of labeling nucleotides, the building blocks of DNA, which are also added in the PCR process, with fluorescent tags. This would become important later on, although initial tests using STRs separated the fragments using gel electrophoresis and used silver staining or a special type of green dye. In 2000 the FBI and other laboratories abandoned the RFLP technique altogether in favor of multiplex STR analysis. Because of their short size, 2 to 6 base pairs (bp) with repeats of total length less than 400 bp, STRs using PCR allowed for typing samples that were not only too minute for RFLP methods but also typing of samples that were degraded such that the lengthy pieces that made up RFLPs or other VNTRs were damaged or lost. It should be noted that while the FBI and some other labs ran the gamut of test techniques, many labs conducted only RFLP testing until STRs were fully accepted and then switched over, as late as the early 2000s. Multiplex PCR and STR analysis is the gold standard in forensic DNA testing at the time of this writing, although notable advances are emerging and will be later discussed.

"Rome has spoken; the case is concluded."
—St. Augustine, *Sermons,* Book 1

The above describes how scientific advances resulted in DNA typing becoming the method of choice to type body fluids. However, before any new technique can enter routine use, it must face its own version of trial by fire, namely it must establish its acceptability through case law. In the United States, the Privacy Act of 1974 laid the groundwork for maintaining records on individuals. Recognizing a need, the FBI formed the Technical Working Group on DNA Analysis Methods (TWGDAM) in 1989. Scientists from federal, state, and local laboratories and the academic community met to address implementation of the new technologies. TWGDAM established, and in 1990 published, a set of quality assurance guidelines to which courts often referred when making admissibility decisions.

People v. Castro (1989) was the first case in which the admissibility of a DNA profile was challenged. The court held that "DNA identification theory and practice are generally accepted among the scientific community." However, DNA testing that showed that the blood on the defendant's watch was not his was allowed into evidence, but tests were not permitted that could demonstrate that the blood belonged to one of the victims.

The Timothy Wilson Spencer case mentioned previously was the first in which DNA evidence was used to convict a person resulting in a death sentence. Convicted in 1989, Spencer was executed in 1994. Also in 1989, Gary Dotson in Illinois became the first person to be exonerated and have his conviction overturned based on DNA testing, and Australia saw its first court case involving DNA evidence. Desmond Applebee in Australia was pinpointed by DNA from his blood sample, which matched that of blood and semen on the victim's clothing. As much attention has been directed at exoneration of innocents as to conviction over the decade and a half since 1990. However, DNA overall

has not been as readily accepted in the courts as some of its forensic science counterparts, such as drug chemistry or toxicology.

Where were you when the O. J. Simpson verdict was announced? Most recall the June 1994 murders of Nicole Brown Simpson and Ronald Goldman, which set off a media frenzy and resulted in what some characterize as the "Trial of the Century" propelling forensic DNA analysis into mainstream conversation. On the evening of Sunday, June 12, 1994, Nicole Brown Simpson and her friend Ron Goldman were brutally murdered on the grounds of her condominium in Brentwood, California. Later that evening, Orenthal James (O. J.) Simpson flew from Los Angeles to Chicago. However, in the early hours of June 13, 1994, Nicole's neighbors, led by her pet Akita dog, found the gruesome scene and notified police. Before sunrise, police went to O. J.'s estate to advise him of the murders but received no answer, although lights were on in the house and vehicles in the driveway. Police detectives noted an apparent bloodstain near the door of O. J.'s white Ford Bronco and other drops on the ground. They also found a glove matching one found at the crime scene. The police were given access to the house by Arnelle Simpson, O. J.'s daughter from his first marriage, who was staying in the guest house. Although the house and grounds were not thoroughly searched at that time, criminalists were called to secure the area, and O. J. was located at a hotel in Chicago. He checked out and returned home at approximately 11:00 a.m. on the thirteenth, less than twelve hours after his flight left for Chicago. A search warrant had been issued prior to his arrival back at the California estate, but items were not seized from the home until the following day, June 14, 1994. This same day, the coroner released a report, and testing began on the evidence.

Samples of bloodstains found at the crime scene were tested for blood type and DNA. Items or areas from which stains were sampled included O. J.'s Bronco, driveway, and house, as well as a pair of socks in his bedroom, the gloves, and the crime scene itself. This blood evidence led to the major point of contingency in the case. Whereas the science and technology of the DNA testing and the match itself were hardly contested, the defense contended that the evidence was contaminated and/or planted. Herein lies a prime example of attacking the scientist and law enforcement, not the science. These accusations were made in lieu of photographs taken throughout the investigation of the bloodstains prior to their being collected or sampled.

Another aspect of this case many recall is the lengthy yet slow and heavily televised vehicle pursuit that ensued on June 17, 1994, when O. J.'s friend A. C. Cowlings drove him throughout the area in the white Ford Bronco. When the two finally returned to O. J.'s estate, Simpson was taken into custody and charged with murder. On July 8, 1994, Judge Kathleen Kennedy-Powell ruled there was sufficient evidence for Simpson to stand trial on the two first-degree murder counts. Judge Lance A. Ito was assigned to hear the case on July 22, 1994, when Simpson pled "absolutely 100 percent not guilty" to the charges.

Pertinent to this discussion, the defense waived a hearing to challenge the prosecution's DNA evidence on January 4, 1995. The trial opened later that month and continued until October 3, 1995, when a not-guilty verdict was returned. During the months of testimony, various statements related to the DNA evidence were heard. The defense, headed up by the "Dream Team" of Robert Shapiro, Johnnie Cochran, F. Lee Bailey, and Alan Dershowitz, began its case by accusing Detective Mark Fuhrman of being racist and involved in a conspiracy plot to frame Simpson for the murders. Attorney Barry Scheck, who had in 1992 first developed the Innocence Project as a nonprofit legal and criminal justice resource center concentrating heavily on exoneration of wrongfully convicted persons using postconviction DNA testing, became an increasingly important tool to the defense as the trial proceeded, as did famed

forensic scientist Dr. Henry Lee. Scheck particularly used Lee as a tool to attack the collection of evidence, for example advising that the two socks collected from O. J.'s house should have been packaged separately and not in the same envelope.

Problems with the blood evidence included the identification of the chemical additive EDTA in some of the samples, as EDTA would only be expected in a blood sample from a test tube containing the additive and not from human circulation. However, this was largely refuted by a second toxicologist's testimony. Also, some of the items had been collected while still wet or damp and placed into paper bags, where some transfer of material from the items to the bags occurred. Some controversy also surrounded the storage of the evidence, as a warm and/or damp environment is conducive to bacterial growth, which can degrade DNA evidence. It is important to note, however, that although improper storage conditions may destroy DNA evidence, they cannot change a DNA profile or test result. Much of this discussion surrounded the work of criminalist Dennis Fung of the Los Angeles Police Department. The trial was also riddled with problems with jurors, and several were dismissed and replaced. At one point, Cochran expressed concern that prodefense jurors were being eliminated from the fold. Contrary to the January proceedings, the defense attempted to challenge the admissibility of DNA evidence in early April, although this was rejected by Judge Ito.

In May Gregory Matheson, chief forensic chemist at the LAPD crime lab, gave testimony regarding serological testing in the case. He provided probabilities based on blood and enzyme typing. Matheson advised the court that mistakes made in collection were technicalities. Matheson also explained the absence of a volume of blood from the vial used to hold O. J.'s reference standard sample drawn for testing by saying it was used by chemists in the laboratory, rather than sprinkled about the scene by LAPD employees as was proposed by the defense. Testimony related directly to DNA testing began the second week of May, including testimony by Robin Cotton of Orchid Cellmark on RFLP testing. As part of the testimony, Cotton indicated that the odds that blood collected from the Brentwood scene could have come from someone other than O. J. were approximately 1 in 170 million and that blood from one of the socks from O. J.'s estate was consistent with Nicole's with an accompanying probability of 1 in 9.7 billion Caucasians. These are just a few of the results presented by DNA experts. Defense attorney Peter Neufeld cross-examined Cotton and others in attempts to belittle the DNA evidence, attacking it from contamination, procedural, and statistical standpoints. The DNA match evidence would later be reaffirmed through testimony of Gary Sims, chief DNA analyst at the California Department of Justice Crime Laboratory. Statistician Bruce Weir also factored into the trial when he apologized for mistakes in computer programming that led to overestimation of match statistics he provided. However, results still supported the prosecution.

Later attacks were directed at the physician who conducted the autopsies, Dr. Irwin Golden. Los Angeles County Medical Examiner Dr. Lakshmanan Sathyavagiswaran provided testimony as to some mistakes made by Golden but also advised on the inability of physicians to determine how many people were involved in conducting the murder, the relative uncertainty of the exact type of murder weapon, and the positioning of the bodies.

In a case where racial tensions ran high and some characters demonstrated quite a professional low, this writing is clearly not indicative of all that occurred throughout the trial. However, one can see just how intertwined the forensic science became with the actions of the police, its effects on the jury, and the myriad interpretations associated with the blood evidence. Plenty of discussion followed the presentation of other types of physical evidence as well, including the infamous leather gloves and Bruno Magli shoe prints. The trial,

touted as the most publicized in U.S. history, was the longest ever held in California, costing over $20 million and producing 50,000 pages of transcript. After hearing from 150 witnesses and being sequestered approximately nine months, in the end, jurors determined that prosecutors Marcia Clark and Christopher Darden failed to prove O. J. Simpson guilty beyond reasonable doubt. Simpson was acquitted and the not-guilty verdict announced on October 3, 1995, after the jury deliberated only four hours on October 2, 1995, to come to its conclusion.

Yet another case that fascinated many involved the President of the United States, William (Bill) J. Clinton. In the summer of 1995, Monica Lewinsky began an unpaid intern position at the White House. She claimed a sexual relationship began with the president that November. In the summer of 1996, Lewinsky began work at the Pentagon, where she met coworker Linda Tripp and began confiding in her. Tripp began taping her conversations with Lewinsky, but Lewinsky soon left the Pentagon seeking work elsewhere. However, by this time, Tripp had consulted *Newsweek* magazine about the taped discussions and possible presidential impropriety. In December 1997 Lewinsky was subpoenaed by attorneys for Paula Jones, who was suing the president on charges of sexual harassment.

In January 1998 two important events occurred. Lewinsky filed an affidavit in the Jones case stating that she never had a sexual relationship with President Clinton, and Tripp contacted Ken Starr, who had served as independent counsel in the Clinton Whitewater scandal. Later that month, although *Newsweek* had postponed running any related stories, the Internet was rife with rumors of the affair. Again, this writing is not intended to detail the entire case, but rather indicate the role of forensic science among the cast of players, which included Lewinsky and Clinton and numerous others such as the president's personal secretary, friends, and members of the FBI and Secret Service. In the summer of

1998, Lewinsky handed over the infamous blue dress, said to contain physical evidence of a sexual encounter with the president. A sample of the president's DNA was taken for comparison. Ultimately, DNA from the president's blood was demonstrated to be consistent with DNA taken from a semen stain on the dress. This provided Starr with physical evidence of the contact between the president and Lewinsky and was key to the resulting Starr report, which was delivered to the House of Representatives in September 1998.

In an interesting turn of events, Ricky McGinn of Texas received a temporary stay of execution in 2000 in order for experts to analyze evidence from a 1994 rape-murder, for which he was convicted and sentenced to death in 1995. At the time of collection, a pubic hair found on his slain twelve-year-old stepdaughter was not suitable for DNA testing. However, advances in technology made this possible in 2000. Although no one can determine exactly why he would call for DNA testing when truly guilty, DNA from the hair indeed matched McGinn, supporting the conviction, and he was executed.

The notorious Green River Killer can be traced to the early 1980s when he began a killing spree that took the lives of at least forty-eight women over the course of twenty years. Gary Ridgeway was first interviewed in 1984 as a potential suspect, and at one point in 1987 was arrested but released as one of many suspects against whom there was little hard evidence. Although he was released, biological samples were taken from him in 1987, and in 2001—aided by modern forensic DNA analysis techniques—DNA from some of the victims was identified as consistent with Ridgeway's DNA. Faced with the death penalty, Ridgeway agreed to cooperate with investigators and confessed to forty-eight killings, for which he was convicted and given consecutive life sentences in a plea deal in November 2003.

In March 2005 DNA from a cigarette butt led investigators to a dead man, Bart Ross, as the suspect in the murder of Federal Judge Joan Humphrey Lefkow's mother and husband. Ross shot himself during a routine traffic stop, and a cigarette butt found in the sink at the crime scene yielded DNA consistent with that of Ross. This evidence and a note found in Ross's van helped investigators quickly solve the high-profile case.

Historical cases involving DNA evidence abound. Underlying these cases is a fascinating set of ever-evolving legislation. In 1994 Congress passed the DNA Identification Act, establishing the DNA Advisory Board (DAB), which is charged with establishing standards and guidelines for forensic DNA testing. The act also recognized the TWGDAM guidelines as an interim standard. It should be noted that in 1992 the National Research Council, an arm of the National Academy of Sciences, had recommended that forensic DNA laboratories establish formal guidelines and seek to implement external review. Based on DAB recommendations, the FBI director issued "Standards for Forensic DNA Testing Laboratories" (1998) and "Standards for Convicted Offender Laboratories" (1999). These standards applied to labs participating in the national DNA database or receiving federal funds. Addressing everything from personnel to facilities to documentation, these documents are still considered the quintessential guidelines for forensic DNA laboratory quality control and operation. Because the DAB completed its mission, it was dissolved, and TWGDAM was designated by the FBI director as the entity responsible for recommendations and amendments to the National Quality Assurance Standards due to technological or other changes. In 1999 TWGDAM changed its name to the Scientific Working Group on DNA Analysis Methods (SWGDAM). SWGDAM has also issued recommendations for analyst training, equipment validation, and data interpretation.

A brief overview of U.S. legislation associated with forensic DNA analysis follows:

- The DNA Identification Act of 1994— established a group to address standards and guidelines in forensic DNA testing, recognized TWGDAM guidelines as an interim standard, and formalized FBI authority to establish CODIS.

- The Violent Offender DNA Identification Act of 1999—modified earlier acts and required the FBI director to develop a plan to assist state and local laboratories in conducting analysis of convicted offender samples for expediting entry into CODIS.
- Paul Coverdell National Forensic Science Improvement Act of 2000—authorized millions of dollars in federal funding for improving forensic science services for criminal justice purposes, supporting crime laboratories and medical examiners' offices, to be awarded over six years.
- DNA Analysis Backlog Elimination Act of 2000—authorized millions of dollars for the years 2001 through 2004 for testing samples for inclusion into CODIS, for testing crime scene samples, and for an overall increase in public laboratory capacity.
- Convicted Offenders DNA Index System Support Act (2000)—designed to facilitate exchange of information between law enforcement entities regarding violent offenders and required plan development and assistance of the FBI director and attorney general.
- Convicted Child Sex Offender DNA Index System Support Act (2003)—sought to eliminate backlog in analysis of samples from convicted child sex offenders.
- DNA Database Completion Act of 2003—authorized a grant program for elimination of the backlog nationwide and in obtaining samples from all persons convicted of a qualifying offense.
- Justice Enhancement and Domestic Security Act of 2003—in addition to appropriations for domestic security

and expanding upon the PATRIOT Act, also addressed grants for DNA training and sexual assault justice.
- Debbie Smith Act of 2003/Rape Kits and DNA Evidence Backlog Elimination Act of 2003—concerned with assessing the extent of backlog of rape kit samples, improving investigation and prosecution of sexual assault cases with DNA evidence, and allowing John Doe indictments, that is, indictments where the person is identified by DNA profile and not by name.
- Justice for All Act of 2004—included new provisions for victims' rights and innocence protection and incorporated the Advancing Justice through DNA Technology legislation.

The president's DNA Initiative, represented on the Internet at www.dna.gov, originated from Attorney General Ashcroft's direction to the National Institute of Justice (NIJ) to assess and make recommendations regarding delays in DNA testing. A working group was formed and met twice in 2002. The initiative has come to encompass funding via the Advancing Justice through DNA Technology initiative, and promotes backlog reduction through system infrastructure improvements and addresses training, research and development, postconviction testing, and missing persons cases.

An NIJ report from 1996 entitled "Convicted by Juries, Exonerated by Science: Case Studies in the Use of DNA Evidence to Establish Innocence after Trial" lent credence to the power of DNA testing. Postconviction DNA testing is governed at the state level. According to the Innocence Project, over 150 individuals have been exonerated or had their innocence proven since 1989 based on postconviction testing. At the time of this writing, thirty-eight states provide access to forensic DNA testing for previously convicted persons. States must enact statutes to provide for postconviction DNA review and make postconviction DNA testing available in order to qualify for funding under the Justice for All Act.

Postconviction testing has also brought to light the phenomenon of false confessions and errors in eyewitness identifications. The former is illustrated by the Central Park Jogger attack case, in which five teenagers confessed to the rape of Trisha Meili when she was jogging in Central park, in 1989. However, Matias Reyes, a convicted serial rapist and murderer, confessed to the crime in 2002, and DNA analysis corroborated his confession. The potential lack of reliability of eyewitness identification was demonstrated by the wrongful conviction of Arthur Lee Whitfield, who served twenty-two years in prison after two women identified him as their rapist. Some of the stories, including the sheer number of early identified cases involving false confessions or admissions, are so compelling that Governor George Ryan of Illinois declared a moratorium on executions in his state in 2000.

One of the crimes for which Timothy Spencer was ultimately blamed in Virginia was originally pinned on David Vasquez based on hair evidence and a confession. However, a pardon was secured for Vasquez in 1989 (just after Gary Dotson's exoneration) after DNA testing of similar crimes demonstrated they were committed by Spencer.

Kirk Bloodsworth of Texas was the first person exonerated from death row due to postconviction DNA testing. Convicted in a 1985 child rape and murder case based on witness accounts that he was seen with the girl, it was determined in 1993 that he could not have contributed to the semen on the victim's underwear.

Forensic DNA analysis is not limited to human testing. In fact, the first conviction involving plant DNA evidence stems from a 1992 Arizona murder (*State v. Bogan,* 905 P.2d 515). A pager found at the scene led to a subject who indicated he had been robbed and did not dump the body. However, DNA extracted from Palo Verde tree pods found in the defendant's pickup truck matched those of a damaged tree near the body.

Preliminary studies indicated that each tree has a unique DNA blueprint, and a *Frye* hearing in the Arizona court held that RFLP testing was generally accepted in the forensic community and allowed admission of the plant DNA evidence and accompanying statistical statements.

A 1994 murder was solved by the RCMP aided by cat DNA. A jacket bearing the victim's blood and white cat hairs was located during a search for the victim's body. Scientists demonstrated that DNA extracted from those cat hairs matched that of the suspect's parents' white cat, Snowball. A jury convicted the defendant, Douglas Beamish, of second-degree murder in his estranged wife's death and laid the groundwork for many subsequent cases involving animal DNA.

Of course not all cases involving DNA evidence make it to the court system. For example, in 1989 in Victoria, Australia, George Kaufman confessed to raping sixteen women after being confronted with DNA evidence. A series of brutal rapes of elderly women in North Carolina in the early 1990s was solved by a cold hit in the state database. When confronted with the DNA evidence more than a decade later, a subject, who was not previously considered a suspect, confessed. It should be noted, however, that DNA evidence is but a piece of the investigative puzzle. Often plausible reasons for the presence of an individual's DNA at a scene exist, and the investigative and prosecution team is charged with the burden of proof in criminal proceedings. In both the United States and Australia, it has been proposed that pleas resulting from DNA evidence have reduced the number of homicide cases going to court. Numerous instances demonstrate that guilty offenders may be more inclined to confess when confronted with DNA evidence.

"I have called this principle, by which each slight variation, if useful, is preserved, by the term of Natural Selection."
—Charles Darwin, *The Origin of the Species,* Chapter 3

Only one-tenth of one percent of the human genetic code differs between individuals. As the number of locations examined and found to match increases, the likelihood that the profile came from someone other than the subject in question becomes progressively smaller. At the time of this writing, the human population of the world is estimated to be greater than 6.4 billion. However, because obtaining the DNA profile of every individual in the world, or even a single country, is not feasible, scientists and legal practitioners depend on databases of known profiles. A profile is composed of a list of a person's type or allele designation at each location examined. The expected frequency of certain types or alleles in a population can be estimated with a great deal of certainty based on a sample of the population. If a particular allele is found 25 percent of the time in a sample of the population, it is generally accepted that the same allele can be expected to occur approximately 25 percent of the time in the whole population. Considering the thirteen locations commonly examined by forensic DNA analysts in the United States, the likelihood that any two unrelated individuals would match at each location examined or have a completely matching profile averages less than one in a trillion. Often, this likelihood is reported as one in three trillion, because it is unlikely that an individual would possess all of the most commonly encountered alleles. This is true even for populations with some minor degree of inbreeding. In forensic cases, partial profiles are often obtained. This means that the analyst was able to obtain the type or allele designation for some, but not all, of the locations examined. Even so, it is not uncommon to hear probabilities in the one in a quadrillion range provided in reports or court proceedings.

Databases allow for estimations of probability for forensic cases, serve as repositories for profiles of known offenders, and provide an arena for comparing unknown profiles such as those collected at more than one crime scene. Databases operating in the United Kingdom, Australia, and the United States will be discussed further.

In 1995 comprehensive legislation enacted in the United Kingdom allowed forensic scientists to set up a national DNA database, the first of its kind, to hold both personal DNA profiles and unknown profiles obtained from crime scenes. The world's first DNA database formally began operation on April 10, 1995. The National DNA Database, as it is known, is under the custodial care of the Forensic Science Society (FSS), which manages the database on behalf of the Association of Chief Police Officers.

In a famous mistaken identity case, the United Kingdom recognized a false cold hit in 1999. An individual was identified as the contributor of a crime-scene stain based on a match at six loci. It was not until a test was conducted looking at ten loci that the individual was absolved of wrongdoing.

In the annual report on the National DNA Database issued in November 2004, the FSS announced having made 584,549 suspect-to-scene matches and 38,417 scene-to-scene matches since 1995 through what most consider the most sophisticated and effective system in operation. As of early 2005, the database reportedly contains more than 2.7 million criminal justice samples and 243,627 unknown crime-scene profiles. The plan in the United Kingdom is essentially to obtain a profile from the entire criminally active population.

In 1997 Australian police services endorsed the establishment of a DNA database and formed a national working party. Victoria, the capital of which is Melbourne, was the first to enact legislation. The Australian federal government committed $50 million (Australian) to establish CrimTrac in 1998. CrimTrac also includes fingerprint and criminal justice records, but DNA is considered the core of the databases.

The CrimTrac DNA database holds convicted offender profiles, allowing for comparison to unknown crime-scene profiles. It also offers a forum for potentially matching DNA profiles from unsolved crime scenes that may or may not be seemingly related. In some states or territories DNA profiles are legally collected from charged suspects in addition to convicted individuals.

The Victorian police obtained the first cold hit from a DNA database in 1999, when the DNA profile of Wallid Haggag, a convicted thief, matched that obtained from blood found in a car used for a burglary. He had not previously been considered a suspect. The database became fully operational in mid-2001. In 2005, Australian laboratories examined or typed alleles at nine locations in addition to sex determination. After the 2002 bombing in Bali, Indonesia, legislation was brought forth to add a Disaster Victim Identification component to the database. Further consideration in 2004 incorporated this use of the database pursuant to antiterrorism legislation. Overall, the CrimTrac system and database has gained recognition worldwide as a highly effective forensic and investigative tool.

In the United States, the Combined DNA Index System (CODIS) began as a pilot project for sharing criminal justice information between a dozen or so laboratories in the early 1990s. The DNA Identification Act of 1994 gave the FBI authority to establish a DNA database for law enforcement purposes.

The Combined DNA Index System (CODIS), through forensic DNA and computer technologies, allows for electronic exchange and comparison of DNA profiles among qualified crime laboratories. CODIS is composed of three tiers and two indices. The tiers are hierarchical, at the local, state, and national levels. The tiered composition allows agencies to participate according to their own legal requirements and constraints, because criminal law is governed at a local and state level. For example, qualifying offenses for required submission of an offender profile vary by state. The local level is referred to as LDIS, the state as SDIS, and the national index as NDIS, with NDIS being governed by the FBI. The FBI's

NDIS became operational in October 1998. At the national level, CODIS operates two indices, the Forensic Index and the Offender Index.

The Forensic Index contains unknown DNA profiles or those obtained from crime-scene evidence for which there is no identified contributor. For example, a profile determined by a crime laboratory analyst in Tampa, Florida, by examination of a rape kit taken from a victim who did not know his/her attacker may not generate any investigative leads. However, the profile could be uploaded into LDIS, SDIS, and finally NDIS where it would reside in the Forensic Index. Matches obtained within the Forensic Index may serve to link crimes together, signify a serial offender, and allow contributing law enforcement agencies to share information and leads. If a profile in the Forensic Index matches that of an individual whose profile is held in the Offender Index, analysts at the two agencies are notified. This potential match identified through CODIS is commonly referred to as a *hit,* and analysts at the laboratories involved then contact each other to confirm or refute the match. As of February 2005 the NDIS Forensic Index contained 99,338 crime-scene profiles. Also as of February 2005 overall, it was reported on the FBI CODIS home page that CODIS had yielded over 20,200 hits with more than 22,100 investigations aided (http://www.fbi.gov/hq/lab/codis/clickmap.htm).

The Offender Index of CODIS contains profiles of convicted offenders, with 2,176,610 being held as of February 2005. Given the likelihood of criminal recidivism, maintaining these profiles allows for speedier investigations and may act as a deterrent to the criminal community. On occasion, an offender's profile is uploaded into the Offender Index by a participating state agency only to set off hits in the Forensic Index for unrelated crimes. Conversely, entry of crime-scene profiles in the Forensic Index may quickly lead to identification of suspects, if their profiles already exist in the Offender

Index. One can easily see the utility of both indices. California, Florida, and Virginia lead the way in offender profile submissions, at the time of this writing with 248,828, 235,810, and 208,527 respectively. Although many states have contributed substantial numbers of profiles, other states have contributed only a few hundred to a few thousand.

In 2004 all fifty states considered both sex crimes and murder as qualifying offenses for submission to the database. Forty-seven states collected samples for typing and submission for those convicted of any violent crime, and thirty-seven states were operating under "all felons" laws. The trend over the past few years has been toward passing legislation allowing states to submit profiles from all persons convicted of felonies. One very interesting fact surrounds the submission of profiles of convicted burglars, which was in force in forty-six states as of 2004. Although traditionally considered a nonviolent or property crime, burglary is often noted as a sort of stepping-stone crime. This supposition has proven valuable in many states. For example, both Florida and Virginia have formally reported very high hit rates for crimes such as rape and robbery against profiles submitted for convicted burglars. Several murders have also been solved across the United States after hits developed against profiles obtained due to convictions of drug or property crimes. Resultant to such success, at the time of this writing, more than half of the states collect samples from those convicted of some misdemeanors for submission to NDIS. Although not retained at a national level, more than half of the states also maintain juvenile offender profiles in their SDIS databases. The U.S. Army, the FBI, and Puerto Rico also participate in CODIS.

Participation in CODIS is limited to public laboratories that meet the Quality Assurance Standards issued by the FBI director, which were devised to ensure that all participating laboratories would be compliant with respect to quality and integrity of data and overall laboratory competency. In some states, this means only one laboratory is a CODIS and/or NDIS participant. Other states have multiple CODIS laboratories in operation at local and/or state levels. For example, California, Florida, and New York have eleven, ten, and nine, respectively. Arizona has seven CODIS labs, while Michigan, Virginia, and Alabama each have four; New Jersey, Utah, and Nevada each have one. DNA profiles cannot be submitted directly from private laboratories. Documents outlining these standards were mentioned previously. Accreditation demonstrates compliance with these standards. Two accrediting bodies exist, the American Society of Crime Laboratory Directors-Laboratory Accreditation Board and the Forensic Quality Services. These entities audit laboratories against a document created by the FBI to check for compliance with the Quality Assurance Standards.

Additional information regarding CODIS and a brochure may be obtained from the FBI's website at http://www.fbi.gov.

"We must all indeed hang together . . ."
—Benjamin Franklin

The words of the great inventor and patriot Benjamin Franklin to John Hancock on the occasion of the signing of the Declaration of Independence are a fitting heading to the concluding section of this history. The story of forensic science is fundamentally the story of the application of science in a highly demanding environment, namely as a part of legal or quasi-legal determinations of fact. To understand forensic science we must understand the forensic as well as the science. This concluding section will treat these separately, showing how we arrived at where we are today.

Science

The essential foundations of forensic science are found in the discoveries and inventions of the Industrial Revolution. These gave the tools for the advancement of physics, chemistry, and biology, and were adopted by inquiring minds to apply to physical evidence and contribute to questions such as: What is this? What happened? How and when did it happen? A visitor to a large forensic science laboratory today will see mass spectrometers no larger than a television set, high-pressure liquid chromatographs, comparison microscopes with integrated digital imaging processors, capillary electrophoresis units, Fourier transform infrared spectrometers, and a range of data-processing stations, unimagined either at all or in their current form just twenty-five years ago. The visitor will also see advanced chemical treatment stations for developing latent fingerprints and traps to collect projectiles from firearms. The visit will probably be arranged by discipline, with the largest areas in the laboratory dedicated to analysis of drugs, fingerprints, firearms, and DNA. A few laboratories will have a small area given to questioned document examination.

A visit made between twenty-five and a hundred years ago would have been quite different. There would have been no DNA section, the equipment in the drug section would have been much more rudimentary and limited in its ability to confirm molecular structure, the latent print laboratory would have been recognizable in comparison to today's equivalent but without the ability to enhance stains that is possible with techniques like superglue fuming. The firearms section, if we ignore the contribution of the IBIS automated analysis and database unit, would have been the most familiar to a present-day forensic s cientist. However, the largest section visited in the retrospective tour is pretty well missing in today's laboratory. There would have been a physical evidence unit, with microscopes and serological testing apparatus. This would have been the cutting-edge section, with scientists comparing glass, paint, soil, hairs, and fibers in samples from the scene, victim, and accused. The serologists would have been identifying species of origin and

blood type of blood and semen stains and excitedly sharing results that showed the stain was only found in one in fifty of the general population.

Today's laboratory reflects the environment in which forensic science operates. Continuing growth in drug abuse with newer synthetic substances sold to abusers means that more types must be added to the range that the laboratory can identify. There is a growing emphasis on intelligence from testing of samples to assist identification of origin. The intelligence is used by agencies to identify and so to control the sources of illicit drugs from imported heroin to domestically produced methamphetamine. The use of firearms in crimes of violence shows no sign of abating. The NIBIN agreement (see **National Integrated Ballistic Information Network**) has extended the work of the section to populating and using databases to link crimes through the weapons used and in some cases to identify the history of the firearm and identify the possible owner.

Latent fingerprints and firearms make an interesting couplet. We have seen that the idea of the fingerprint as something that was truly unique to the individual goes back a long way. A flurry of activity in the last quarter of the nineteenth century established fingerprinting as an integral part of law enforcement investigations. It was the first example of Paul Kirk's dictum of forensic science being about identity. It long predated the development of the exchange principle named after Edmond Locard. It was the first application of forensic science to use a form of database to match evidence samples with reference material from known criminals. Firearms examination also has a long history, with scattered individual contributions to the development of the discipline throughout the late nineteenth century and then their consolidation in the early to mid-twentieth century through the work of Goddard and others. They share the property of being the only significant parts of forensic science that belong to forensic science rather than being a niche offshoot of some other and much larger scientific discipline.

Drug chemistry is chemistry, and DNA testing is molecular biology. Forensic science borrowed the discoveries of the wider scientific areas and adapted them for forensic applications. Sometimes they did not do it very well, at least in the beginning. For example, the work in the *Castro* case discussed above received the most damning of all comments from the molecular biologists who reviewed it and declared it not to be of a standard fit for publication.

The role of trace evidence has diminished almost to the point of extinction. Advocates point out that the Locard principle applies to virtually all associations and that the chemical and microscopy talents of the experienced trace examiner can be applied to most of them. They then argue that if traditional trace evidence capacity and capability is lost, then so is the ability to exonerate or corroborate in many serious cases, particularly those in which there is no biological evidence. The contra and prevailing view is that conventional trace examinations are highly labor intensive, take far too long to complete, and leave us with evidence that is unquantifiable as to its significance. The story of Joyce Gilchrist supports that position. Many of the cases that she was criticized for were hair cases. DNA testing of the hairs gave clear-cut information as to origin and the weight of which could be evaluated from databases. Conventional trace examination, even if conducted without flaws, could not approach the objectivity available from DNA.

Any doubt about the correctness of concentrating resources on DNA testing has been dispelled by the incredible effectiveness of national databases coupled with the almost unbelievable sensitivity of the techniques now used. The case for DNA used to hinge on the fact that it produced high-quality evidence in crimes of violence where blood and semen were transferred between victim and assailant. Proponents conceded that there are many other crimes where there is no such transfer. However, today DNA profiles can be developed from what are called "touch traces." The burglar leaves traces on objects touched during entry, the armed robber leaves traces on the gun, and even a smudged fingerprint with insufficient details for conventional ridge analysis can yield a DNA profile.

The outstanding success of DNA to solve crimes and exonerate the wrongfully convicted should mean that forensic science is universally regarded as a valuable tool for the public good. It is not, and to find out why, we must take a look at the second issue identified in the opening paragraph of this section, before ending with a synthesis of the two parts, science and law.

Law

Any judicial or quasi-judicial hearing is a determination of fact. The hearing only happens after someone complains that a law or rule or regulation has been broken, and after some inquiry into the complaint. The proceedings may be criminal, with investigation by law enforcement officers and a hearing at a criminal trial. They may be civil, with each side subjecting the other's case to scrutiny before proceeding to a civil trial. Or they may be regulatory, such as an infringement of rules for drug administration in sport.

In each case, the participants are contesting to have the hearing accept their version of events. If a scientist is involved, then two things happen: (1) The scientist is generally held to be independent of the partisan conflict, and (2) the science itself is regarded as somehow absolute and exact. The two are joined in the concept of the "expert witness," generally recognized in Western legal systems as a witness who, because of education, training, and experience, possesses knowledge beyond the ken of the layperson, and who is therefore permitted (or in the view of the Scottish Appeal Court in Preece, required) to give an interpretation to the court of the meaning of the material presented. In the context of this work, the expert is the forensic scientist, and includes crime-scene, fingerprint, questioned document, and firearms examiners who may not have a university degree but are trained and experienced. In practice, the scope is wider, and translators, for example, are a group considered as expert witnesses.

The two concepts are both wrong: The scientists cannot possibly be independent of the conflict and science is not exact.

The reason why the scientists cannot possibly be independent of the conflict is that they are instructed by one of the adversarial parties. Lawyers will not relinquish what they regard as their right to have their own expert testing and their own opinion on the evidence, so the obvious solution of court-appointed experts will not happen. The information that the scientists are given and the tests that they are asked to conduct are specified by the instructing party. The presentation above of the *Chamberlain* case briefly contrasted the inductive and deductive approaches to forensic science. The potential coloring of work in the deductive approach is clear: The instructing side says, "here is the evidence, what can you tell me from it that shows my client is correct and the other side is not?" The frame of reference for the scientist is thus constrained and ultimately biased from the start. It is important that it is understood that "biased" is not the same as "flawed"—the work can be biased, but accurate. *Webster's* definitions of bias include: "an inclination of temperament or outlook: such prepossession with some object or point of view that the mind does not respond impartially to anything related to this object or point of view."

The inductive approach is just as bad. Everyone with an interest in forensic science should read David Macauley's excellent book *Motel of the Mysteries* (Houghton Mifflin, New York, 1979). Set in the year 4022, the book chronicles the adventures of amateur archeologist Howard Carson as he explores the lost continent of Usa, where he comes upon the remains of a motel. With absolutely no knowledge of what a motel was, or any of the room incidentals, Carson classifies the site as a burial chamber of the ancients. It is humorous and, to a forensic scientist, chilling in its absolutely logical unfolding of events leading to a conclusion on the part of Carson that is completely incorrect. There may be no question that the mind is entirely impartial but it can be totally wrong.

The idea that science is exact is just as off the mark as the idea that isolation from knowledge of the circumstances will somehow ensure the purity and relevance of the investigation. No science is exact and no scientist should claim that it is. Fundamental principles and data from applications tell us so. Actually, the concept of science itself is not exactly exact. There is general agreement about something known as the scientific method and that has been with us since the time of Francis Bacon, if not earlier. But asking someone to define what makes something scientific will illicit a range of responses. Most will include some form of Popper's concept of falsification. Pure science is about advancing our knowledge and understanding of nature by the process of formulating and challenging laws of the way that the universe works. Scientific laws are predictive statements that have stood the test of time and have been arrived at from observations, that is, inductively. The key to Popper's view is "stood the test of time." He says that no amount of testing can prove something is an absolute law; however, the results of just one well-designed experiment can disprove it, hence "falsification." And so it is with associative evidence. No amount of testing can prove the association did take place, but just one well-designed test can disprove it. Even fingerprints and DNA tests fall into this category because the hypothesis being challenged is the story of the complainant. The DNA could have arisen from consensual sex, the fingerprint may have been deposited by a legitimate activity at a different time. In other words, there are alternative explanations for the forensic science test results, just as there are for the experimental findings in the test of the scientific law.

Thomas Kuhn's *Structure of Scientific Revolutions* (University of Chicago Press, 1996) presents a different view of science but one of considerable relevance to forensic science. Kuhn argues that science is identified by the behavior of scientists, who follow unwritten consensus rules until there is a compelling reason to shift. Perhaps the time is coming for a revolution in the unwritten rules of forensic science.

"He was a gentleman on whom I built an absolute trust."
—Shakespeare, *Macbeth,* I, iv, 7

So said Duncan of Macbeth. Something similar might be said by many about forensic science today, and the question of the reliability of forensic science must be addressed. The concept that it is not an exact science must not be confused with the belief that forensic science is therefore unreliable. The best illustration of inexact in the sense of applied testing is that of blood alcohol. Many factors come into play in the analysis. The exact volume of blood used, the settings of the instrument, the exact amount of ethanol in the standard, whether the operator is tired or alert. All these add up to the fact that no two tests on the same sample will give exactly the same result. The size of the variability can be measured and reasonable decisions made as to the acceptable maximum or minimum result. Typically most of the results from repeated analysis of a single sample in a blood-alcohol test will fall in a relatively narrow range. If we define the value of the average of all the tests as 100 units, then very few individual results will be greater than 106 or less than 94 units. In the case of measurement of tetrahydrocannabinol (THC), the active ingredient of cannabis, in the blood, the individual results will be much more spread out. THC is present at much smaller concentrations in blood, and the nature of the chemical makes the test much more complex than that for the relatively high levels of the relatively simple chemical ethanol.

Here the spread of results could easily be between 25 and 175 units.

The issue for the opponents in the hearing is not whether the scientific test result is exact, but rather whether it is reliable and to what degree. The answer lies in the concept of objective test, defined as one "which having been documented and validated is under control so that it can be demonstrated that all appropriately trained staff will obtain the same results within defined limits" (International Laboratory Accreditation Cooperation Guide 19:2002 Guidelines for Forensic Science Laboratories, available at http://www.ilac.org/). Factors that contribute to "control" include the validation of the test by a range of techniques. The "objective test" therefore meets the needs of users for reliability and acceptability.

Isaac Newton, one of the greats of science, modestly said, "If I have seen further it is by standing on ye shoulders of Giants." The tale of forensic science has its share of giants: Orfila, Bertillon, Locard, Goddard, Vollmer, and others. Unfortunately, it has had more than its share of pseudo giants, too. Mention was made earlier of Michael Saks and his thoughtful work on bias in forensic science. Saks, like many other lawyers, is highly critical of the *ipse dixit* (literally "he himself said it," meaning an unsupported assertion usually made by a person of standing) mentality of some forensic scientists. We first saw that in Orfila, who, although an outstanding scientist of his time, became sought-after not for his ability but for his status. Cameron in the dingo baby case and Zain as perceived by the Texas and West Virginia police are other examples.

If forensic science has done anything in recent years to assure the public and users about its quality, that thing is the move away from the individual "expert" to a more controlled systems approach. There are well-established and internationally respected processes for control of the reliability of laboratory testing. They have widespread applicability in fields from aircraft construction to food safety. The same principles are now being introduced into forensic practice. The benchmark is accreditation programs that are based on the requirements of ISO/IEC 17025, General Requirements for the Competence of Testing and Calibration Laboratories published by the International Organization for Standardization. These programs incorporate standards for the training, education, and competency of personnel; for controlling the quality of testing procedures; and for continuing internal and external reviews of operations. Cases such as *Preece* and *Chamberlain* and individual problems such as with Zain should be prevented by conformity to ISO 17025 requirements.

" . . . all the hopes of future years."
—Henry Wadsworth Longfellow, The Building of the Ship (1849)

Testing will move from the laboratory to the crime scene. The United States led the way with the adoption of the "breathalyzer," and the technical reasons for not having an equivalent portable device for the identification of drugs, fingerprints, and DNA are being solved. Drugs will probably be first, with field instruments to identify controlled substances in someone's possession, and detect from a saliva sample whether or not their levels in blood constitute an impairment to driving. The prospect of a crime-scene examiner passing a sensor over a semen stain at a scene and getting a report within ten minutes of the name and last-known address of the person it came from may not be just around the corner, but is well within the realm of science fact rather than science fiction. DNA testing will also produce an anthropological sketch of the source—sex, race, height, and hair and eye color, for example.

Based on current trends, most future advances are going to be driven by security needs. Foolproof and ultrasensitive sniffers to identify explosives on the clothing of suicide bombers or in car bombs are needed. The recovery and analysis of computer data will need to keep pace with the ability of those who wish it to be kept secret to hide digital evidence. Biometric data systems to permit and prohibit passage through checkpoints such as airline and airport security and immigration screening will be commonplace. It may be possible to implement a reliable lie detector by 2023 and overturn a century of case law based on *Frye*, as well as simplify criminal investigation and trials.

Beyond that, all that can be said is that forensic science has an excellent record of latching onto advances in science and adapting them for investigative and evidentiary use and will probably continue to do so.

Further Reading: Books and Journal Articles

Andreasson, R., and A. Jones. "Historical anecdote related to chemical tests for intoxication." *Journal of Analytical Toxicology* 20 (1996): 207–208. An interesting historical note, referred to in the text.

Borkenstein, R. F., R. F. Crowther, R. P. Shumate, W. B. Ziel, and R. Zylman. "Report on the Grand Rapids Survey." Department of Police Adminstration, Indiana University, 1964. This is a classic publication that is out of print, but is widely referenced in works on drinking and driving. It should be available from libraries.

Conan Doyle, Sir Arthur. *The New Annotated Sherlock Holmes: The Complete Short Stories.* Edited by J. Lecarre and L. Klinger. New York: W. W. Norton, 2004.

DiMaio, V. J. M., and D. DiMaio. *Forensic Pathology.* 2nd ed. Boca Raton, FL: CRC, 2001. One of the standard U.S. textbooks on forensic pathology.

Electronic Crime Scene Investigation: A Guide for First Responders. Washington, DC: U.S. National Institute of Justice, 2001.

Exonerated by Science: Case Studies in the Use of DNA Evidence to Establish Innocence after Trial. Washington, DC: U.S. National Institute of Justice, 1996.

Fay, S., L. Chester, and M. Linklater. *Hoax: The Inside Story of the Howard Hughes–Clifford Irving Affair.* New York: Viking, 1972.

Fisher, B. A. J. *Techniques of Crime Scene Investigation.* Boca Raton, FL: CRC, 2000.

Forensic Examination of Digital Evidence: A Guide for Law Enforcement. Washington, DC: U.S. National Institute of Justice, 2004.

Golan, T. *Laws of Men and Laws of Nature: The History of Scientific Expert Testimony in England and America.* Cambridge, MA: Harvard University Press, 2004. One of the very best books available for those interested in the interaction of science and law.

Gross, H. *Criminal Investigation.* 4th ed. Edited by Ronald Martin Howe. London: Sweet and Maxwell, 1949. This book is long out of print but copies may be available from libraries.

Hamby, J. E., and J. W. Thorpe. "The History of Firearm and Toolmark Identification." *Association of Firearm and Tool Mark Examiners Journal,* 30th Anniversary Issue. Vol. 31, no. 3 (1999).

Helmer, W. J., and A. J. Bilek. *The St. Valentine's Day Massacre: The Untold Story of the Gangland Bloodbath That Brought Down Al Capone.* Nashville, TN: Cumberland House, 2004.

Hilton, O. *Scientific Examination of Questioned Documents.* Rev. ed. New York: CRC, 1992. This is the only one of the three classic books on questioned document examination (the others are Albert S. Osborn's *Questioned Documents* and Wilson Harrison's *Suspect Documents*) that is currently available.

Houck, M. M. *Trace Evidence Analysis: More Cases in Mute Witness.* Burlington, MA: Elsevier, 2004.

Jasanoff, S. *Science at the Bar: Law, Science, and Technology in America.* Cambridge, MA: Harvard University Press, 1997. Certain to irritate any forensic scientist reading it, but well-regarded by attorneys.

Kakis, F. J. *Drugs: Facts and Fictions.* New York: Franklin Watts, 1982.

Kind, S. S. *The Scientific Investigation of Crime.* London: Forensic Science Services, 1987. This is the report of an internally commissioned study and is no longer available for purchase. It may be available from libraries. The publication's ISBN is 0–9512584–0–0. Those interested in how scientific principles can be used to investigate crime should try to obtain a copy. Many of Stuart's ideas form the basis of modern "cold case" investigations.

Knight, B., and P. Saukko. *Forensic Pathology.* London: Hodder Arnold, 2004. Bernard Knight's *Forensic Pathology* is the standard text in Britain. This is the most recent edition, updated by his coeditor, Dr. Saukko.

Kuhn, T. S. *The Structure of Scientific Revolutions.* Chicago: University of Chicago Press, 1996. Not an easy read for the nonscientist but a book that influenced business and scientific thinking—this work is where the popular "paradigm shift" concept came from.

Lee, H., and J. Labriola. *Famous Crimes Revisited.* Southington, CT: Strong, 2001. Henry Lee is well known for his many appearances in the media giving expert commentary on forensic science cases. This book deals with the Sacco-Vanzetti case, O. J. Simpson, the Lindbergh kidnapping, Sam Sheppard, and JonBenet Ramsey, among others.

Macaulay, David. *Motel of the Mysteries.* Boston: Houghton Mifflin, 1979. A "must read" to see how apparently logical conclusions can be horribly wrong. Lay readers interested in science will enjoy other books by the same author.

Popper, K. *The Logic of Scientific Discovery.* Routledge Classics. Routledge, London and New York, 2004. Popper is the father figure in modern scientific philosophy. His concept of falsification is a powerful tool to distinguish good science from junk science. This paperback edition of his *Logic* is a must-read for those who struggle with the principles behind the *Daubert* decision and for those truly concerned about keeping poor science out of the courtroom.

Reference Manual on Scientific Evidence. Washington, DC: Federal Judicial Center, 2000. This is the definitive reference on science and the law.

Royal Commission of Inquiry into Chamberlain Convictions. "Report of the Commissioner the Hon. Mr. Justice T. R. Morling." Government Printer of the Northern Territory, 1987.

Rudin, N., and K. Inman. *An Introduction to Forensic DNA Analysis.* 2nd ed. Boca Raton, FL: CRC, 2003.

Saferstein, R. *Criminalistics.* 8th ed. Upper Saddle River, NJ: Pearson Education, 2004. This book is so well accepted as the basic forensic science text in the United States that it is known simply as "Saferstein."

Saks, M. J. *The Daubert / Kumho Implications of Observer Effects in Forensic Science: Hidden Problems of Expectation and Suggestion with Risinger, Rosenthal & Thompson.* 90 CAL. L. REV. 1 (2002). Saks is one of the leading writers challenging the basic way in which crime laboratories operate. This paper is one of many and gives an insight into his perspective.

Smith, S. *Mostly Murder.* New York: Dorset, 1989. Read this to learn how the "expert" sees himself.

Twain, Mark. *Mississippi Writings: Tom Sawyer, Life on the Mississippi, Huckleberry Finn, Pudd'nhead Wilson.* New York: Library of America, 1982.

U.S. Department of Justice, Office of the Inspector General. "Status of IDENT/IAFIS Integration" Report No. I–2002–003, December 7, 2001.

Wambaugh, J. *The Blooding.* New York: William Morrow, 1989. One of the first books to jump on the DNA "bandwagon" but a most enjoyable read.

Wecht, C., M. Curriden, and B. Wecht. *Cause of Death.* New York: E. P. Dutton, 1993.

Wilson, Colin. *Written in Blood: The Trail and the Hunt.* New York: Warner Books, 1989. Could well have been titled "how science solved crime before DNA," covers interesting cases and easy to follow.

Further Reading: Cases

Daubert v. Merrill Dow Pharmaceuticals, Inc., 509 U.S. 579, 596 (1993).

Frye v. United States, 293 F. 1013 (D.C. Cir. 1923).

Opinion of the Supreme Court of Appeals of West Virginia, filed in November 1993 (No. 21973 "In the matter of an investigation of the West Virginia State Police Crime Laboratory, Serology Division"). http://www.state.wv.us/wvsca/DOCS/FALL93/21973.htm

People v. Castro, 545 N.Y.S. 2d 985 (Sup. Ct. 1989).

People v. Jennings, 96 N.E. 1077 (1911).

Regina v. Buckely, 143 SJ LB 158 (1999).

State v. Bogan, 905 P.2d 515 (1992).

United States v. Havvard, 117 F. Supp. 2d 848 (D. Ind. 2000).

United States v. Llera Plaza, 181 F. Supp. 2d 414 (E.D. Pa. 2002).

Further Reading: Websites

Although websites do not have the advantages of permanency and prepublication review possessed by most books and journals, there are some that are worth visiting, either because of their official standing or because they have displayed a degree of longevity. Some are cited below.

www.DNA.gov is a recent initiative from the U.S. Department of Justice's National Institute of Justice (NIJ). It is being established as a "one-stop shop" on forensic DNA.

www.ojp.usdoj.gov/nij/ is the homepage of the NIJ.

www.ascld.org is the website of the American Society of Crime Laboratory Directors. The society's newsletter can be accessed from the site. Although it is generally directed to professional and technical matters, the newsletter does contain some articles of more general interest.

http://forensic.to/forensic.html will take you to Zeno's Forensic Web Site. This is a reliable site that has been in operation for many years and provides comprehensive and up-to-date information.

http://www.forensic.gov.uk/forensic/news/press_releases/2003/NDNAD_Annual_Report_02–03.pdf is the source for the FSS Annual Report on the National DNA Data Base, only available in electronic form.

http://www.fbi.gov/hq/lab/codis/clickmap.htm provides statistics for the U.S. National DNA Database

A

ABO Blood Groups

There are many thousands of inherited characteristics in blood. Identifying these can be used in forensic science to associate a stain of body fluid with a possible source. However, in everyday language, *blood group* refers to the characters that determine whether blood transfusions will be compatible between a donor and recipient.

Physically, blood is a suspension of red cells in a fluid called serum (strictly speaking, in the body the fluid is plasma, which is converted to serum by removal of fibrinogen and platelets during the clotting process). Karl Landsteiner established the basis of transfusion compatibility at the turn of the twentieth century. He took blood from a number of donors and separated the cells and serum, then mixed them. He found that some serum-cell combinations resulted in clumping of the cells.

Landsteiner identified four basic blood types. In type A, the cells have a chemical on their surface—an antigen—that reacts with an antibody in serum from type B people. In type B, the antigen on the cells reacts with antibody in serum from type A people. Group O people have neither of these antigens but both types of antibody. Group AB people have both antigens but neither antibody. Thus group A people

will respond to transfusions of type B blood by their antibody destroying the transfused cells.

The frequency of the various ABO blood groups in the population has been measured and is shown in the table below. These are approximate values for the population as a whole—the exact frequencies vary by race and ethnic group.

ABO blood types can also be detected in other body fluids such as semen and saliva in more than 80 percent of the population.

The ABO blood type of the donor of a blood or semen stain can be shown by very sensitive testing techniques. Provided that the stain is dry and has not putrefied, the antigens can be preserved for many years.

ABO typing was therefore one of the main methods used to characterize body fluid stains before the widespread adoption of DNA typing.

Table 2 Blood Type

Type	Cells	Serum	% of Population
A	A	anti-B	41
B	B	anti-A	10
O	none	anti-A and anti-B	45
AB	A and B	none	4

See also Blood; Blood Grouping; Saliva; Semen Identification

References

De Forest, P., R. E. Gaensslen, and H. C. Lee. *Forensic Science: An Introduction to Criminalistics.* New York: McGraw-Hill, 1983.

James, S. H, and J. J. Nordby. *Forensic Science: An Introduction to Scientific and Investigative Techniques.* Boca Raton, FL: Taylor and Francis, 2005.

Saferstein, R. *Criminalistics.* 8th ed. Upper Saddle River, NJ: Pearson Education, 2004.

White, P. *Crime Scene to Court: The Essentials of Forensic Science.* Cambridge: Royal Society of Chemistry, 1998.

Abortion

The forensic investigation of criminal abortion relates to identification of fetal remains and association with a putative mother or the investigation of maternal death. Maternal death is very rare in countries where abortion is legal. However, it does occur and requires medical investigation.

Criminal abortion can result from physical violence, from the use of abortifacient drugs or chemicals, and the use of instruments. Physical violence can be self-inflicted, accidental, or deliberate but not necessarily intended to result in abortion—domestic abuse for example. Drugs such as pennyroyal and oil of turpentine that can induce menstrual flow can result in abortion. Other drugs that cause contraction of the wall of the uterus are also encountered. Most effective abortifacient drugs are controlled by law and not readily accessible. By contrast, instruments used to procure abortion by physical interference, with or without introduction of such substances as soap solutions or slippery elm, are readily available. They are also dangerous. Death can result from shock, air embolism, instrument injury, or sepsis.

Where illegal abortion is suspected, therefore, the examination of the deceased should include the physical condition of the uterus and contents, such as perforations, signs of hemorrhage, and foul-smelling products. Signs of abortifacients such as soapy fluids or foreign bodies should be looked for. Evidence of air embolism in pregnancies of less than twenty-four weeks duration is also an indication that death may have been related to a failed illegal abortion. Examination of fetal remains consists of physical and histological measurements of the fetus and organs.

Abortion is an issue in domestic abuse, where pregnancy may be a stimulus to, and object of, physical violence by the male domestic partner.

References

Byard, R., T. Corey, C. Henderson, and J. Payne-James. *Encyclopedia of Forensic and Legal Medicine.* London: Elsevier Academic, 2005.

Knight, B. *Simpson's Forensic Medicine.* London and New York: Oxford University Press, 1997.

Payne-James, J., A. Busuttil, and W. Smock. *Forensic Medicine: Clinical and Pathological Aspects.* London: Greenwich Medical Media, 2003.

Accelerant Residues

Scientific examination of fire scenes is a special area of forensic science. The main role of the testing is to determine how the fire started—that is, whether it was the result of an accident or arson—but seldom provides evidence of the individuals who may have been involved. However, this is not always so. For example, if an unusual accelerant is identified, the information may lead to development of a suspect based on information about the purchase of the material. Evidence of an accelerant in debris from the scene is usually a good indication of a deliberately set fire.

Examination of the fire scene is a two-stage process. It begins with the scene investigator looking for physical evidence of the ignition source, such as a burned out electrical appliance. Where the source is localized but not associated with a likely cause, arson using an accelerant must be suspected.

Even after a fierce fire, traces of the accelerant will remain. The traces can be recovered from the debris and identified in the laboratory. Trained dogs can be used at the scene to indicate the location of possible accelerant residues. In the laboratory, techniques such as gas chromatography and (preferably) gas

Firefighters examining evidence from a house fire. Evidence collected from the scene of a fire can help distinguish an accident from arson. (Michael Donne / Photo Researchers, Inc.)

chromatography–mass spectrometry (GC-MS) are used. The assay consists of extraction of residue from the debris, followed by its characterization by chemical analysis. Several extraction methods have been used over the years, but most laboratories now use some form of adsorption-desorption process. Headspace (the container gas space above the sample) from the debris is collected on an adsorptive strip such as activated charcoal. Accelerant traces are then desorbed chemically or by heating. Assay is best conducted by GC-MS. Ignitable liquid residues detected are characterized according to a scheme published by the American Society for Testing and Materials (ASTM).

Accelerants commonly used to set fires include gasoline, lighter fuels, and alcohols. Interpretation of results requires knowledge of the effects of the fire on the composition of the accelerant. Ignitable liquid accelerants, such as gasoline, are usually complex mixtures. The heat of the fire will distort the relative proportions of the constituents.

The more volatile components are lost to a greater extent than those with a higher boiling point and so the apparent composition of the residue from gasoline, for example, will not be exactly the same as that of the original liquid. The distortion is referred to as *weathering*. Interpretation also requires awareness of the effects of the material burned. For example, pinewood contains chemicals of the same class as the constituents of turpentine.

See also Arson; Gas Chromatography; Mass Spectrometry

References

De Forest, P., R. E. Gaensslen, and H. C. Lee. *Forensic Science: An Introduction to Criminalistics.* New York: McGraw-Hill, 1983.

James, S. H., and J. J. Nordby. *Forensic Science: An Introduction to Scientific and Investigative Techniques.* Boca Raton, FL: Taylor and Francis, 2005.

Saferstein, R. *Criminalistics.* 8th ed. Upper Saddle River, NJ: Pearson Education, 2004.

White, P. *Crime Scene to Court: The Essentials of Forensic Science.* Cambridge: Royal Society of Chemistry, 1998.

Accreditation

Laboratory accreditation is the process of an independent competent authority inspecting policies, practices, and procedures for compliance with a credible set of standards. It provides the public and users of the services with an objective measure of reliability of the laboratory's operations.

There are two main accreditation programs applicable to forensic laboratories. The first is the one provided by the American Society of Crime Laboratory Directors Laboratory Accreditation Board (ASCLD/LAB). This has been in operation since 1982, and at the start of 2003, 237 of the approximately 350 crime laboratories in the United States were ASCLD/LAB accredited.

There are also laboratories in Australia, Canada, Hong Kong, New Zealand, and Singapore that are ASCLD/LAB accredited. However, the main accreditation programs for testing laboratories worldwide are based on ISO Standard 17025. This has been used as the basis of forensic programs in Canada, Australia, England, and Holland.

The ASCLD/LAB program recognizes controlled substances, biology (including DNA), trace evidence, toxicology, latent print development and comparison, questioned documents, firearms and tool marks, and crime scene examination as areas of testing or disciplines in forensic science. In addition to crime scene examination, a laboratory must submit itself for evaluation in all disciplines in which it is active and cannot elect to become accredited in a selected subset of a discipline. The evaluation consists of a compliance audit against a published set of standards, expressed as criteria. The criteria are classified as "essential," "important," and "desirable." To become accredited, a laboratory must meet all essential criteria that apply, 75 percent of important criteria, and 50 percent of desirable criteria.

The ISO-based programs differ somewhat in operation. The main international standard provides a set of clauses describing quality management and technical requirements that must be met. These same standards apply to all testing laboratories seeking ISO accreditation, no matter what the testing area is. The program is tailored to specific testing areas, such as forensic science, by means of amplification documents that provide guidance on acceptable interpretations of clauses in the field of application. There is no list of disciplines, and laboratories are not required to seek accreditation in every area in which they conduct testing. However, consensus operation of ISO programs requires that the laboratory and its accrediting body develop and publish a scope of accreditation that describes the tests and materials covered by the accreditation. Forensic Quality Services (FQS) was formed by the National Forensic Science Technology Center (NFSTC) and provides ISO-based accreditations to forensic science laboratories in the United States.

There is, de facto, a third accreditation program in forensic testing in the United States. This is for forensic DNA-testing and DNA-databasing laboratories. Laboratories seeking to produce data for inclusion in the national DNA database (see **Combined DNA Index System [CODIS]**) or to receive federal government financial support for such work are required to show compliance with national quality assurance standards. Such compliance can be demonstrated by accreditation through the ASCLD/LAB or FQS accreditation programs that use the national standards as field specific guidelines. Direct accreditation can also be obtained through FQS.

There is no program that establishes compulsory certification of forensic laboratory testing personnel (see **American Board of Criminalistics** for details on a voluntary program). However, all three accreditation programs contain requirements for establishing and maintaining the competency of personnel. Typically, these are requirements for training, professional development, and proficiency testing.

Training requirements establish the knowledge base and practical skills required before being permitted to conduct unsupervised casework. Professional development establishes pathways whereby analysts continue to improve their theoretical and practical abilities. Proficiency testing provides a process for monitoring the overall competency of systems and individuals by presenting them with samples the composition of which is known (or can be determined) by the testing agency but not by the analyst.

See also American Society of Crime Laboratory Directors/Laboratory Accreditation Board (ASCLD/LAB)

References
American Society of Crime Laboratory Directors/Laboratory Accreditation Board; www.ascld-lab.org (Referenced July 2005).
Forensic Quality Services; www.forquality.org (Referenced July 2005).

Acid Phosphatase

Acid phosphatase (AP) is an enzyme found in many body tissues, including the prostate, that is used as a screening technique for semen. Human seminal fluid contains concentrations of the enzyme several hundred times higher than any of the other tissues in which it can be detected, including vaginal secretions. Detection of AP cannot therefore be used as proof of the presence of semen (it is found in other tissues, especially vaginal) but can be used as a reliable screening procedure (it is found in especially high levels in semen).

The enzyme promotes the removal of phosphate groups from organic molecules and its activity is greatest under acid conditions. This contrasts with alkaline phosphatase, an enzyme found at high levels in other tissues, including bone. The test usually performed in screening for semen applies an acid solution of sodium α-naphthyl phosphate together with a dye called Fast Blue B to the suspect stain. If AP is present, the phosphate bond is hydrolyzed and a purple diazo compound is formed within thirty seconds.

The usual next step is examination of stain extracts for human spermatozoa, which provides a specific confirmation of human semen. Acid phosphatase as a screening test is gradually being replaced by methods based on detection of a prostate specific protein called P30.

See also Ejaculate; Semen Identification

References
Butler, J. *Forensic DNA Typing.* 2nd ed. Burlington, MA: Elsevier, 2005.
De Forest, P., R. E. Gaensslen, and H. C. Lee. *Forensic Science: An Introduction to Criminalistics.* New York: McGraw-Hill, 1983.
James, S. H., and J. J. Nordby. *Forensic Science: An Introduction to Scientific and Investigative Techniques.* Boca Raton, FL: Taylor and Francis, 2005.
Saferstein, R. *Criminalistics.* 8th ed. Upper Saddle River, NJ: Pearson Education, 2004.
White, P. *Crime Scene to Court: The Essentials of Forensic Science.* Cambridge: Royal Society of Chemistry, 1998.

Admissibility of Scientific Evidence

There are many sources (case law, rules of evidence, statutory requirements) governing the admissibility of scientific or expert evidence in a court of law. The most well-established is the *Frye* case of 1923, which set a standard requiring that the basis of the scientific examination at issue had to be generally accepted in the field. This is a conservative approach that has been criticized for denying the court useful information that could be gleaned from novel scientific tests. In *Frye,* the court ruled against the admissibility of polygraph (lie detector) test results on the grounds that it was a novel method and not generally accepted in the field.

The Federal Rules of Evidence (Rule 702) are more liberal and are based on a determination by the judge as to whether or not the evidence will assist the trier of fact to understand or determine a fact or issue.

More recently, Supreme Court decisions in *Daubert v. Merrill Dow Pharmaceuticals* (1993) and *Kumho Tire Co. v. Carmichael* (1999) have confirmed that the gatekeeper for admissibility is the trial judge, but have offered a framework for the decision-making process. The framework consists of a number of questions, each of which attempts to address whether the evidence is scientific.

The *Daubert* decision covered many aspects of what would permit a judge to accept evidence as scientific. They included whether the evidentiary method at issue was taught at a university, whether it was the subject of research published in peer-reviewed literature, whether error rates for the method were known, and whether the hypothesis upon which it was based could be falsified. Falsification is part of the deductive-inductive cyclical process of the scientific method. A possible scientific law is deduced from observation and expressed as a hypothesis. To have validity, the hypothesis must forbid something. Experiments are then planned to test the hypothesis by searching for the occurrence of the forbidden attribute. This is the process of falsification, usually associated in scientific philosophy with the writings of Karl Popper. It is a very simple and powerful concept, yet one that puzzled a minority of the judges on the *Daubert* Court.

In contrast to "scientific" evidence, the acceptance of an "expert" is a pragmatic decision that the witness possesses skills or knowledge beyond those of a layperson and that these skills or knowledge can assist the court. *Daubert* thus soon came under challenge in regard to the admissibility of well-established evidence such as fingerprint comparisons, questioned document examinations, and exami-nation of tool marks and footwear impressions. This seems to have been resolved (for example in *Kumho Tire*) in a confirmation of the admissibility of expert evidence provided that the witness can show that the work was performed in a manner that resulted in reliable findings and conclusions.

However, *Daubert* (and *Kumho*) are misleading and have led to needless confusion. The lack of understanding of the courts about science is partly forgivable—scientists are confused about science. What is less forgivable is that the *Daubert* Court was really trying to assess the reliability of conflicting scientific evidence, albeit in a very complex case. The reliability of testing is a quite different issue and depends on the application of correctly controlled objective tests.

See also Frye Rule; *Daubert* Ruling; *Kumho Tire* Ruling

References

Daubert v. Merrill Dow Pharmaceuticals. 509 U.S. 579 (1993).

Frye v. United States, 293 F. 1013 (D.C. Cir. 1923).

Kumho Tire Co. v. Carmichael, 526 U.S. 137 (1999).

Age Estimation

Determination of age occurs in forensic examinations for two main reasons: (1) to determine whether a threshold age of legal significance has been achieved, such as the age of consent for sexual intercourse, and (2) in identification of remains. Some factors used are teeth, ossification of bones, and closure of skull sutures. However, most of these are complete by age twenty-five and age determination thereafter is more problematical.

Age determination from teeth is based on the time scale of eruption of teeth in the child and their replacement by permanent teeth. Age determination by ossification and closure of skull sutures depends on the conversion of young cartilage or fibro-membrane-based bones as calcification and hardening occur.

References

Bowers, C. M., and G. Bell. *Manual of Forensic Odontology*. 3rd ed. Saratoga Springs, NY: American Society of Forensic Odontology, 1995.

Byers, S. N. *Introduction to Forensic Anthropology: A Textbook*. Boston: Allyn and Bacon, 2001.

Alcohol

Alcohol is the name for a family of chemicals of the general formula $C_nH_{2n+1}OH$. Ethyl alcohol, also called ethanol, is the common form of potable alcohol and has the formula C_2H_5OH. Other forms, such as methyl alcohol (CH_3OH), are poisonous.

Alcohol in alcoholic beverages is produced from fermentation of sugar by yeasts. In beers and wines the concentration of alcohol is about 5 percent and 13 percent alcohol by volume respectively. The final alcohol concentration is determined by the amount of sugar in the starting material and the duration of the fermentation. There is a natural maximum of about 15 percent as the alcohol kills off the yeast at higher concentration. Stronger alcoholic beverages are produced by distillation of the fermented liquor. Spirits usually have alcohol concentrations in the range 40 to 50 percent.

Alcohol has been imbibed socially for centuries. Alcoholic beverages are appreciated aesthetically, and moderate intake of wine has been associated with reduced risk of heart disease. On the other hand, alcohol is an addictive, depressant drug that is poisonous at high concentrations and results in considerable social harm through its causal role in road traffic accidents and in personal violence. Chronic alcohol abuse can result in liver failure.

The observed pharmacological effects of ingesting alcohol depend on the sequence in which it depresses functions controlled by the brain. The sequence from pleasure to death is well-known. The first effect of alcohol is on the brain area responsible for inhibitions. The sequence continues through speech control (slurring and unattenuated volume), to motor control (staggering), to coma (passing out). If enough is taken, death due to depression of the center controlling breathing will result.

However, many people show marked tolerance to the effects of alcohol. A chronic drinker will be much less affected than a moderate one. For example, most people will be unconscious at blood-alcohol levels of 0.350 percent, but such readings are encountered regularly in the laboratory in samples from drunk drivers. There are recorded cases of recovery from alcoholic coma with blood levels greater than 1 percent.

Alcohol and Drinking

Consumption of alcoholic beverages is a widely accepted social activity. Unfortunately, there is no doubt that there is a causal relationship between alcohol intake and the likelihood of causing a vehicle accident. Because driving is a major part of everyday life, alcohol cases—specifically drunk driving—probably affect a greater number of citizens than any other aspect of forensic science.

People vary in their sensitivity to the effects of alcohol intake and to the relationship between intake and blood-alcohol level. However, the physiological processes controlling the relationship are well understood and the variation is a normal biological phenomenon.

When an alcoholic beverage is consumed, it first enters the stomach. From there, some of it is taken up into the bloodstream and passes to the liver (which is also the main organ by which it is detoxified and thus eliminated from the body), and so to the body as a whole, including the brain (where it exerts its pharmacological activity). Any unabsorbed alcohol continues its passage into the small intestine where absorption is completed.

The rate at which the alcohol enters the bloodstream depends on the local conditions in the stomach and small intestine. These include the rate at which the stomach empties into the small intestine, the extent to which the alcohol acts on the lining of the gut to increase its blood flow, and the effects of foodstuffs consumed at the same time.

Absorbed alcohol is dissolved in the body's water. This means that the same amount of alcohol consumed will result in a blood-alcohol level inversely proportional to lean body size. Lean body mass is about 70 percent of the weight of a man of normal build and about 65 percent for a woman. Together with their lower average weight, this means that the same amount of alcoholic beverage will result in higher blood-alcohol levels in women than in men.

Alcohol is metabolized mostly by the liver, by chemical conversion by the enzyme alcohol dehydrogenase. Some is eliminated unchanged in urine and a tiny amount in breath.

The relationship between alcohol intake and blood alcohol is thus a balance between the intake (amount and rate of drinking, and local effects on absorption in the stomach and intestines), distribution in the body (lean body size), and elimination (by the liver and in urine).

It so happens that the usual volumes of alcoholic beverages and the concentration of alcohol in them are such that there is approximately the same amount of alcohol in a shot of liquor (whisky, vodka, rum, etc.) as in a glass of wine or a glass of beer.

For an average person, drinking in social circumstances, one drink will result in a maximum blood alcohol of about 0.015 percent about one half hour after drinking. There is a slight relationship between the nature of the drink and the resulting blood alcohol level, with carbonated drinks of moderate concentration being the most rapidly absorbed. Examples include vodka and coke, champagne, and gin and tonic. The body eliminates the equivalent of one average drink each hour no matter what the nature of the alcoholic beverage. Although many urban legends persist as to sobering cocktails, nothing other than time will have any significant effect on blood-alcohol levels.

It is emphasized that the normal variation in size, body composition, elimination rates, and absorption rates all come together to result in a very wide range of possible blood-alcohol levels per drink and in the time to eliminate the alcohol from the blood. There is thus no "safe" rate of drinking in terms of blood (or breath) alcohol levels.

Alcohol and Driving

There is no doubt that drinking and driving do not mix. For example, it is generally accepted that alcohol is involved in approximately 40 percent of fatal road traffic accidents, within the United States, with the total economic cost for alcohol related crashes exceeding $50 billion each year (National Commission Against Drunk Driving).

There is a well-established relationship between blood-alcohol levels and the likelihood of causing an accident. The benchmark study was performed in the 1970s in Grand Rapids, Michigan. The blood-alcohol levels of drivers who had caused accidents were compared to those in a similar group who had not been involved in an accident. The data can be used to describe the relative risk, from the proportion of the driving population with a given blood-alcohol concentration to that in drivers who have caused an accident. The relative risk rises as blood alcohol reaches about 0.04 percent. The risk rises steeply and disproportionately as blood alcohol continues to increase. Someone with a blood-alcohol level of 0.100 percent is more than 10 times more likely than normal to cause an accident, and one at 0.15 some 40 times more likely (Borkenstein et al. 1964; Borkenstein et al. 1974).

Many people claim that their ability to drive is not affected at lower blood-alcohol levels. However, the Grand Rapids study showing an increased risk even at low levels has been supported in so many other surveys that there is no justification at all for leniency. A British study offered some insight into why accident rates rise at levels where handling skills are probably not affected.

Professional bus drivers were taken to an off-road site. They were divided into three groups: those given no alcohol, those given one drink, and those given three drinks. They were then asked to set gates to a gap through which they believed they could drive their vehicles. They were then allowed to drive through the gates, and their performance was measured.

Actual performance was as expected. The small amount of alcohol had no effect on the average gap that they could clear (but some individuals did show marked impairment). The larger amount was associated with impairment, the group requiring an increase in the gate width to pass safely through. However, even the low-level group showed clear impairment of judgment as to the minimum gap that they estimated they could clear with their vehicles.

We thus see that alcohol will impair judgment before motor skills. A low blood-alcohol level can indeed leave the drinker perfectly able to perform the manipulations required for driving. However, the ability of the driver to respond to a situation requiring judgment will be impaired.

Alcohol Detection in Blood and Breath

Measuring the concentration of alcohol in blood is one of the most straightforward and reliable tests performed in forensic science. The procedure used universally is that of gas chromatography with an internal standard.

The internal standard is another member of the alcohol family that behaves very like ethanol in the assay but is distinguishable from it. A fixed amount of the internal standard is added to the blood at the start of the process. Any variation in each step is corrected for as the ethanol and internal standard are affected in the same way.

A set of samples of known alcohol concentrations is run. The ratio of the detector response for the ethanol to that of the response for the internal standard is calculated and a standard curve prepared. Thus the ratio of ethanol to internal standard in the unknown blood sample can be used to measure the concentration of alcohol in the sample.

There is some variation in the configuration of the test chromatographs. Some are set to measure alcohol in liquid extracts of the blood samples; others measure alcohol in the vapor headspace in a sealed vial of sample and internal standard.

The physics of partition between liquid and gas that allows headspace analysis of alcohol in blood samples also permits estimation of blood alcohol by breath testing. Alcohol in the blood is in equilibrium with the concentration of alcohol in the gases in the lungs. The ratio of concentrations is highly in favor of the blood levels, at 2,100 to 1. The measured breath level is converted to a blood equivalent using the blood to air ratio.

There are many devices used in breath testing. The earliest widespread instrument was the Breathalyzer brand. The original Breathalyzer depended on the development of a color due to oxidation of the alcohol in the sample. Some criticism was levied at earlier breath devices because the detection system was not absolutely specific for ethanol, and substances that could be present in the breath of diabetics could contribute to the reading.

Some of the earlier devices were also susceptible to regurgitation effects. The concentration of alcohol in the beverage that enters the stomach is around 10 percent, the concentration in blood is around 0.015 percent, and the blood to breath ratio is 2,100 to 1. If there is any regurgitation ("burping") of stomach content when the breath test is conducted, it will have a substantial false positive effect on the breath reading.

Modern instruments use a range of detection systems, including infrared spectrometry, electrochemical cells, and miniature gas chromatographs. They have the advantage that they are much more sensitive, specific, and rapid than the Breathalyzer was. This means that operators can take more than one sample for greater accuracy, the instruments allow for rapid repeat testing and can detect regurgitation effects and negate readings so affected, and potential interferences from natural sources, such as acetone in diabetics, can be allowed for.

Alcohol Laws

The clearly established relationship between blood alcohol and accidents has led to most jurisdictions adopting the so-called per se law, in which a blood-alcohol reading over a stated level is taken as evidence of intoxication. Until recently, the level was 0.100 percent in most U.S. jurisdictions, but the federal government is promoting a lower level of 0.08 percent. Other countries generally have lower limits (zero in Scandinavia, 0.05 percent in Australia). The United Kingdom has a direct breath-alcohol limit of 35 μg percent, which is a little under 0.08 percent equivalent in blood.

See also Field Sobriety Tests

References

Borkenstein, R.F., F.R. Crowther, R.P. Shumate, W.B. Ziel, and R. Zylman. The role of the drinking driver in traffic accidents. Department of Police Administration, Indiana University, 1964.

————. The role of the drinking driver in traffic accidents. The Grand Rapids Study. *Blutalkohol*, 11, Supplement 1, 1974.

De Forest, P., R. E. Gaensslen, and H. C. Lee. *Forensic Science: An Introduction to Criminalistics.* New York: McGraw-Hill, 1983.

James, S. H., and J. J. Nordby. *Forensic Science: An Introduction to Scientific and Investigative Techniques.* Boca Raton, FL: Taylor and Francis, 2005.

Levine, B. *Principles of Forensic Toxicology.* Washington, DC: American Association for Clinical Chemistry, 1999.

National Commission Against Drunk Driving; http://www.ncadd.com/ (Referenced December 2005).

Saferstein, R. *Criminalistics.* 8th ed. Upper Saddle River, NJ: Pearson Education, 2004.

White, P. *Crime Scene to Court: The Essentials of Forensic Science.* Cambridge: Royal Society of Chemistry, 1998

Amelogenin

The amelogenin gene is found on the X and Y chromosomes. The gene can be detected in body tissues using the polymerase chain reaction (PCR) procedure. The X- and Y-specific products are of different size, and so sex can be determined, as material of female origin will produce only a single product, but that of male origin will produce both. This process can thus be used to determine gender.

See also DNA in Forensic Science; Polymerase Chain Reaction (PCR)

Reference
Butler, J. *Forensic DNA Typing.* 2nd ed. Burlington, MA: Elsevier, 2005.

American Board of Criminalistics (ABC)

Forensic science consists of many disciplines. Paul Kirk coined the term *criminalistics* to describe the general area of scientific examination of materials for associative or inceptive evidence. In recent years there has been an upsurge of interest in formalized demonstration of competency of forensic practitioners. The American Board of Criminalistics was established in 1995 to create and maintain a program for certification of individual criminalists.

The program has been very successful, and by 2000 there were 450 people certified as diplomates and 45 as fellows.

Diplomate status is attained by passing an extensive theory test covering all the disciplines within forensic science. The objective is to ensure that there are analysts with a sufficiently broad awareness to be able to recognize, preserve, and collect potential physical evidence.

Fellow status recognizes that there are areas in which the testing and interpretation of evidence requires many years of experience and specialized study. Certification as a fellow requires demonstration of continuing competency through proficiency testing and completion of continuing education.

Reference
American Board of Criminalistics; http://www.criminalistics.com/ (Referenced July 2005).

American Society of Crime Laboratory Directors (ASCLD)

The American Society of Crime Laboratory Directors (ASCLD) is a nonprofit professional society that strives to improve the operation of crime laboratories through improved communications among crime laboratory directors, promoting and encouraging high standards in the field as well as promoting the development of management techniques among its membership. This is achieved through an annual meeting of its membership where training is provided in current issues relating to crime laboratory management.

Reference
American Society of Crime Laboratory Directors; www.ascld.org (Referenced July 2005).

American Society of Crime Laboratory Directors/Laboratory Accreditation Board (ASCLD/LAB)

In 1982 the ASCLD recognized the need to set some minimum standards for operation of crime laboratories and to conduct an objective evaluation of their performance. It therefore set up the American Society of Crime Laboratory Directors Laboratory Accreditation Board (ASCLD/LAB).

The ASCLD/LAB operates two accreditation programs covering laboratory management and operations, personnel qualifications, and physical plant. The first is their traditional "legacy" program. In 2004, ASCLD/LAB became one of two accrediting bodies in the United States to offer an ISO based accreditation program to forensic science laboratories. ASCLD/LAB has called this program "ASCLD/LAB—International." As well as being ISO 17025 based, this program also incorporates key components of their legacy program and ILAC G–19 standards.

See also Accreditation; ASCLD; Forensic Quality Services
References
American Society of Crime Laboratory Directors/Laboratory Accreditation Board; www.ascld-lab.org (Referenced July 2005).

Ammonium Nitrate–Based Explosives

Mixtures of ammonium nitrate and fuels provide stable, low-cost explosives. Commercial preparations include water gels in which ammonium and sodium nitrate are gelled with a gum and include fuel such as aluminum. Ammonium nitrate mixed with fuel oil creates an explosive known as ANFO.

Ammonium nitrate is readily available as fertilizer. Mixed with fuel oil and contained in metal milk churns, the resulting "fertilizer bombs" are cheap, easy, and deadly weapons for terrorism.

See also Explosions and Explosives
References
De Forest, P., R. E. Gaensslen, and H. C. Lee. *Forensic Science: An Introduction to Criminalistics*. New York: McGraw-Hill, 1983.
Saferstein, R. *Criminalistics*. 8th ed. Upper Saddle River, NJ: Pearson Education, 2004.
White, P. *Crime Scene to Court: The Essentials of Forensic Science*. Cambridge: Royal Society of Chemistry, 1998.

Ammunition

A live round of ammunition is properly referred to as a *cartridge*. The bullet is the projectile that is fired from the barrel of the firearm. The cartridge case holds all the components of the cartridge—the propellant or gunpowder charge and the primer—and is crimped at the mouth to hold the bullet in place.

Modern ammunition contains smokeless powder, which is made up largely of nitrocellulose. Before smokeless powder, black powder was used. The fast-burning gunpowder is what creates the force that propels the bullet down the barrel. The primer is a mixture of chemicals detonated by the impact of the firing pin, which in turn ignite the gunpowder charge.

Rim-fire cartridges (primer located at the rim as opposed to the center of the cartridge) were once common but are now exclusively 0.22 caliber. They are not reloadable. Centerfire cartridges have the primer located in the center of the cartridge case. Almost all calibers other than 0.22 are centerfire. The primers can be removed after firing and replaced, making this type of cartridge reloadable.

Examination of ammunition requires a facility for test firing. Test firing must be done in a manner that prevents damage to the test bullets. This is usually accomplished by test firing into water, although the older method of firing into cotton waste is still in use in some laboratories.

Traditionally, a comparison microscope is used for side-by-side viewing of tests and evidence. Oblique, reflected lighting is used to highlight individual characteristics. Test specimens are first compared with each other to determine if the weapon produces reproducible markings each time it is fired. Today, automated imaging systems such as IBIS are used for entry of markings into a database and also to produce and record the features used in individual cases.

Bullet Ammunitions

The projectiles in handgun or rifle ammunition are called *bullets*. Shotgun ammunition fires *pellets*.

Examination of ammunition is carried out to compare fired rounds with each other and with possible sources. The examinations depend on comparison of markings on the bullets and cases.

Individual characteristics are the striation-type tool marks impressed onto the bullet as it travels through the barrel. These markings result from microscopic imperfections transferred from the tools used to create the barrel. These markings are unique to each barrel, and look similar to a barcode when viewed under a microscope. Any manufactured physical object (not just bullets) will show individual characteristics that may be used for individualization. They may also arise from random wearing patterns. These characteristics can be altered by extensive use, cleaning, or abuse.

In addition to determining the type of weapon that may have fired a bullet, the examiner can also check to see if a bullet bears sufficient individual characteristics for comparison, either with other bullets from the same or different crime scene, as well as test bullets fired from a specific weapon.

Cartridge cases can also be marked during the firing process with individual characteristics that are formed during the manufacture of various parts located in the breech area. The negative impression of the firing pin is pressed into the soft metal of the primer. As the gunpowder charge ignites, the pressure forces the base of the cartridge case backward against the breech face, leaving impressed markings on the base of the primer and/or cartridge case.

The expanding gases from the burning gunpowder create tremendous pressure that forces the sides of the cartridge case into tight-fitting contact with the inside surfaces of the chamber. Markings from irregularities in the chamber can be scratched into the sides of the cartridge case when it is extracted from the chamber.

Markings caused by pressure, scratching, and scraping on the sides of the cartridge case by the magazine or loading mechanism may also bear individual characteristics. The clawlike extractor, which grips the base of the cartridge case and pulls it out of the chamber, may also leave striated toolmarks. In the same manner, the ejector may leave its mark when it strikes the base of the cartridge case, throwing it out of the weapon.

Shotgun Cartridges

Shot pellets are usually lead spheres, although pellets used for hunting waterfowl must be made of materials other than lead, such as steel, bismuth, or tungsten. Pellets are surrounded by wadding and are classified by number sizes; the larger the number, the smaller the pellet. Birdshot pellets are smaller than buckshot pellets, ranging in size from 0.05 to 0.17 inch in diameter. Buckshot pellets range from 0.24 to 0.36 inch. In addition to pellets, elongated, hollow, lead rifled slugs may be used. Pellets can be sized by weight and diameter if they are not badly deformed.

Shotgun wadding helps to cushion and protect the shot from the hot gases produced by the burning gunpowder, as well as to keep the shot together as it exits the barrel. On close-range shots, the wadding may enter the wound along with the shot. Wadding may be made of paper, felt, plastic, or plastic granules. The wadding usually helps to identify the ammunition manufacturer. In some instances, plastic wadding can receive markings when fired from shotgun barrels that have been sawed off, have adjustable chokes, or have other gross defects. The type of wadding used is often characteristic of a particular ammunition manufacturer.

Plastic wadding may be marked with identifying markings, particularly when fired from sawed-off barrels with rough edges, or when the wadding scrapes against adjustable chokes or front sights that protrude into the barrel.

See also Firearms; Integrated Ballistic Identification System (IBIS)

References
De Forest, P., R. E. Gaensslen, and H. C. Lee. *Forensic Science: An Introduction to Criminalistics.* New York: McGraw-Hill, 1983.

James, S. H., and J. J. Nordby. *Forensic Science: An Introduction to Scientific and Investigative Techniques.* Boca Raton, FL: Taylor and Francis, 2005.

Saferstein, R. *Criminalistics.* 8th ed. Upper Saddle River, NJ: Pearson Education, 2004.

Warlow, T. *Firearms, the Law, and Forensic Ballistics.* 2nd ed. London and Bristol, PA: Taylor and Francis, 1996.

White, P. *Crime Scene to Court: The Essentials of Forensic Science.* Cambridge: Royal Society of Chemistry, 1998.

Amphetamines

Amphetamine and methamphetamine are frequently abused stimulant drugs. As the name implies, they exhibit a stimulating effect on the central nervous system. They cause a sense of well-being, increased alertness, decreased fatigue, and a loss of appetite. They can lead to a strong psychological dependence.

Most amphetamine and methamphetamine are illegitimately produced in clandestine laboratories. Production requires little training, a small amount of equipment, and relatively inexpensive chemicals. They are white to tan powders that are snorted, injected, smoked, or ingested. "Ice" is a very pure smokeable form of methamphetamine. Their effects last from four to twelve hours.

See also Drugs; Methamphetamine
References
Christian, D. R. *Forensic Investigation of Clandestine Laboratories.* Boca Raton, FL: CRC, 2004.

De Forest, P., R. E. Gaensslen, and H. C. Lee. *Forensic Science: An Introduction to Criminalistics.* New York: McGraw-Hill, 1983.

Drummer, O. H. *The Forensic Pharmacology of Drugs of Abuse.* New York: Oxford University Press, 2001.

James, S. H., and J. J. Nordby. *Forensic Science: An Introduction to Scientific and Investigative Techniques.* Boca Raton, FL: Taylor and Francis, 2005.

Levine, B. *Principles of Forensic Toxicology.* Washington, DC: American Association for Clinical Chemistry, 1999.

Saferstein, R. *Criminalistics.* 8th ed. Upper Saddle River, NJ: Pearson Education, 2004.

U.S. Drug Enforcement Administration, Drug Descriptions, Methamphetamine/Amphetamines; http://www.usdoj.gov/dea/concern/amphetamines.html (Referenced July 2005).

White, P. *Crime Scene to Court: The Essentials of Forensic Science.* Cambridge: Royal Society of Chemistry, 1998.

Andrews, Tommie Lee

This case presents the first U.S. trial in which DNA evidence was admissible in court. The case began in Orlando, Florida, in 1986, when a young woman named Nancy Hodge was raped at knifepoint. She briefly saw her attacker's face, but most of the time he held his hand over her face so that she could not see him. Over the following months, more than twenty women were raped, each time the attacker covering their faces so that they could not see him. Following an attack in 1987, police found two fingerprints on the window screen and finally, on March 1, 1987, police were alerted to the scene of an attack and arrested Tommie Lee Andrews following a two-mile car chase. His fingerprints were found to match those found on the window screen, and he was charged with the rape of that young woman. Andrews was also identified by Nancy Hodge as being the man who raped her. However, this case was unlikely to be successful if based only on identification by the victim.

In order to show that Andrews was, in fact, a serial rapist, police turned to DNA analysis. The DNA of semen from the rapist was compared with that from a blood sample from Tommie Lee Andrews, and the profiles were found to be identical.

DNA evidence had not yet been accepted as admissible evidence in court, and so a pretrial hearing was necessary to show that the technique was scientifically sound in its theory, practice, and interpretation. Following the extensive pretrial hearing, the evidence was admitted into court. However, when the evidence was submitted by the prosecution, it was stated that there was only a 1 in 10 million chance that Andrews could be falsely accused, and when this was challenged by the defense, it could not be substantiated. The jury was split and the case declared a mistrial. Andrews was then tried for the second rape, for which fingerprint evidence was found, and found guilty. For this he was sentenced to twenty-two years in prison. The Hodge case was then retried a few months later, when the DNA evidence was better represented. Andrews was found guilty of serial rape and his prison term extended to 115 years.

See also Admissibility of Scientific Evidence; DNA in Forensic Science
References
Court TV's Crime Library, DNA in Court; http://www.crimelibrary.com/criminal_mind/ forensics/dna/6.html?sect=21 (Referenced July 2005).
Dr. George Johnson's Backgrounders, DNA Fingerprinting; http://www.txtwriter.com/ Backgrounders/Genetech/GEpage14.html (Referenced July 2005).

Anthropology

Anthropology is the study of the origins, physical characteristics, and social institutions of mankind. It is the part of anthropology dealing with physical characteristics that is employed in forensic science. The applications are devoted to identification of remains. These may be bodies found accidentally, mass disaster victims, war grave remains, or mass homicide victims. Remains come upon accidentally could be of homicidal, suicidal, accidental, or natural causes.

The objectives of the anthropological investigation are to identify the species, gender, race, stature, and age of remains. These identifications are made from known data about the human skeleton and its anatomy. However, the investigation can also provide information on health and the manner and cause of death. Some features may even result in discovering individual identity. For example, pipe smoking can result in characteristic deformation of teeth that, if allied to data on age and stature, could result in a reliable assignment of identity comparing remains with a list of possible persons.

Features measured include the cranium and its lines of ossification, teeth, long bones, pelvic structure, rib cage, and the overall skeletal configuration. Ancillary material such as hair and nails are also of value. For example, examination of the nails for Beau's lines (grooves on the fingernails often found following disease) can indicate the recent health history of the individual, as periods of severe illness result in cessation of nail growth. Hair left at a scene, even when the body has undergone severe changes and decomposition, may be used to determine the hair color of an individual.

Finally, today's technology can examine DNA recovered from hair or bones. As well as offering individualization of recent remains by comparison with known or family samples, some DNA characteristics can provide an indication of race.

See also Mass Disaster Victim Identification
References
Byers, S.N. *Introduction to Forensic Anthropology: A Textbook.* Boston: Allyn and Bacon, 2001.
Maples, W. R., and M. Browning. *Dead Men Do Tell Tales: The Strange and Fascinating Cases of a Forensic Anthropologist.* New York: Doubleday, 1995.

Antibody

An antibody is a naturally occurring protein found in blood serum. It contains sites that will bind with specific chemical groupings in antigens, thereby rendering them inactive. The proteins are termed *immunoglobulins.* Different classes of immunoglobulins react in different ways with the antigens. Some combine with an antigen in a way that results in formation of a large polymer, which is insoluble and precipitates. Others cross-link with antigens on cell surfaces, causing them to clump together, or agglutinate.

Challenge with an antigen (usually a protein foreign to the host organism) results in a range of antibodies. Different regions of the antigen produce antibodies of different specificity, and the same region can produce antibodies of different immunoglobulin classes.

The immunoglobulins are produced by white cells. When the antibody production to the antigen challenge is at a maximum, the white cells can be harvested and cell lines grown in culture. The antibody from each line (clones) can be examined and highly specific antibodies obtained. These are defined as monoclonal antibodies.

Many of the techniques of traditional forensic biology depend on antibody-antigen reactions. There are many examples, ranging from the traditional blood-group serology tests—typing blood as group A, B or O, for example—to confirmatory tests still in routine use. For example, the identification of species of origin of blood or other body fluids is usually based on reaction of an extract of the stain with a species-specific antiserum.

Other tests are based on chemical detection of tags used to measure the binding of antigen to antibody. These include a large range of drug screens.

See also Antigen; Blood Grouping; Serology; Species Identification
References
Antibody Resource Page, The;
    http://www.antibodyresource.com/ (Referenced July 2005).
Butler, J. *Forensic DNA Typing.* 2nd ed. Burlington, MA: Elsevier, 2005.
Saferstein, R. *Criminalistics.* 8th ed. Upper Saddle River, NJ: Pearson Education, 2004.

Antigen

An antigen is a chemical substance that will cause the production of antibodies when it is introduced into the body. Antigens must be foreign to the body and be of a nature that will cause the body's defense system to respond. They are usually proteins or chemicals attached to a protein backbone. The immune response to the challenge results in production of a cocktail of antibodies. Each has been formed in response to a tiny part of the antigen and each will combine with that site in an antigen-antibody reaction.

See also Antibody
References
Antigen Presentation;
    http://users.rcn.com/jkimball.ma.ultranet/BiologyPages/A/AntigenPresentation.html (Referenced July 2005).

Butler, J. *Forensic DNA Typing.* 2nd ed. Burlington, MA: Elsevier, 2005.
Saferstein, R. *Criminalistics.* 8th ed. Upper Saddle River, NJ: Pearson Education, 2004.

Antiserum

An antiserum is serum from an animal containing antibodies to a defined source. The antiserum may be natural, for example produced by the body in response to infection, or may be provoked by deliberate challenge with antigen.

Each antibody in the antiserum will combine specifically with the antigen site that produced it, like a lock and key. This is the principle that makes immunization an effective treatment against infectious diseases such as polio. The immunization causes production of antibodies. The cells that manufacture the antibodies remember the specific antigens and are triggered into producing the antibody if challenged again. The antibodies bind with the antigens and so neutralize them.

Most antisera reactions used in forensic science depend on the fact that the antibody immunoglobulin molecules are a dimer of two identical chains: There are therefore two identical binding sites on each immunoglobulin. In this way, the antibodies can cross-link to each other. It is this cross-linking that produces the agglutinations first seen by Karl Landsteiner when studying blood types, and the precipitin reactions used in species identification.

See also Species Identification
Reference
Butler, J. *Forensic DNA Typing.* 2nd ed. Burlington, MA: Elsevier, 2005.

Arson

Arson is the offense of setting fire to property. The main task of the forensic scientist in arson investigation is the examination of the scene and materials removed from it to determine whether the fire was deliberately caused. This is an instance in which forensic science is used to identify the occurrence of a crime and not to associate an individual with a scene or another person. This is known as inceptive evidence.

As well as detection of accelerant residues, the forensic scientist may examine appliances for integrity and the remains of possible initiation devices, such as fuses. Physical examination can yield good evidence. For example, if a fire is caused by the burn-out of an electric motor, the inside of the appliance will be sooted. If the appliance was burned during the fire, the inside will often be soot free.

Detector dogs have been shown to be effective at pointing to sites of accelerant residues at arson scenes.

See also Accelerant Residues

References
Almirall, J., and K. Furton. *Analysis and Interpretation of Fire Scene Evidence.* Boca Raton, FL: CRC, 2004.
De Forest, P., R. E. Gaensslen, and H. C. Lee. *Forensic Science: An Introduction to Criminalistics.* New York: McGraw-Hill, 1983.
James, S. H., and J. J. Nordby. *Forensic Science: An Introduction to Scientific and Investigative Techniques.* Boca Raton, FL: Taylor and Francis, 2005.
Nic Daeid, N. *Fire Investigation.* Boca Raton, FL: CRC, 2004.
Saferstein, R. *Criminalistics.* 8th ed. Upper Saddle River, NJ: Pearson Education, 2004.
White, P. *Crime Scene to Court: The Essentials of Forensic Science.* Cambridge: Royal Society of Chemistry, 1998.

Arson and Explosives Incidents System (AEXIS)

By their nature, arson and explosive sites present the scientist with substantial challenges not found in other crime scenes. The fire or explosion will have produced a site of devastation and destruction and consumed much of the potential evidence. The goal of any first responders on the scene should be to ensure safety and to minimize and control further damage by extinguishing the fire or securing the physical site. Only then can the scene investigator enter and begin work. Unfortunately, onlookers (including senior law enforcement personnel who should know better) are often present and further complicate matters. By then, the fire and damage

control may have destroyed much of the available evidence.

The Bureau of Alcohol, Tobacco, and Firearms (ATF) has built up considerable experience in dealing with major fire and explosive scenes and has established the Arson and Explosives Incidents System (AEXIS) to provide investigators with a source of information relating to arson and explosives incidents. This information includes details on the total numbers of explosives and bombing incidents grouped according to the type of target; the total number of explosives incidents for each U.S. state; the total number of bombing incident fatalities for each type of target, including information on deaths, injuries, and property damage; as well as the total number of bombing and arson incidents grouped by motive, the nature of the explosive device, and the related injuries and damage.

See also Arson

Asphyxia

Asphyxia, or suffocation, can be the cause of death in an accident, a suicide, or a homicide. Examples of accidental asphyxia include burial in avalanches and autoerotic experiences that go wrong. Suicidal asphyxia often results from hanging. Homicidal asphyxia includes strangulation and smothering.

Classical signs of asphyxia include cyanosis, congestion, petechial hemorrhages, and bleeding from the nose and mouth. Cyanosis occurs whenever there is a lack of oxygen, which makes the blood darken in color, taking on a bluish tint, when seen through the fingernail bed for instance. However, the presence and the absence of cyanosis are not reliable guides to whether or not death was due to asphyxia. Interference with venous return and the effect of lack of oxygen on capillary blood vessels can cause congestion and edema of the face. However, one of the most reliable signs of asphyxia is the presence of tiny, pinpoint lesions in the conjunctiva and the skin of the eyelids. These petechial hemorrhages are also found in other circumstances, including insulin overdose and heart attacks. Bleeding from the nose and mouth may occur in strangulation asphyxia.

Autopsy findings include intracranial bleeding and edema of the lungs and brain. Very characteristic hemorrhages will be found in the viscera. They are named Tardieu's spots after their discoverer.

References
Byard, R., T. Corey, C. Henderson, and J. Payne-James. *Encyclopedia of Forensic and Legal Medicine.* London: Elsevier Academic, 2005.
Knight, B. *Simpson's Forensic Medicine.* London and New York: Oxford University Press, 1997.
Payne-James, J., A. Busuttil, and W. Smock. *Forensic Medicine: Clinical and Pathological Aspects.* London: Greenwich Medical Media, 2003.
Saukko, P., and B. Knight. *Knight's Forensic Pathology.* 3rd ed. London: Arnold, 2004.

Assault

Investigation of the victims of violent attack requires careful documentation and thorough physical examination. This is true irrespective of the nature of the assault: deliberate personal attack, terrorist action, domestic violence, or rape, for example. Factors to be determined begin with determining that there is indeed physical evidence of an assault. The examination will afford information as to the nature of the injury; the influence of any preexisting illness; whether there is a history of injury and treatment; the time of injury; and perhaps information on weapons used. Involvement of medicines and illicit drugs can be elicited by testing blood and urine samples.

The history of the victim must be documented carefully; for example, sports injuries and some diseases such as psoriasis may result in misleading physical signs.

Documentation can consist of notes, photographs, and copies of x-rays and other laboratory tests. The notes are discoverable in any litigation.

The physical examination should bear in mind that injuries can result from deliberate self-harm and from defensive actions. In the case of self-inflicted injuries, the injury will be on a body area that is accessible, such as limbs or the abdomen, and will probably demonstrate handedness. These injuries are typically minor, multiple, and regular, and there may be signs of older similar injuries. Defensive injuries arise from the actions of the victim to avoid or minimize damage. They include cuts to the palms in attempts to fend off knife attacks and bruising to the arms in attempts to ward off beating injury to the head.

Custody and arrest injuries warrant particular attention. Handcuff injuries include neuropathy from damage to nerves, abrasions, and bruising. Custody injuries may result from use of restraints or use of force in attempting to restrain a violent prisoner.

Differentiation of accidental causes from deliberate injuries presents significant challenges to the examining physician. One of the more common differential situations is that of distinguishing falls from assault. The most serious example is where death could have resulted from a head injury. A simple fall, caused, for example, by tripping or alcohol impairment, in which the head strikes a concrete floor, can certainly produce a head injury sufficient to result in death. There will not necessarily be any characteristic features to permit distinguishing between accidental and deliberate causation. Some features worth exploring are the presence of other injuries that could have resulted from a fight, blood-alcohol levels, medical history, and the nature of the head injuries—for example evidence of multiple blows.

References
Byard, R., T. Corey, C. Henderson, and J. Payne-James. *Encyclopedia of Forensic and Legal Medicine.* London: Elsevier Academic, 2005.
Knight, B. *Simpson's Forensic Medicine.* London and New York: Oxford University Press, 1997.
Payne-James, J., A. Busuttil, and W. Smock. *Forensic Medicine: Clinical and Pathological Aspects.* London: Greenwich Medical Media, 2003.

Associative Evidence

Associative evidence seeks to establish an association between people or between people and places. For example, the results of typing of semen from the vagina of a rape victim are compared to those in a nominated suspect or in an offender database. Where the types in the evidence sample differ from those in the nominated suspect, the suspect is eliminated as a possible source of the semen. If there are no differences in type, the suspect is not eliminated. However, the apparent association could be because the suspect was indeed the source of the semen or could be because the suspect had the same type by chance.

In the example of typing a semen sample in a rape case, if DNA typing was used, the types detected can be very rare—often rarer than one person in a million of the general population. DNA typing methods often develop figures of one in a billion of the population. That being so, it is not very likely that the suspect would happen by chance to have the same DNA type. However, before DNA typing became commonplace, ABO blood typing was used in these cases. The population frequencies of the different ABO types vary from about 3 percent (type AB) to about 50 percent (type O). The strength of the association there would be much less.

The most compelling associative evidence is fingerprinting. No two people (not even identical twins) have the same fingerprints. The finding of a matching print at a scene or on a weapon is accepted as conclusive proof of association between the person and the object. How the print got there is another matter. There is no reliable way to tell the age of a print, and it is possible to plant fingerprints.

Other associative evidence includes comparison of fabric fibers. The concept is that contact between the fabrics will result in transfer of fibers from one to the other. Thus, if a clothed body is transported in the carpeted trunk of a car, fibers will be transferred from the shirt of the body to the carpet and from the carpet to the shirt. This concept of transfer is often described as "every contact leaves a trace" and is attributed to Edmund Locard.

There are many problems in evaluating the strength of the association postulated by the finding of transferred fibers. Fabrics and their constituent fibers are not unique. Denim jeans, for example, are very common and there is nothing sufficiently characteristic about one pair of jeans to provide the basis for a unique association.

The transferred fibers are not retained indefinitely by the recipient fabric and the weight of a finding of a few fibers, indistinguishable from the targeted source, has been challenged many times.

At best, associative evidence is often described more correctly as corroborative, which means that it has to be weighed alongside other evidence and not taken as having substantial weight in its own right.

See also DNA in Forensic Science; Fibers; Fingerprints

References

De Forest, P., R. E. Gaensslen, and H. C. Lee. *Forensic Science: An Introduction to Criminalistics.* New York: McGraw-Hill, 1983.

James, S. H., and J. J. Nordby. *Forensic Science: An Introduction to Scientific and Investigative Techniques.* Boca Raton, FL: Taylor and Francis, 2005.

Saferstein, R. *Criminalistics.* 8th ed. Upper Saddle River, NJ: Pearson Education, 2004.

White, P. *Crime Scene to Court: The Essentials of Forensic Science.* Cambridge: Royal Society of Chemistry, 1998.

Automated Fingerprint Identification System (AFIS)

Fingerprints are unique to an individual. Fingerprint patterns can be left at a scene. The print, or latent, can be made visible and its features compared with reference samples. The basic principles of identification through fingerprints were established more than a century ago. Police forces began to preserve data banks of fingerprints and compared latents with those in the data bank using a classification system established by Sir Edward Richard Henry in 1897. The procedure is labor intensive and time-consuming, and too cumbersome for effective use with large databases.

A relatively recent and effective development has been the introduction of computer-based systems for storing and comparing fingerprint images. These automated fingerprint identification systems (AFIS) provide rapid and accurate comparisons even with large databases.

AFIS systems use computers to establish and compare patterns in minutiae in the recovered and referenced fingerprints. They therefore offer a faster and more objective means of identifying the source of a latent print. They also make the use of extensive databases feasible in a way that would not be possible using conventional classification systems.

See also Fingerprints
References
James, S.H., and J. J. Nordby. *Forensic Science: An Introduction to Scientific and Investigative Techniques.* Boca Raton, FL: Taylor and Francis, 2005.
Komarinski, P. *Automated Fingerprint Identification Systems (AFIS).* Amsterdam and Boston: Elsevier, 2005.
Saferstein, R. *Criminalistics.* 8th ed. Upper Saddle River, NJ: 2004.

Automobile Examination

Automobiles can be stolen, involved in hit-and-run incidents, or used to deliver bombs. Vehicles have unique identifiers (VINs) in many places throughout their structure. Plates are affixed to the body in several places and numbers are ground into parts such as the engine and transmission. Window glass bears stamps identifying the make and date of manufacture.

Identification of a stolen vehicle relies on inspection of the VIN information. This may involve restoration of ground-out, stamped numbers. Identification of a vehicle from parts in the debris relies on the same principle.

In a hit-and-run incident, paint is often transferred between vehicles or from vehicle to victim. Identification of a vehicle from transferred materials thus usually depends on examination of the paint. Surface color can be compared to manufacturers' charts. If flakes or chips of paint are left, the layer sequence and sometimes the physical shape of the chip can provide good evidence when compared to those in a possible source vehicle.

Other valuable evidence can be obtained from tire prints, especially if there is any characteristic wear pattern. Oil residues and chips of plastic or glass from broken lights are also sources of good associative evidence.

See also Glass; Paint
Reference
James, S.H., and J. J. Nordby. *Forensic Science: An Introduction to Scientific and Investigative Techniques.* Boca Raton, FL: Taylor and Francis, 2005.

Autopsy

The autopsy, also sometimes referred to as the postmortem examination or necropsy, is the examination of a body to ascertain the cause of death and the character and extent of change caused by disease. The three main reasons for conducting an autopsy are therefore clinical, statutory, and investigative.

Statutory autopsy examinations are conducted whenever the circumstances of death prevent an authorized person from signing a death certificate. The definition of "authorized person" and the circumstances that lead

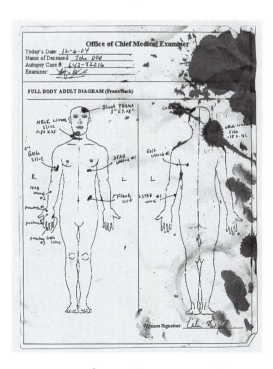

An autopsy report from a stabbing victim. (iStockphoto.com)

to his or her being unable to sign vary from jurisdiction to jurisdiction. However, generally the person is a medical practitioner and the circumstances are violent or unexplained death.

Investigative autopsies focus on determining the nature and cause of death, and possibly the time of death. Samples that should be taken at the autopsy include blood, head hair and pubic hair, fingernail scrapings, swabs from orifices such as the vagina and rectum, and stomach contents. Many of these are routine and samples are included in standard examination kits. The purpose of the sampling is to provide material for laboratory examination for poisoning (blood, urine, and stomach contents) and physical contact (blood, semen on the swabs, and hair).

Blood from at least two different sites should be taken from vehicle accident victims to screen for alcohol. Postmortem production of alcohol is a well-established problem artifact. It is also a very localized phenomenon and differing levels in each of the samples will show that there has been postmortem conversion of glucose to alcohol.

Clothing should be examined before removal for stains and evidence of weapon damage such as knife cuts and bullet trajectories. If the clothing has to be cut off the body, care must be taken to avoid interfering with any evidentiary marks or damage. The clothing itself should be packaged in a way that prevents contamination and degradation of evidence; sealed paper bags are best.

The outer surfaces of the body should be examined for injuries and scars. If death due to administration of a drug or poison is suspected, look for injection sites. These can be vital evidence in cases such as insulin poisoning in which the drug itself is rapidly removed from the blood but residues of overdose may be detected at the site of administration. The internal organ examination is conducted only after completion of the careful external scrutiny. Organ appearance and weights are recorded and samples taken for histology.

The forensic autopsy has many serious aspects. For example, if there is no obvious cause of death, decisions must be made about the extent of the examinations to be conducted and the time that they will take. The pathologist must be aware that the family of the deceased will be awaiting release of the body so that their grieving process can be closed. A balanced approach is required, and some situations are more fraught than others. For example, microscopy of the brain takes considerable time and the brain is an organ of considerable emotional significance to the family. Yet this can be an invaluable test in some cases. In one such case, a car ran a red light, crashed into a motorcycle, and the car driver and the cyclist were both declared dead at the scene. On the face of it, the car driver was culpable, but examination of the brain revealed that he had suffered a brain hemorrhage and would have been dead or unconscious before the accident.

References

Byard, R., T. Corey, C. Henderson, and J. Payne-James. *Encyclopedia of Forensic and Legal Medicine.* London: Elsevier Academic, 2005.

DiMaio, V. J. M., and D. DiMaio. *Forensic Pathology.* 2nd ed. Boca Raton, FL: CRC, 2001.

Dolinak, D., E. Matshes, and E. O. Lew. *Forensic Pathology: Principles and Practice.* Amsterdam and Boston: Elsevier Academic, 2005.

Knight, B. *Simpson's Forensic Medicine.* London and New York: Oxford University Press, 1997.

Payne-James, J., A. Busuttil, and W. Smock. *Forensic Medicine: Clinical and Pathological Aspects.* London: Greenwich Medical Media, 2003.

Saukko, P., and B. Knight. *Knight's Forensic Pathology.* 3rd ed. London: Arnold, 2004.

B

Barbiturates

Barbiturates are depressant drugs used as sleeping tablets. There are several members in the family, each with different properties. Some have a rapid onset but short duration of action (for example amobarbital) and some have a more delayed onset but longer duration of action (for example phenobarbital). Phenobarbital also has clinical use as an antiepileptic. Chronic barbiturate users can develop a dependency.

The drugs are toxic in overdose. They are also abused as "downers."

See also Drugs
References
Drummer, O. H. *The Forensic Pharmacology of Drugs of Abuse.* New York: Oxford University Press, 2001.
Levine, B. *Principles of Forensic Toxicology.* Washington, DC: American Association for Clinical Chemistry, 1999.
U.S. Drug Enforcement Administration, Drug Descriptions, Barbiturates; *http://www.usdoj.gov/dea/concern/barbiturates.html* (Referenced July 2005).

Battered Baby Syndrome

Physical child abuse, sadly, is a worldwide problem. It is usually identified as repeated willful injury inflicted by parents or caregivers on children. The cause can be stress on the part of emotionally inadequate parents or an excess of what the parent considers to be reasonable discipline, or the child may become enveloped in an environment of domestic violence.

Features encountered in the battered baby included repeated injuries from rough handling, multiple fractures, delay in seeking treatment for the child's injuries, and misleading explanations to account for the injuries. Accidents do happen, and children do injure themselves in the course of play. However, warning signs include any injury in children under twelve months of age and inappropriate locations of the injuries, such as small rounded bruises on the front upper chest caused by finger pressure.

References
Byard, R., T. Corey, C. Henderson, and J. Payne-James. *Encyclopedia of Forensic and Legal Medicine.* London: Elsevier Academic, 2005.
Knight, B. *Simpson's Forensic Medicine.* London and New York: Oxford University Press, 1997.
Payne-James, J., A. Busuttil, and W. Smock. *Forensic Medicine: Clinical and Pathological Aspects.* London: Greenwich Medical Media, 2003.

Benzidine

Benzidine is a chemical that can be used to detect blood. The benzidine test is one of the most sensitive available, capable of producing a positive response to less than one part per thousand of blood. Unfortunately benzidine is toxic, capable of inducing cancer. It is seldom used and has been replaced by safer alternates such as phenolphthalein. Attempts have been made to alter benzidine to retain the sensitivity but without the safety problems. Tetramethyl benzidine is one such safer chemical derivative. A different approach is used in the commercial preparation known as Hemastix, in which the benzidine is immobilized on a matrix and protected by a covering membrane.

See also Blood; Color Tests
References
Butler, J. *Forensic DNA Typing.* 2nd ed. Burlington, MA: Elsevier, 2005.
De Forest, P., R. E. Gaensslen, and H. C. Lee. *Forensic Science: An Introduction to Criminalistics.* New York: McGraw-Hill, 1983.
Saferstein, R. *Criminalistics.* 8th ed. Upper Saddle River, NJ: Pearson Education, 2004.

Bite Marks

Comparison of marks is the basis of many tests in forensic science. Ammunition, tools, footwear, tire tracks, and even fingerprints are examples. The basis of such tests is that the object examined contains unique physical features caused by random processes such as wear or manufacturing defects.

Bite marks are also in this category. Teeth vary from person to person due to natural variation and to the consequences of dental treatments. Bite marks can be left on foodstuffs and on skin.

However, comparison of bite marks is much less reliable than the comparisons in the other areas mentioned. This is because there is seldom a clear impression with sufficient individualizing characteristics available. Bite marks, unlike full dental impressions or x-rays, seldom possess sufficient individuality and the media—especially skin—do not preserve a good enough impression.

See also Odontology
References
Bowers, C. M. *Forensic Dental Evidence: An Investigator's Handbook.* Amsterdam and Boston: Academic, 2004.

Bowers, C. M., and G. Bell. *Manual of Forensic Odontology.* 3rd ed. Saratoga Springs, NY: American Society of Forensic Odontology, 1995.
Dorion, R. B. J. *Bitemark Evidence.* New York: Marcel Dekker, 2004.

Black Powder

Black powder is a mixture of potassium nitrate, charcoal, and sulfur. Known as an explosive for many centuries, black powder has been used to propel rockets, to fire ammunition, to act as a detonator, and to make bombs.

See also Explosions and Explosives

Blood

Blood is the fluid that carries nutrients to body tissues and waste materials away from them. About 40 percent of the volume of blood is made up from red cells, which give it its color. The cells derive the color from a chemical, hemoglobin. The heart pumps blood through the lungs and body tissues. In the lungs, the hemoglobin in the red cells forms a chemical complex with oxygen in the breathed air. The oxygen is transferred to tissue where it is taken up during metabolism, and the waste gas carbon dioxide is carried back to the lungs to be eliminated in the air we breathe out.

Blood contains other cells—the white blood cells. These play an important role in the body's defense against infections. The fluid that carries the red and white cells is called *plasma*. Plasma is a solution of salts, proteins, and other chemicals. Blood responds to cuts and other wounds by clotting and so prevents the injured from bleeding to death. The clotting process converts one of the proteins in plasma, fibrinogen, to fibrin, which forms the matrix for the clot to form. The fluid left after the clot has formed is called serum. Blood removed from the body, for example for blood grouping, will always clot unless treated with an anticoagulant.

Nutrients, hormones, drugs, and other substances are carried around the body either in solution or chemically bound to some of the plasma proteins.

See also ABO Blood Groups; Blood Grouping; Blood Spatters; Bloodstain Identification; Color Tests

References
Butler, J. *Forensic DNA Typing.* 2nd ed. Burlington, MA: Elsevier, 2005.
Saferstein, R. *Criminalistics.* 8th ed. Upper Saddle River, NJ: Pearson Education, 2004.
White, P. *Crime Scene to Court: The Essentials of Forensic Science.* Cambridge: Royal Society of Chemistry, 1998.

Blood Alcohol Analysis

See *Alcohol*

Blood Alcohol and Driving While Intoxicated

See *Alcohol*

Blood Grouping

Blood grouping is the process of characterizing blood, whether in a stain or as liquid sample, by identifying inherited characteristics in it. The clinically important ABO system is the oldest of the many hundreds of grouping classifications (see **ABO Blood Groups**). To be of use in forensic science, a characteristic substance must show variation between people (sometimes expressed as "discriminating power"), not change with time in an individual, be stable in stains, and be capable of reliable detection in stains or other evidence samples. Because most cases in which body fluids are typed in the laboratory are sexual offenses, the group should be found in semen and other body fluids as well as in blood.

The overwhelming advantages of DNA typing as a grouping system are clear, and there are very few places still routinely using other systems. The advantages of DNA work out in many different circumstances: for example, cold cases in which there is no usable eyewitness testimony, and cases in which there are no other leads. Each of these can be illustrated with examples.

The largest group of cold case examples comes from the Innocence Project, where DNA testing has been used to exonerate many of the falsely convicted. Testing of the most minute traces of evidence remaining after exhaustive testing more than a decade previously has resulted in DNA profiles being developed that prove that the material did not originate from the convicted persons. Only DNA has the combination of stability, sensitivity, and discriminating power to be successful in these circumstances of traces of material remaining after many years.

DNA databases are producing evidence associating individuals with samples from crime scenes. Even where there is no database reference material, DNA testing can be a powerful adjunct to careful police investigation, as for example in the Jarrett case in South Australia. Briefly, a widow in her eighties, living alone, was found dead in her home. She had been raped and died during the assault. Various features of the scene led the police to conclude that her assailant was known to her. They therefore interviewed family members and men (such as workmen) who would be well enough known to the deceased that she would allow them into the house. Each was asked to donate a sample for DNA testing. The sixteenth person interviewed, David Jarrett, was a youth who had repaired her roof a few months earlier. His DNA matched that of the semen on the body. He was subsequently convicted of the murder. One of the features of this case—and of those like it—was that the DNA types are so rare that they constitute compelling evidence of association.

See also ABO Blood Groups; Blood; DNA in Forensic Science

References
Butler, J. *Forensic DNA Typing.* 2nd ed. Burlington, MA: Elsevier, 2005.
Saferstein, R. *Criminalistics.* 8th ed. Upper Saddle River, NJ: Pearson Education, 2004.
White, P. *Crime Scene to Court: The Essentials of Forensic Science.* Cambridge: Royal Society of Chemistry, 1998.

Blood Spatters

The physical nature of bloodstains can contain valuable information. Blood shed from any violent impact, such as a blow with a club, is broken up into small droplets. The higher the energy of the impact, the smaller the droplet size. As the blood makes contact with a solid surface, it leaves elongated stain patterns, the linear direction of the stain pointing back to the point of impact. Blood cast off from the club can be distinguished from that originating from the point of impact by the larger drop size and the pattern of the stains. Blood spatters can also be created by secondary energy transfer such as running through a pool of blood.

The value of blood spatter examination is seen in distinguishing between blood transferred during an attack and that transferred to someone rendering assistance. The first will take the form of small spatters. However, blood transferred from wet stains or bleeding injuries will take the form of streaks and smears.

Investigations in the Bajada case in Malta illustrate the power of blood spatter examination. Police were called to a scene on a quiet road late one night. There they found Angelo Bajada standing outside of his car with a wound to his left arm and his wife in the passenger seat dead from a bullet wound to the right temple (cars are right-hand drive and drive on the left of the road in Malta). Bajada told the police that he and his wife were returning home after dining at a restaurant when their car was flagged down by a pedestrian who subsequently pulled a gun, demanded money, and shot Bajada and his wife when they did not comply. Bajada was ill with blood loss as the bullet to his arm had penetrated his chest and ruptured a blood vessel. He was taken to hospital by ambulance. There were sufficient inconsistencies in the story presented at subsequent interviews that the police began to doubt his story and suspect that he had killed his wife and shot himself as a cover-up. For example, he claimed his wife had been shot from outside of the car, yet the wound was to her right temple and there was a spent cartridge case caught in the folds of her dress.

The dash of the car was covered with a fine spray of blood, and the laboratory tried to group it to ascertain whether it came from Mr. or Mrs. Bajada. They were unsuccessful in developing sufficient markers. However, the bloodstains showed directional characteristics, and a reconstruction was instituted using cords to align with the splashes. It was discovered that there were two points of origin. One was from inside the car at the level of the dash and just to the left of center. The other was high on the right side near the driver's door. This information, along with the distribution of cartridge cases, was consistent with Mrs. Bajada being pulled down toward the driver and shot inside the car, and then Mr. Bajada shooting himself in the arm. Angelo was subsequently convicted of the murder of his wife.

See also Blood

References

Bevel, T., and R. M. Gardner. *Bloodstain Pattern Analysis: With an Introduction to Crime Scene Reconstruction.* 2nd ed. Boca Raton, FL: CRC, 2001.

James, S. H., and W. G. Eckert. *Interpretation of Bloodstain Evidence at Crime Scenes.* 2nd ed. Boca Raton, FL: CRC, 1998.

Wonder, A.Y. *Blood Dynamics.* San Diego, CA: Academic, 2001.

Bloodstain Identification

When initially shed, blood is red in color. As it dries into a stain, the color darkens to red-brown and then to brown. The color is used to identify possible blood during the visual examination of evidence items or crime scenes. Thereafter, blood identification proceeds in four stages. These are:

- A screening, or presumptive, test to see if the material could be blood
- A confirmatory test to prove that it is blood
- A test to prove that it is human blood
- A typing test to investigate the origin of the blood

The screening test uses the peroxidase activity of the heme component of the hemoglobin in red cells to catalyze the conversion of a colorless chemical to a colored oxidized form. Other substances such as vegetable material can also catalyze the conversion. The timing of the reaction is critical and the test should only be called as a positive if the color is developed within thirty seconds of treating with reagent.

Although it is almost unknown for a brown stain giving a positive reaction to the screening test to be anything other than blood, the lack of specificity of the test means that it must always be regarded as a screen only. Phenolphthalein (Kastle-Meyer test) and benzidine are examples of screening tests. There are several confirmatory tests. The most well established is the hemochromogen or Takayama test in which characteristic crystal forms are produced when the stain extract is mixed with a pyridine reagent.

The third step traditionally used a specific antiserum to human serum. The stain extract and the antiserum are allowed to diffuse through a gel. When the two solutions come into contact, an antigen-antibody complex is formed that precipitates out of solution and forms a visible mark in the gel. This is known as a precipitin reaction, and the technique of allowing the solutions to diffuse through the gel is known as the Ouchterlony method. A variant is to drive the serum and antiserum together using an electric field.

The steps of confirming the presence of blood and proving human origin can be combined using antihuman hemoglobin rather than antihuman serum. Some DNA markers are primate specific and so give species information as well as typing. However, strictly speaking, material that gives a positive screening test and a DNA result in a system that is primate specific should be described as "bloodlike" rather than blood.

The fourth step is blood typing. Modern DNA typing includes a step that will only respond to human (or in some cases primate) DNA. Thus, today many laboratories go straight to DNA testing after obtaining a positive screening result.

See also Blood; Blood Grouping; Color Tests; DNA in Forensic Science; Species Identification

References

Butler, J. *Forensic DNA Typing.* 2nd ed. Burlington, MA: Elsevier, 2005.

De Forest, P., R. E. Gaensslen, and H. C. Lee. *Forensic Science: An Introduction to Criminalistics.* New York: McGraw-Hill, 1983.

James, S. H., and J. J. Nordby. *Forensic Science: An Introduction to Scientific and Investigative Techniques.* Boca Raton, FL: Taylor and Francis, 2005.

Saferstein, R. *Criminalistics.* 8th ed. Upper Saddle River, NJ: Pearson Education, 2004.

White, P. *Crime Scene to Court: The Essentials of Forensic Science.* Cambridge: Royal Society of Chemistry, 1998.

Bombs

Four things are needed to produce a bomb: (1) an energy source, (2) a material that will undergo a rapid exothermic reaction to produce a large volume of gas when ignited by the energy source, (3) a method of containment so that the energy of the phase transition is not dissipated harmlessly, and (4) a method of initiation so that the bomb explodes where and when intended.

The fragments of the container impart damage and can cause serious injury to anyone in the vicinity of the exploding bomb. However, it is the very fast shockwave from the explosion that is responsible for most of the damage to property. (See **Ammonium Nitrate–Based Explosives; Explosions and Explosives.**)

A pipe bomb typically contains black powder in a pipe with ends closed by threaded caps. Detonation from within the pipe causes a huge volume of gas that expands almost instantaneously into the surrounding air when the pipe bursts. The effect is like a gale, but one traveling at up to 7,000 miles per hour. In contrast, disposal of a bomb by exploding it from outside results in a harmless phase change, as there is no compression release.

See also Ammonium Nitrate–Based Explosives; Arson and Explosives Incident System (AEXIS); Explosions and Explosives
References
De Forest, P., R. E. Gaensslen, and H. C. Lee. *Forensic Science: An Introduction to Criminalistics.* New York: McGraw-Hill, 1983.
Saferstein, R. *Criminalistics.* 8th ed. Upper Saddle River, NJ: Pearson Education, 2004.
White, P. *Crime Scene to Court: The Essentials of Forensic Science.* Cambridge: Royal Society of Chemistry, 1998.

Bradley, Stephen

Shortly after Bazil and Freda Thorne of Australia won the lottery, their son Graeme was kidnapped while on his way home from school. The Thornes received a phone call from a man with a heavy accent demanding $25,000. Following the call, however, they heard nothing more.

Five weeks later, the body of the boy was found wrapped in a rug approximately ten miles from home. It was determined that he had been suffocated and then clubbed to death. On the boy's clothing was found a pink granular substance together with human and animal hairs, and traces of mold were found on his socks. Several types of plant material were also found with the body. The mold suggested that the boy had been dead for some time and so was probably killed soon after he was taken. The animal hairs were examined and identified as being from a Pekinese dog. The pink substance was found to be a type of mortar used in house building, and examination of the plant material revealed the presence of rare cypress seeds that did not grow in the area in which the body was found. Given these findings, police then made a public request to be informed of any house that contained both the pink mortar and the rare cypress. A mailman suggested a house in Clontarf, Australia, where police found the pink mortar and the rare cypress, as well as other plants that had been found with the body. The previous tenant of this house was found to be Stephen Bradley, who moved out of the house on the day the boy was kidnapped

and was reported to speak with a heavy accent. An old photograph was found in the house of Bradley and his family sitting on the rug in which the boy's body had been wrapped, and a missing tassel from the rug was actually found in the house. It was found out that Bradley had recently sold a blue Ford Customline, and when this was located at a local dealership, the same pink granular material was found in the trunk. It was then discovered that Bradley and his family had booked passage to sail to England aboard the *Himalaya,* which was already en route to Colombo, Ceylon, and his Pekinese dog was found at a local veterinary hospital with instructions left to send it to England. Detectives waited for Bradley in Colombo and he was taken back to Australia. Stephen Bradley was found guilty of the murder and sentenced to life imprisonment.

See also Fibers; Hair
References
Evans, C. *The Casebook of Forensic Detection: How Science Solved 100 of the World's Most Baffling Crimes.* New York: Wiley, 1998.
Owen, D. *Hidden Evidence: Forty True Crimes and How Forensic Science Helped Solve Them.* Willowdale, Ontario: Firefly, 2000.

Breath Alcohol

See *Alcohol*

Bulbs (Automobile, Examination of in Accidents)

When a vehicle is involved in an accident after dark, whether or not its lights were illuminated can be an issue. This can often be resolved by examination of the bulb. A lightbulb consists of a filament contained in an airtight envelope filled with inert gases under low pressure. The filament glows when an electric current passes through it and the inert atmosphere prolongs the life of the bulb by preventing oxidation. The filament also heats up when the bulb is on. The effects of an impact on the filament can differ depending on whether it is on or off.

When the bulb is off, the filament may experience a cold break. This is a clean disruption of the wire and there are no other effects. When the bulb is on, the hot wire will be stretched by the force of the impact, rather than breaking. If the glass envelope is broken, the filament will be covered with a coating of fused glass dust.

See also Glass; Headlight Filaments
References
Caddy, B. *Forensic Examination of Glass and Paint: Analysis and Interpretation.* London and New York: Taylor and Francis, 2001.
De Forest, P., R. E. Gaensslen, and H. C. Lee. *Forensic Science: An Introduction to Criminalistics.* New York: McGraw-Hill, 1983.
Saferstein, R. *Criminalistics.* 8th ed. Upper Saddle River, NJ: Pearson Education, 2004.
White, P. *Crime Scene to Court: The Essentials of Forensic Science.* Cambridge: Royal Society of Chemistry, 1998.

Bullets

Bullets are the projectiles in a round of ammunition. They are generally made of lead and may be encased partially or entirely with a metal—typically copper, aluminum, or steel. Lead bullets, especially 0.22 caliber, frequently have a coating of copper or brass, which reduces fouling of the bore and allows the projectile to travel through the barrel more easily. Bullet styles typically include round nose, wadcutter (for target shooting), semiwadcutter, full metal jacket, soft point, hollow point, and semi-jacketed hollow point. Metal-jacketed bullets are typically used in semiautomatic weapons. Their harder surfaces are less likely to jam when feeding out of the magazine and into the chamber.

Bullets are marked as they pass through the barrel of the gun. There are two types of marks imparted. Class characteristics are typical of the type of gun. They include the bullet caliber and the number and direction of twist of the lands and grooves left by the rifling of the barrel. For example, a 0.38 caliber bullet with 5 lands and grooves and a right twist is consistent with bullets fired from Smith and Wesson, Taurus, Ruger, and I.N.A. revolvers.

The bullet will also bear individualizing characteristics, in the form of fine striations caused by imperfections made during the barrel rifling process. The individualizing marks are unique to the individual gun.

The gross physical characteristics of a bullet can be determined by examination under very low-power magnification. The finer individualizing markings require a higher magnification. Typically, a comparison microscope is used so that the evidence bullet and one from the suspect gun can be mounted and the images viewed side by side. The bullets can be rotated and the markings aligned. A photographic record can be made to illustrate the agreement in detail.

Today, many laboratories use digital image capture. The information can be stored in a database and an evidence bullet compared with those from other shootings or known guns. The main difficulty is that bullets can be deformed by the impact, and cartridge cases are therefore more widely used in examinations.

The automated search and database comparison systems currently used in the United States and in most other countries use the IBIS system developed by Forensic Technology Incorporated. The system can develop images from cartridges and bullets. It uses an automated algorithm to search for possible matches and displays candidates for review by an experienced operator.

See also Ammunition; Firearms
References
De Forest, P., R. E. Gaensslen, and H. C. Lee. *Forensic Science: An Introduction to Criminalistics.* New York: McGraw-Hill, 1983.
Saferstein, R. *Criminalistics.* 8th ed. Upper Saddle River, NJ: Pearson Education, 2004.
Warlow, T. *Firearms, the Law, and Forensic Ballistics.* 2nd ed. London and Bristol, PA: Taylor and Francis, 1996.
White, P. *Crime Scene to Court: The Essentials of Forensic Science.* Cambridge: Royal Society of Chemistry, 1998.

Bundy, Ted

The Bundy case is of forensic significance due to its reliance on bite-mark evidence.

During the period of 1969 to 1975, a series of killings occurred in California, Oregon, Washington, Utah, and Colorado. The dozens of victims all looked alike— young females, most with long hair parted in the middle—and they were all attacked in the evening. At first it was difficult to realize the similarities between the murders because they occurred in multiple states, but it soon became clear that one link among the murders was that a young law student, Ted Bundy, was present in the vicinity of each of the crimes. In each case, a man fitting Bundy's description was seen around the area with a cast on an arm or leg asking female passersby for assistance. A tan-colored VW beetle was also seen in many of the areas.

In 1974 a young woman was attacked by Ted Bundy in his VW Beetle, but managed to fight back and escape, and was later able to identify Bundy as her attacker. However, Bundy escaped prison, was recaptured eight days later, and then somehow a few months later escaped again.

In January 1978 a resident of Chi Omega sorority house at Florida State University (FSU) in Tallahassee returned home to find the front door open. On entering, she heard someone running upstairs, followed by footsteps coming downstairs. Assuming the footsteps were that of a burglar, she hid until they were gone and then went upstairs to alert the other residents. However, what she found was two

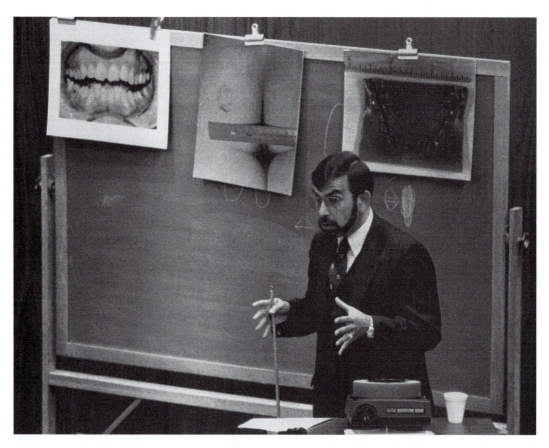

Dr. Lowell J. Levine, a New York forensic odontologist, testified that the bite marks found on the buttock of Florida State University coed Lisa Levy reflect characteristics of Ted Bundy's teeth. Photos of Ted Bundy's teeth and bite marks are displayed on the chalkboard behind Dr. Levine. ca. 1979, Tallahassee, Florida. (Bettmann/Corbis)

sisters alive and another two dead in their beds. All had been beaten and sexually assaulted. One of the girls who had been found dead, Lisa Levy, was found to have bite marks on her legs, buttocks, and one of her nipples. When pictures were taken of these bite marks, a ruler was included in the picture. It was these bite marks that provided the most compelling evidence in the conviction of Ted Bundy for the murders. Initially, Bundy refused to allow moulds of his teeth to be made, but a search warrant was obtained and investigators were authorized to obtain the moulds they needed by force if necessary. During his trial for the FSU murders, an overlay was shown of Bundy's teeth over the bite marks, and the fit was determined to be exact. Ted Bundy was found guilty and was executed by electrocution. Shortly before his execution, Bundy indicated that he had killed between forty and fifty young women.

See also Bite Marks; Odontology

References
BBC Website, Crime Case Closed, Infamous Criminals; http://www.bbc.co.uk/crime/caseclosed/tedbundy1.shtml (Referenced July 2005).
Court TV's Crime Library, Criminal Minds and Methods; http://www.crimelibrary.com/bundy/attack.htm (Referenced July 2005).
Evans, C. *The Casebook of Forensic Detection: How Science Solved 100 of the World's Most Baffling Crimes*. New York: Wiley, 1998.
Owen, D. *Hidden Evidence: Forty True Crimes and How Forensic Science Helped Solve Them*. Willowdale, Ontario: Firefly, 2000.

Bureau of Alcohol, Tobacco, and Firearms (BATF)

The Bureau of Alcohol, Tobacco, and Firearms (BATF) is a federal government agency that was part of the Department of the Treasury until the creation of the Office of Homeland Security in 2002. Its responsibilities include laboratory examinations in tax law cases concerning alcohol and tobacco products, and examination of explosive devices and weapons in cases covered by federal law. The bureau also examines material from the scenes of crimes involving the above and so has developed substantial expertise in areas such as examination of fire debris for accelerant and explosive traces. The laboratory headquarters are in Rockville, Maryland.

Reference
ATF Online, Bureau of Alcohol, Tobacco, Firearms and Explosives, U.S. Department of Justice; http://www.atf.treas.gov/ (Referenced July 2005).

Burns

Burns are well-recognized, painful injuries caused by excessive heat or some chemicals such as acids and caustics. In forensic science, examination of burns is required when there is suspicion that the burn was nonaccidental or when the victim has died.

As is the case with most injuries, it can be difficult to differentiate accidental and deliberate burn injury. Often the history of the victim, the testimony of witnesses, and the specific nature of the burn will be important. For example, severe burning from immersion in hot water requires the heat to be applied for more than the transient exposure resulting from reflex withdrawal of the exposed part. Face burns from acid or alkyl sprays are also unlikely to be due to accident.

Examination of fire victims is directed to ascertaining the cause of death. The main factor is usually inhalation of smoke and poisoning from carbon monoxide or toxic products of combustion of materials such as polyurethane and nylon (which produce cyanide). There is always the rare possibility that the deceased may be a homicide victim and that the fire was an attempt to cover up the true cause of death.

The effects of heat cause certain artifacts. One of the first to develop is the body taking up what is referred to as the "pugilistic position" in which heat causes muscle shrinkage and results in the hands being raised in front of the face, similar to the stance of a boxer. This must not be taken as an indication that the deceased was defending against attack.

Determination of the timing of injury and death, relative to the events caused by the fire, depends on several factors. Someone alive when the fire started will inhale smoke; someone who was dead will not. The presence of soot in the airways is therefore a good indicator. The question of whether some of the burns themselves may have been inflicted before the fire is less clear. Antemortem (before death) burns will usually have signs associated with injury and the subsequent healing response, whereas those caused by the fire will not. Typical signs include a red rim to the injury and blistering. However, these are not infallible indicators and occasionally occur in postmortem burn injuries.

Death can result from exposure to heat well below the degree encountered in fires or that will induce burns. Technically, any environmental situation that causes the body temperature to rise above 41° C (105.8° F) can cause death from hyperthermia. Risk factors include severe heatwaves, exercise in hot weather, and improper use of saunas. Most deaths from environmental heat are related to preexisting conditions such as heart disease, obesity, and alcoholism.

References

Byard, R., T. Corey, C. Henderson, and J. Payne-James. *Encyclopedia of Forensic and Legal Medicine.* London: Elsevier Academic, 2005.

Knight, B. *Simpson's Forensic Medicine.* London and New York: Oxford University Press, 1997.

Payne-James, J., A. Busuttil, and W. Smock. *Forensic Medicine: Clinical and Pathological Aspects.* London: Greenwich Medical Media, 2003.

C

California Association of Criminalists (CAC)

There are seven regional forensic science societies in the United States. The California Association of Criminalists (CAC) is the oldest and the one with the most well-established format. The CAC holds well-attended scientific meetings twice a year, the proceedings of which are published by the internationally acknowledged journal *Science and Justice*.

Reference
California Association of Criminalists; http://www.cacnews.org/ (Referenced July 2005).

Cannabis

Cannabis preparations are obtained from the perennial plant *Cannabis sativa,* which is thought to have originated somewhere in central Asia. The leaves of this plant are distinctive—they are long and slender with serrated edges. There are both male and female plants (it is dioecious)—the female plants are tall and bushy, while the male plant is smaller and not so bushy. In order for the female plant to become fertilized and produce seeds, a sticky resin is produced from the flowering top of the female plant to allow it to collect airborne pollen from the male flower.

The main active ingredient of cannabis is Δ9-tetrahydrocannabinol (THC). However, cannabis contains an entire family of constituents known as the cannabinoids, of which over sixty have been identified.

Preparations

Marijuana refers to the dried leaves and flowers of the cannabis plant. It is usually smoked in a cigarette, cigar, or pipe, or may be baked in cookies or brownies.

Bhang refers to dried leaves of female plants from which the resin has been removed. Bhang does not normally have a great deal of psychotropic activity.

Ganja is made from the tops of female plants from which the resin has not been removed. This is three or four times more potent than bhang.

- Note that the distinction between bhang and ganja is only really made in India.
- In the West Indies, the term *ganja* is used in place of *marijuana* because their cannabis was largely introduced from India.
- In Jamaica, *ganja* refers to the entire cannabis plant.

Hashish (or *charas* in India) refers to the dried resin from the top of the female plant. This resin is a pale yellow color when harvested, but turns almost black when dried.

A police officer examines a marijuana plant during a raid on a marijuana grower. (Phil Schermeister / Corbis)

This can be smoked alone or mixed with tobacco. It can also be baked in cookies or brownies. This preparation typically contains 6 to 10 percent cannabinoids.

Hash oil is a more concentrated version of hashish. It is made by boiling the hashish in a solvent such as alcohol, filtering the residue, and then allowing the solvent to evaporate. This oil can be dropped onto a regular cigarette and then smoked, or can be dropped onto hot aluminum foil, and the smoke inhaled. This oil is very potent and may contain up to 60 percent cannabinoids.

Cannabis has been shown to have legitimate clinical uses in the treatment of nausea and vomiting in cancer patients undergoing chemotherapy and in stimulation of appetite in AIDS and cancer patients. (However, its use in AIDS patients is controversial because cannabis may further suppress the immune system.)

Despite the widespread belief that cannabis is a relatively harmless drug, it has a number of important physiological effects that may be experienced by users, including:

- Tachycardia—heart rate may increase by 20 to 50 percent.
- Increased blood pressure (however, note that when a user stands up after lying down, he or she may experience a drop in heart rate and blood pressure).
- Decreased body temperature.
- Dry mouth and throat.
- Reddening of the conjunctivae of the eyes.
- Decreased intraocular pressure.
- Decreased size of pupils.
- Hunger.

Note that because cannabis is often smoked, many of the adverse effects associated with smoking may also be experienced. These include lung disease, and wheezing and coughing. It is often difficult to distinguish which effects are due to cannabis use and which are due to tobacco products because many cannabis users use both.

Cannabis also has a number of behavioral effects that tend to be dependent not only on the dose and form of cannabis taken, but on the state of mind, mood, and expectations of the individual prior to use of the substance, as well as the atmosphere and setting. At moderate doses these effects include euphoria, heightening of the senses, altered sense of time (time appears to pass much more slowly), and short-term memory impairment. At higher doses, other more serious effects may be (though are not usually) experienced, including anxiety, confusion, aggressiveness, hallucinations, nausea, and vomiting.

Cannabis alters the ability of a user to perform skilled tasks, and this includes the ability to drive. It is illegal to drive while under the influence of cannabis.

One of the major concerns of cannabis use is the likelihood that it will lead to use of other, more harmful drugs such as cocaine and heroin. Studies have found that if cannabis use starts at a young age, there is more likelihood that the user will go on to use other drugs such as these.

See also Drugs; Hashish; Marijuana

References
De Forest, P., R. E. Gaensslen, and H. C. Lee. *Forensic Science: An Introduction to Criminalistics*. New York: McGraw-Hill, 1983.
Drummer, O.H. *The Forensic Pharmacology of Drugs of Abuse*. New York: Oxford University Press, 2001.
James, S. H., and J. J. Nordby. *Forensic Science: An Introduction to Scientific and Investigative Techniques*. Boca Raton, FL: Taylor and Francis, 2005.
Levine, B. *Principles of Forensic Toxicology*. Washington, DC: American Association for Clinical Chemistry, 1999.
Office of National Drug Control Policy, Drug Facts, Marijuana; http://www.whitehousedrugpolicy.gov/drugfact/marijuana/index.html (Referenced July 2005).
Saferstein, R. *Criminalistics*. 8th ed. Upper Saddle River, NJ: Pearson Education, 2004.
U.S. Drug Enforcement Administration, Drug Descriptions, Marijuana; http://www.dea.gov/concern/marijuana.html (Referenced July 2005).
White, P. *Crime Scene to Court: The Essentials of Forensic Science*. Cambridge: Royal Society of Chemistry, 1998.

Carbon Monoxide

Carbon monoxide is formed when fuels such as propane or butane burn in an oxygen-deficient atmosphere. It is also a product of gasoline combustion in engines. It is a colorless, odorless gas that is very poisonous. It combines with the sites on hemoglobin in blood and prevents oxygen uptake and transportation. The effects depend on the degree to which the carbon monoxide replaces the oxygen, usually expressed as percent saturation. Carbon monoxide poisoning is associated with a cherry-red coloration of the skin and muscles.

Normal, healthy people can have carbon monoxide present in their blood at concentrations of up to 5 percent, depending on whether they live in the country or city and whether or not they smoke. Levels between 10 and 50 percent are associated with breathlessness and headache. Levels over 50 percent are associated with coma and death. Like all toxic chemical effects, these are average figures for healthy people and actual responses will vary between individuals. For example, a frail elderly person could die due to carbon monoxide poisoning with a level of the order of 25 percent.

Death in suicide by inhalation of automobile exhaust fumes is due to the carbon monoxide in the exhaust gases. Accidental carbon monoxide deaths occur in winter in cold climates due to incomplete combustion in propane-burning stoves in poorly ventilated areas. Sealing the room to prevent drafts also prevents air renewal, reduces the oxygen supply to the fire, and prevents dissipation of the products of incomplete combustion. Because the gas is odorless, there is little or no warning to the room's occupants of the buildup of carbon monoxide.

See also Toxicology
References
Goldfrank, L., N. Flomenbaum, N. Lewin, M. A. Howland, R. Hoffman, and L. Nelson. *Goldfrank's Toxicologic Emergencies.* 7th ed. New York: McGraw-Hill Professional, 2002.

Klaassen, C. *Casarett and Doull's Toxicology: The Basic Science of Poisons.* 6th ed. New York: McGraw-Hill Professional, 2001.

Levine, B. *Principles of Forensic Toxicology.* Washington, DC: American Association for Clinical Chemistry, 1999.

Casts

Impressions left by tires, footwear, and tool marks can show individual characteristics due to wear and manufacturing defects. These can be used to provide a link between source and scene as good as that of a fingerprint. However, the impression must be collected so that its characteristics and dimensions can be recorded and compared. For samples such as tire tracks in mud, footprints in snow, and crowbar impressions in wood, the mark has to be collected by casting. Plaster and silicone are used. Casting may also be used in forensic odontology to cast bite marks for comparison to potential suspects.

See also Fingerprints; Footwear; Tire Tracks; Tool Marks
References
De Forest, P., R. E. Gaensslen, and H. C. Lee. *Forensic Science: An Introduction to Criminalistics.* New York: McGraw-Hill, 1983.

James, S. H., and J. J. Nordby. *Forensic Science: An Introduction to Scientific and Investigative Techniques.* Boca Raton, FL: Taylor and Francis, 2005.

Saferstein, R. *Criminalistics.* 8th ed. Upper Saddle River, NJ: Pearson Education, 2004.

White, P. *Crime Scene to Court: The Essentials of Forensic Science.* Cambridge: Royal Society of Chemistry, 1998.

Cause of Death

Identifying the cause of death is a recurring issue in forensic investigations. Was death by accident, suicide, or murder? What was the manner of death? If accidental, can the information be used to prevent others suffering in the same way? If homicidal, can the information assist in identifying the perpetrator?

However, many of the situations that present themselves are not amenable to specific assignment of cause. There is a spectrum of possible causes that merges into probable causes, and mostly the forensic evidence is corroborative—supportive of other evidence but not in itself clear-cut and compelling.

Even when clear presentation of facts is possible, different doctors can provide different interpretations.

References
Byard, R., T. Corey, C. Henderson, and J. Payne-James. *Encyclopedia of Forensic and Legal Medicine.* London: Elsevier Academic, 2005.

Knight, B. *Simpson's Forensic Medicine.* London and New York: Oxford University Press, 1997.

Payne-James, J., A. Busuttil, and W. Smock. *Forensic Medicine: Clinical and Pathological Aspects.* London: Greenwich Medical Media, 2003.

Chain of Custody

There must be no doubt about the integrity of objects admitted as evidence. The usual way to establish integrity is through what is known as the chain of custody. The chain of custody, through written or secure electronic records, is a way to show the history of the object from the time it was first collected until it is admitted in court as evidence. The chain is an unbroken audit trail, which permits identification of all of the people who have handled the item and the transfers of custody that they make. The item itself is identified by some unique marking affixed directly to it or to its proximal container.

References
De Forest, P., R. E. Gaensslen, and H. C. Lee. *Forensic Science: An Introduction to Criminalistics.* New York: McGraw-Hill, 1983.

James, S. H., and J. J. Nordby. *Forensic Science: An Introduction to Scientific and Investigative Techniques.* Boca Raton, FL: Taylor and Francis, 2005.

Saferstein, R. *Criminalistics.* 8th ed. Upper Saddle River, NJ: Pearson Education, 2004.

White, P. *Crime Scene to Court: The Essentials of Forensic Science.* Cambridge: Royal Society of Chemistry, 1998.

Charred Documents

It is sometimes possible to read writing or printing on documents that have been damaged by fire. Provided that the surface is not completely destroyed, examination using infrared or reflected light will sometimes reveal the original text.

See also Document Examination
References
Hilton, O. *Scientific Examination of Questioned Documents.* Boca Raton, FL: CRC, 1993.
Saferstein, R. *Criminalistics.* 8th ed. Upper Saddle River, NJ: Pearson Education, 2004.

Clandestine Drug Laboratories

Some illicit drugs, such as cocaine, opiates, and marijuana, have their origins in plant materials. Others, like speed, ecstasy, and LSD, are manufactured in the laboratory. Even those of natural origin are usually laboratory processed to make the final preparation sold on the street. The laboratories can be anything from a small setup in the kitchen of a house to a fully equipped facility. Whatever the size and complexity, clandestine drug laboratories present challenges to the scientific and enforcement arms of illicit drug control.

Coca- and opium-processing laboratories require little by way of special chemical apparatus and are easy to conceal. Laboratories producing amphetamine analogs such as ecstasy are more difficult to hide (some of the processes make use of intermediates with a strong and unpleasant odor). All can be potentially dangerous to the enforcement agencies, either through the hazards from the chemicals, booby trapping, or the danger of serious intoxication from exposure to the highly potent product.

Investigation of a clandestine laboratory is thus a task best left to properly trained technical experts. As well as the safety issues, identification of apparatus and precursor chemicals can reveal valuable information about the drugs being manufactured.

See also Amphetamines; Drugs; Methamphetamine; Lysergic Acid Diethylamide (LSD)

References
Christian, D. R. *Forensic Investigation of Clandestine Laboratories.* Boca Raton, FL: CRC, 2004.
James, S. H., and J. J. Nordby. *Forensic Science: An Introduction to Scientific and Investigative Techniques.* Boca Raton, FL: Taylor and Francis, 2005.
Lee, H. C., T. Palmbach, and M. T. Miller. *Henry Lee's Crime Scene Handbook.* London and San Diego, CA: Academic, 2001.

Class Characteristics

Markings on physical objects such as ammunition, door frames forced by screwdrivers or crowbars, and footwear and tire tracks can contain information about the source. For example, a sneaker print may be of a pattern that reveals it was made by one of the Nike Air Jordan series. However, these patterns are designated class characteristics and do not contain any individualizing features that would permit the impression to be linked to a specific item.

See also Individual Characteristics
References
De Forest, P., R. E. Gaensslen, and H. C. Lee. *Forensic Science: An Introduction to Criminalistics.* New York: McGraw-Hill, 1983.
James, S. H., and J. J. Nordby. *Forensic Science: An Introduction to Scientific and Investigative Techniques.* Boca Raton, FL: Taylor and Francis, 2005.
Saferstein, R. *Criminalistics.* 8th ed. Upper Saddle River, NJ: Pearson Education, 2004.
White, P. *Crime Scene to Court: The Essentials of Forensic Science.* Cambridge: Royal Society of Chemistry, 1998.

Cocaine

Cocaine is a powerfully addictive stimulant that directly affects the brain. Pure cocaine was first extracted from the leaf of the *Erythroxylon coca* bush, which grows primarily in Peru and Bolivia, in the mid-nineteenth century. In the early 1900s, it became the main stimulant drug used in most of the tonics and elixirs that were developed to treat a wide variety of illnesses. Today, cocaine is a Schedule II drug, meaning that it has high potential for abuse, but can be administered by a doctor for legitimate medical uses, such as a local anesthetic for some eye, ear, and throat surgeries.

There are basically two chemical forms of cocaine: the hydrochloride salt and the freebase. The hydrochloride salt, or powdered form of cocaine, dissolves in water and, when abused, can be taken intravenously (by vein) or intranasally (in the nose). *Freebase* refers to a compound that has not been neutralized by an acid to make the hydrochloride salt. The freebase form of cocaine is smokable.

Cocaine is generally sold on the street as a fine, white, crystalline powder, known as "coke," "C," "snow," "flake," or "blow." Street dealers generally dilute it with such inert substances as cornstarch, talcum powder, and/or sugar, or with such active drugs as procaine (a chemically related local anesthetic) or with such other stimulants as amphetamines.

Common cutting agents are similar to those found in heroin and include inert sugars (glucose, mannitol, lactose) as well as other drugs, including caffeine, codeine, and amphetamines.

Cocaine was previously recommended for use in eye surgery because it has local anesthetic properties as well as a vasoconstrictor action that limits hemorrhage. It is no longer the agent of choice for this purpose because of damage to the eye, but it is still used in the surgery of ear, nose, and throat.

Cocaine activates the brain's pleasure centers, which results in euphoria, increased motor activity, and psychotic symptoms. It is also a potent vasoconstrictor and produces increased heart rate, a combination that can lead to sudden changes in blood pressure. Cocaine-related deaths are typically associated with cardiac failure or cerebral hemorrhage. Cocaine use can lead to severe psychological dependence, but there is no evidence that it leads to physical dependence. Cocaine consumption is a major health and social problem and a major revenue generator for the traffickers.

Stimulant drugs have a very strong psychological dependence associated with them—users exhibit extreme drug-seeking behavior

to the detriment of their health and lives in general.

The effects sought by users of cocaine include feelings of euphoria and increased energy and alertness, as well as appetite-suppressant effects.

See also Crack Cocaine; Drugs

References

De Forest, P., R. E. Gaensslen, and H. C. Lee. *Forensic Science: An Introduction to Criminalistics.* New York: McGraw-Hill, 1983.

Drummer, O. H. *The Forensic Pharmacology of Drugs of Abuse.* New York: Oxford University Press, 2001.

James, S. H., and J. J. Nordby. *Forensic Science: An Introduction to Scientific and Investigative Techniques.* Boca Raton, FL: Taylor and Francis, 2005.

Levine, B. *Principles of Forensic Toxicology.* Washington, DC: American Association for Clinical Chemistry, 1999.

Office of National Drug Control Policy, Drug Facts, Cocaine; http://www.whitehousedrugpolicy.gov/drugfact/cocaine/index.html (Referenced July 2005).

Saferstein, R. *Criminalistics.* 8th ed. Upper Saddle River, NJ: Pearson Education, 2004.

U.S. Drug Enforcement Administration, Drug Descriptions, Cocaine; http://www.usdoj.gov/dea/concern/cocaine.html (Referenced July 2005).

White, P. *Crime Scene to Court: The Essentials of Forensic Science.* Cambridge: Royal Society of Chemistry, 1998.

CODIS

See *Combined DNA Index System*

Color Tests

Color or spot tests are widely used for preliminary screening in the forensic laboratory. They share the property that a target substance will produce a color when mixed with a test reagent. Color tests are not chemically specific and cannot be used as proof of the presence of the material. Examples include the Kastle-Meyer test for blood, the Marquis test for opiates and amphetamines, the Duquenois-Levine test for marijuana, the Scot test for cocaine, and the Griess test for explosive residues.

Principles of Color Tests

A reagent is added to the unknown substance, and any color change is noted. There may be no change, one color change, or a series of color changes. Spot tests are commonly used in the preliminary identification of drug substances. The reagent used may be specific for a compound under investigation, but is generally more indicative of a certain class of drugs. The resulting color can usually be attributed to a particular aspect of the drug's structure and so can be indicative of particular functional groups (components of chemical structure that determine reactivity), or groups of drugs. In general, these tests have wide applicability and provide useful and diagnostic information. The color of each reaction may vary depending on the conditions of the test, drug concentration, salt form, pH, and the presence of extraneous material in the sample. A negative result for a color test may be used to rule out a drug. Some of these tests are also used as sprays or locating agents for thin-layer chromatography. As well as being used for the preliminary identification of chemical substances, with minor modifications these tests may be used in the preliminary identification of chemical substances in biological fluids, tissues, and stomach contents.

The advantages of color tests are that they are sensitive and fairly quick. The disadvantages are that they are nonspecific, destructive, and may be affected by the conditions of the test.

Screening tests for body fluids depend on detection of constituent chemicals. Presumptive blood tests are rapid chemical tests that may be used to locate and differentiate between blood and other similarly colored stains. They are sensitive, easy to perform, and allow minute traces of blood to be located when they otherwise might not be easily noticed. They are sensitive enough to apply to areas of scenes that may have been cleaned in an attempt to hide the evidence.

Most screening tests for blood depend on the peroxidase activity of hemoglobin. The most common ones depend on the oxidation of colorless reduced indicators, many of which are unconjugated systems and are known or suspected carcinogens.

Color tests for semen are based on the hydrolysis of phosphate esters and detection of the liberated organic moiety by production of a color complex. Acid phosphatase, present in seminal fluid, reacts with sodium alpha-napthylphosphate and the dye Fast Blue B to produce a purple-blue coloration.

As with the screening test for blood, a positive result is the rapid formation of the intensely colored product. It is important to note that a number of vegetable and fruit juices can result in the generation of false positive results with this test.

Color tests for organic and inorganic explosives can be conducted on acetone washes of debris. The three main test reagents are Greiss reagent, diphenylamine, and potassium hydroxide. Greiss reagent changes to a pink-red color in the presence of nitrate, nitroglycerin, PETN, RDX, and tetryl. Diphenylamine produces a blue color in the presence of most of the common explosives, but not TNT. Alcoholic potassium hydroxide produces color changes only with TNT and tetryl substances.

See also Blood; Explosions and Explosives; Saliva; Semen Identification

References

Butler, J. *Forensic DNA Typing.* 2nd ed. Burlington, MA: Elsevier, 2005.

De Forest, P., R. E. Gaensslen, and H. C. Lee. *Forensic Science: An Introduction to Criminalistics.* New York: McGraw-Hill, 1983.

James, S. H., and J. J. Nordby. *Forensic Science: An Introduction to Scientific and Investigative Techniques.* Boca Raton, FL: Taylor and Francis, 2005.

Lee, H. C., T. Palmbach, and M. T. Miller. *Henry Lee's Crime Scene Handbook.* London and San Diego, CA: Academic, 2001.

Saferstein, R. *Criminalistics.* 8th ed. Upper Saddle River, NJ: Pearson Education, 2004.

White, P. *Crime Scene to Court: The Essentials of Forensic Science.* Cambridge: Royal Society of Chemistry, 1998.

Combined DNA Index System (CODIS)

The Combined DNA Index System (CODIS) is a national database of DNA profiles. It is administered by the FBI. The value of DNA databases is now thoroughly established by results from CODIS and from the DNA database operated in Britain by the English Home Office Forensic Science Service.

All states have now passed legislation permitting the taking of reference samples from persons convicted of certain offenses. Usually these are the more severe offenses against the person such as homicide and rape. DNA results from the scene or the body of a victim are compared with the reference profiles in the database. There have been many successes, even though the systems have only been operating for a short time and even though not all states are currently contributing reference samples.

Experience with the British system, which accepts data from a wide range of offenses including burglary and traffic violations, shows that 300 to 400 matches per week can be obtained on a regular basis. DNA databases are thus a powerful tool for public safety. However, there remain concerns regarding the impact on civil liberties. For example, there is a fear the data could be used for health screening of job or insurance applicants, if one of the forensic markers turned out to be associated with a genetically determined disease, or even a predisposition to a disease.

See also DNA in Forensic Science
References
Butler, J. *Forensic DNA Typing.* 2nd ed. Burlington, MA: Elsevier, 2005.
Federal Bureau of Investigation, Combined DNA Index System; http://www.fbi.gov/hq/lab/codis/index1.htm (Referenced July 2005).

Comparison Microscope

The comparison microscope is made up of two microscopes joined by an optical bridge. The images of the objects in the light path are viewed side by side in a split screen. The physical characteristics can thus be compared.

The microscopes can be low power, capable of viewing and comparing three-dimensional objects such as bullets or cartridge cases, or high power for viewing the features of fibers or hairs.

The main use of the comparison microscope is exactly what the name suggests—the comparison of microscopic features in the examination of physical evidence.

See also Fibers; Firearms; Hair; Microscope
References
De Forest, P., R. E. Gaensslen, and H. C. Lee. *Forensic Science: An Introduction to Criminalistics.* New York: McGraw-Hill, 1983.
Saferstein, R. *Criminalistics.* 8th ed. Upper Saddle River, NJ: Pearson Education, 2004.

Contact Gunshot Wounds

Scientific testing can sometimes produce information as to the distance between the weapon and the victim of a shooting. There are many variables, but one relatively clear-cut situation is where the gun was in contact with the victim.

When the muzzle of the gun is in firm contact with the body of the victim, the combination of heat and a partial seal of discharge residues results in typical blackening of the skin and clothing. There is searing at the entry wound and there may be an imprint of the muzzle. In the case of a contact wound to the head, the gases blow out between the soft skin of the scalp and the hard bone of the skull. Back pressure can result in a star-shaped, or stellate, entry wound.

Where the gun is not in firm contact with the body, there is still a concentrated blackened discharge pattern, but searing and an imprint of the muzzle will be absent or slight. The patterns that result as the distance of the weapon from the victim's body increases depend on the type of gun and ammunition used, and distance determination becomes more approximate.

See also Firearms
References
Payne-James, J., A. Busuttil, and W. Smock. *Forensic Medicine: Clinical and Pathological Aspects.* London: Greenwich Medical Media, 2003.

Schwoeble, A. J., and D. L. Exline. *Current Methods in Forensic Gunshot Residue Analysis.* Boca Raton, FL: CRC, 2000.

Warlow, T. *Firearms, the Law, and Forensic Ballistics.* 2nd ed. London and Bristol, PA: Taylor and Francis, 1996.

Saferstein, R. *Criminalistics.* 8th ed. Upper Saddle River, NJ: Pearson Education, 2004.

U.S. Drug Enforcement Administration; http://www.usdoj.gov/dea/ (Referenced July 2005).

White, P. *Crime Scene to Court: The Essentials of Forensic Science.* Cambridge: Royal Society of Chemistry, 1998.

Controlled Substances

One of the ways to manage drug abuse is to place the chemicals on a list of controlled substances. In that way, trading or possession of medicines without a prescription or of nonmedical substances becomes a defined legal offense. The principal legislation in the United States is the federal Controlled Substances Act. There are five schedules to the act, controlling the manufacture, distribution, use, and pharmacy dispensing of the listed substances, as well as establishing norms for record keeping and reporting, and penalties for trafficking of the listed drugs.

Schedule I drugs are those with no accepted medical use and a high potential for abuse. Heroin and LSD are in this category, as is marijuana. Schedule II drugs also have a high potential for abuse but with some restricted medical applications. Examples include methadone and amphetamines. Schedule III drugs have an accepted medical use but with a potential for moderate abuse. Anabolic steroids and some codeine preparations are in this schedule. Schedule IV and V drugs are those with more medical and fewer illicit reasons for use. Phenobarbital and valium are in Schedule IV, and some opiate—but non-narcotic—preparations are in Schedule V.

See also Amphetamines; Cannabis; Cocaine; Crack Cocaine; Drugs; Ecstasy; Gamma Hydroxybutyrate (GHB); Hashish; Heroin; Lysergic Acid Diethylamide (LSD); Methadone; Methamphetamine; Morphine; Narcotics; Opium; Phencyclidine, Phenylcyclhexyl, or Piperidine (PCP); Psilocybin
References
De Forest, P., R. E. Gaensslen, and H. C. Lee. *Forensic Science: An Introduction to Criminalistics.* New York: McGraw-Hill, 1983.

James, S. H., and J. J. Nordby. *Forensic Science: An Introduction to Scientific and Investigative Techniques.* Boca Raton, FL: Taylor and Francis, 2005.

Corroborative Evidence

There are very few situations in which forensic evidence is definitive of an offense and the perpetrator. These include driving while intoxicated and drug possession offenses. In other situations, such as trying to link someone to a scene or to contact with someone else, the forensic evidence is less firm. The best instance would be DNA typing of semen recovered from the vagina of a rape victim. There is no doubt about intimate contact or, with modern typing systems, the biological origin of the material. However, the issue may be one of consensual rather than forced intercourse. Fingerprinting is also definitive as to biological origin but does not provide information about the circumstances of the deposition of the print. In both cases, non-matching forensic evidence will eliminate the suspect as the source of the material (semen or fingerprint). Thus the forensic science corroborates some other evidence, lending a degree of weight to it that depends on the nature of the test and the circumstances of the alleged offense.

See also Associative Evidence
Reference
Saferstein, R. *Criminalistics.* 8th ed. Upper Saddle River, NJ: Pearson Education, 2004.

Counterfeit Currency

Fraudulent reproduction of currency, whether coins or notes, is called counterfeiting. Because higher denomination currency is in note form, these are the main target of counterfeiters. Counterfeit currency notes used to require considerable skill to produce, through preparation of accurate engraved printing plates. Color copiers and digital scanners make it easy to

National Basketball Association star Kobe Bryant arrives at the Eagle County Courthouse October 9, 2003, after he attended a preliminary hearing in a case against him by an alleged victim for rape. Evidence was presented at the hearing. Bryant claimed that the sex was consensual and the Lakers guard denied that he raped the woman. The preliminary hearing was extended to October 15, 2003, Eagle County, Colorado. (Corbis)

produce a facsimile of a banknote. Currency now contains security features to prevent successful reproduction. Some currency, for example in Australia, goes so far as to use plastic, rather than paper, for banknotes. Examination of counterfeit currency consists of examination for the presence of security features, examination of the paper, and of the inks.

See also Document Examination
References
Saferstein, R. *Criminalistics.* 8th ed. Upper Saddle River, NJ: Pearson Education, 2004.
U.S. Department of the Treasury; http://www.ustreas.gov/topics/currency/ (Referenced July 2005).
White, P. *Crime Scene to Court: The Essentials of Forensic Science.* Cambridge: Royal Society of Chemistry, 1998.

Crack Cocaine

Crack cocaine is made by heating a solution of cocaine and baking soda, then drying the product and breaking it into small pieces or rocks. It is a more potent form of the drug as it is in the form of the freebase and can be smoked to produce an intense high of rapid onset.

See also Cocaine; Controlled Substances; Drugs
References
Office of National Drug Control Policy, Drug Facts, Cocaine; http://www.whitehousedrugpolicy.gov/drugfact/cocaine/index.html (Referenced July 2005).
Saferstein, R. *Criminalistics.* 8th ed. Upper Saddle River, NJ: Pearson Education, 2004.
U.S. Drug Enforcement Administration, Drug Descriptions, Cocaine; http://www.usdoj.gov/dea/concern/cocaine.html (Referenced July 2005).
White, P. *Crime Scene to Court: The Essentials of Forensic Science.* Cambridge: Royal Society of Chemistry, 1998.

Crime Laboratories

Government forensic science laboratories are often termed "crime laboratories," reflecting the purpose of the examinations conducted. There are some private laboratories conducting forensic examinations (for example companies offering forensic DNA testing) and there are government laboratories doing forensic work not in a criminal context (for example, medical examiners seeking information as to cause of death, and civil insurance investigations).

There is a view that crime laboratories should not be part of police departments, on the grounds that they must be independent of law enforcement and that being part of the police department will lead to some conscious or unconscious bias in the work of the scientists. There is no good objective evidence to support this assertion. On the contrary, there is evidence that it is not so and that the key factor in keeping testing unbiased is the scientific independence of the laboratory director. For example, many of the headline cases of error in crime laboratories in the 1990s involved private, nonpolice laboratories: the dingo baby case in Australia, the several cases associated with Allan Clift in England, and the Guy Paul Morin case in Canada, to name a few. By contrast, one of the most highly regarded laboratories internationally was the Metropolitan Police Forensic Science Laboratory in England. The "Met Lab," as it was known, had a policy of always appointing a distinguished scientist to the post of director.

Crime Scene

Examination of the crime scene is the critical first step in a forensic investigation. The scene must be protected against contamination but still permit access to key investigation personnel such as photographers and pathologists. The crime scene investigator must be able to recognize potential evidence, to collect and preserve it so that it is not lost or degraded, and to establish the chain of custody for the evidence. Different jurisdictions approach crime-scene examination in different ways. Some use police officers with minimal training; others use police officers specially trained in scene examination and evidence collection. Some use laboratory personnel as a support in especially difficult circumstances. There is no evidence that one system is best, as long as the personnel concerned understand the basic principles. One important example is the examination of a clandestine laboratory.

See also Chain of Custody; Clandestine Drug Laboratories

References
De Forest, P., R. E. Gaensslen, and H. C. Lee. *Forensic Science: An Introduction to Criminalistics.* New York: McGraw-Hill, 1983.
James, S. H., and J. J. Nordby. *Forensic Science: An Introduction to Scientific and Investigative Techniques.* Boca Raton, FL: Taylor and Francis, 2005.
Lee, H. C., T. Palmbach, and M. T. Miller. *Henry Lee's Crime Scene Handbook.* London and San Diego, CA: Academic, 2001.
Saferstein, R. *Criminalistics.* 8th ed. Upper Saddle River, NJ: Pearson Education, 2004.
White, P. *Crime Scene to Court: The Essentials of Forensic Science.* Cambridge: Royal Society of Chemistry, 1998.

Criminalistics

The term "criminalistics" was first used by Hans Gross in 1891. It is widely used in the United States, especially in California, but is almost unknown elsewhere. Sometimes used synonymously with *forensic science,* it is coming to have a more restricted meaning, namely the general examination of evidence in the crime laboratory. Criminalists are those who conduct such examinations.

References
De Forest, P., R. E. Gaensslen, and H. C. Lee. *Forensic Science: An Introduction to Criminalistics.* New York: McGraw-Hill, 1983.
James, S. H., and J. J. Nordby. *Forensic Science: An Introduction to Scientific and Investigative Techniques.* Boca Raton, FL: Taylor and Francis, 2005.
Saferstein, R. *Criminalistics.* 8th ed. Upper Saddle River, NJ: Pearson Education, 2004.
White, P. *Crime Scene to Court: The Essentials of Forensic Science.* Cambridge: Royal Society of Chemistry, 1998.

Crystal Tests

See *Microcrystal Tests*

Cyanide

Beloved of mystery writers, cyanide is an extremely toxic chemical that kills by interfering with the metabolic pathways within the cell. Cyanide is used in industries such as electroplating and is a toxic by-product of combustion of certain plastics. It is found naturally in almonds and apricots.

Cyanide poisoning can result from inhalation of cyanide gas, ingestion of solutions of cyanide salts, or from skin contact. It is fast acting—less than thirty minutes—but the effects can be reversed by inhalation of amyl nitrite and oxygen. Intravenous sodium nitrite administration has also been used.

See also Poisoning; Toxicology
References
Goldfrank, L., N. Flomenbaum, N. Lewin, M. A. Howland, R. Hoffman, and L. Nelson. *Goldfrank's Toxicologic Emergencies.* 7th ed. New York: McGraw-Hill Professional, 2002.
Levine, B. *Principles of Forensic Toxicology.* Washington, DC: American Association for Clinical Chemistry, 1999.

Cyclotrimethylenetrinitramine (RDX)

Cyclotrimethylenetrinitramine (RDX) is one of the most powerful high explosives used by the military. RDX, also known as cyclonite or hexogen, is a white crystalline solid usually used in mixtures with other explosives, oils, or waxes. RDX compositions are mixtures of RDX, other explosive ingredients, and desensitizers or plasticizers. When combined with other explosives or inert materials, RDX forms the base for detonator charges in common military explosives, projectiles, rockets, and land mines.

RDX gives a pink to red coloration with Greiss reagent and a blue color with diphenylamine spot test reagent. Analysis for the presence of explosive materials requires the rinsing of blast debris using acetone. The acetone rinses of trace materials can then be analyzed by thin layer chromatography (TLC) and high performance liquid chromatography (HPLC). When sufficient sample is recovered, confirmatory tests can be conducted by infrared spectroscopy or x-ray diffraction.

See also Color Tests; Explosions and Explosives; High Performance Liquid Chromatography (HPLC); Thin Layer Chromatography (TLC)
References
Saferstein, R. *Criminalistics.* 8th ed. Upper Saddle River, NJ: Pearson Education, 2004.
White, P. *Crime Scene to Court: The Essentials of Forensic Science.* Cambridge: Royal Society of Chemistry, 1998.

D

Databases

Identification of a suspect from a fingerprint left at a scene is one of the more usual representations of forensic science in fictional works. In this case, fiction is indeed close to fact. The success of fingerprint identification is a well-established example of a forensic database in action. The reference source is a collection of exemplar prints collected as inked fingerprint cards, and the suspect sample is a print recovered from an item associated with the crime or the scene. Comparison can be made using the inked impressions or digital images. The print recovered from the scene is compared to those in the database and the suspect identified, provided of course that there is a print in the reference set and that the matching has been conducted correctly.

Other significant databases currently used in forensic science deal with DNA testing to compare biological material left at a scene with the DNA of people in a database, and with comparison of fired ammunition to associate shootings with each other or with seized weapons. There is a great potential for enhanced public safety through application of these three database comparisons. The main attractions are that the information is objective, can be fast, and allows crime-scene material to be compared to a range of possible suspects, even where no suspect is identified (for example, a masked rapist). There are smaller databases dealing with drugs and footwear sole patterns.

The largest fingerprint database in the world is that established by the FBI in 1930. Thus, the concept and practice of using databases in forensic science to identify suspects from materials left at the scene is very well established indeed. However, the most recent application, that of DNA databases, has generated some controversy. The controversy is not about the successful applications: More than 100 wrongfully convicted prisoners have been freed in the United States alone, and many victims of long-unsolved serious crimes have had their cases resolved through DNA testing. By September 2005, the CODIS database had produced over 25,900 hits assisting in more than 27,800 investigations since its establishment in 1998 (http://www.fbi.gov/hq/lab/codis/success.htm). In Britain, with a population about one-fifth that of the United States, the DNA database is producing over 1,700 hits per week, providing associations between crime scenes and between scenes and individuals (http://www.ojp.usdoj.gov/nij/pdf/uk_finaldraft.pdf).

The concerns relate to possible misuse of data. Most of the U.S. states applying DNA databasing have legislation that restricts the reference source to samples taken from convicted felons, and often only those convicted of sexual offenses or homicide. In contrast, the highly successful British database permits the taking of samples from suspects and from those convicted of offenses such as house breaking and motor vehicle law infringements. The fear associated with DNA database entries contrasts with the acceptance of fingerprint records. It is hard to see why there is such a difference in acceptability. One possible reason is that the reliability of fingerprinting has been established in the courts and community from decades of use, whereas DNA testing is relatively modern and complex.

Another possibility is that people fear that information about their DNA could be misused—for example, by allowing genetic information about their intelligence, alcoholism, social problems, or degenerative diseases to fall into the hands of employers, insurance companies, or others who might use it to deny the individual benefits or damage his or her reputation.

Three things would all need to happen for this fear to be realized. First, a single gene would have to be identified that played a major role in determining the characteristic. The examples cited are not ascribable to a single gene. Next, one of the markers used in forensic DNA testing would need to be associated with that gene and so act as a marker for it. Finally, there would need to be a breach of the stringent security controls that protect the databases. Taking all of these together, it is just not likely that there would be real risk to personal privacy rights from forensic DNA databases.

In contrast, the real prospect of crime-scene personnel having testing technologies to identify the evidence material and use wireless communication to enter the results into databases would mean that perpetrators could be identified and apprehended before they could flee or commit another offense. It would also mean that innocent people would be spared the trauma of needless investigation and that law enforcement resources would be more efficiently deployed.

There has been less controversy about firearms databases. The National Integrated Ballistic Information Network (NIBIN) program is a network of 16 multistate regions established by the BATF. The regions encompass over 225 sites, each equipped with IBIS imaging technology and a link to the BATF database.

The system has produced many hits, linking ammunition to ammunition and or guns. For example, NIBIN has assisted Chicago police in finding a link between an armed suspect and a shooting with no other investigative leads available. In the first incident, a victim was shot and wounded in Chicago. Although a physical description was given, no suspect was identified. One cartridge case was recovered at the scene and submitted for IBIS entry at the Illinois State Police Department's Chicago laboratory. A few months after this incident, a confidential informant told police that one of his associates had committed a shooting and was in possession of a firearm. After the suspect was arrested, his weapon was seized and submitted for test firing and NIBIN entry. Correlation highlighted a potential link between the arrested suspect and the shooting four months earlier, and this link was confirmed by firearms examiners who compared the original evidence in the cases.

See also Arson and Explosives Incidents System (AEXIS); Automated Fingerprint Identification System (AFIS); Combined DNA Index System (CODIS); DNA Databases; Drugfire; Integrated Ballistic Identification System (IBIS)

References
Bureau of Alcohol, Tobacco, Firearms, and Explosives, National Integrated Ballistic Information Network; http://www.nibin.gov/ (Referenced July 2005).
De Forest, P., R. E. Gaensslen, and H. C. Lee. *Forensic Science: An Introduction to Criminalistics.* New York: McGraw-Hill, 1983.
Federal Bureau of Investigation, Combined DNA Index System; http://www.fbi.gov/hq/lab/odis/index1.htm (Referenced July 2005).

Forensic Technology, IBIS/Ballistic Identification; http://www.fti-ibis.com/en/s_4_1.asp (Referenced July 2005).

James, S. H., and J. J. Nordby. *Forensic Science: An Introduction to Scientific and Investigative Techniques.* Boca Raton, FL: Taylor and Francis, 2005.

Komarinski, P. *Automated Fingerprint Identification Systems (AFIS).* Amsterdam and Boston, MA: Elsevier, 2005.

Saferstein, R. *Criminalistics.* 8th ed. Upper Saddle River, NJ: Pearson Education, 2004.

White, P. *Crime Scene to Court: The Essentials of Forensic Science.* Cambridge: Royal Society of Chemistry, 1998.

Date Rape

Date rape is the popular name for nonconsensual sexual relations that take place when the victim is incapacitated due to the effects of drugs, including alcohol. The nature of the incapacity can be physical inability to resist or an impairment of the faculties required to give a true consent.

Without doubt, the most common date-rape drug is alcohol. It is not generally acknowledged as such, partly because it can be difficult to separate the social environment of alcohol use from one that constitutes clear nonconsensual sexual activity. However, it is the general experience of laboratories that alcohol is found in the blood (or urine) of about half of the samples from rape victims.

Media attention is focused on a few drugs that are more unequivocally implicated in setting the scene for nonconsensual sexual activity. These are drugs with pharmacological actions that facilitate date rape and that are readily available. The relevant properties are that they can be administered without the recipient knowing, and cause loss of inhibitions or consciousness, and amnesia; that is, they can be used to spike drinks, impair the awareness of what is happening, and cause blackouts and memory loss in the recipient. There are three drugs that fit this description: gammahydroxybutyrate (GHB), flunitrazepam (roofies), and ketamine.

GHB is odorless, colorless, and tasteless. It is not available legitimately but is easily obtained. It has been associated with deaths. Flunitrazepam is the active ingredient of a prescription drug, Rohypnol. When taken with alcohol it reduces inhibitions, and like GHB, causes amnesia. It is used as a recreational drug often with alcohol or marijuana to increase the high. The manufacturer has reformulated the medicine form so that is it a bright blue color and thus hard to disguise if used to spike a drink. Ketamine is a veterinary anesthetic that can be used to spike drinks or cigarettes.

Date-rape drugs are short acting. They are removed from the blood rapidly and so it can be difficult to detect their use from analysis of samples taken more than a few hours after the incident.

See also Gammahydroxybutyrate (GHB); Rape; Sexual Offenses; Time Since Intercourse

References

Goldfrank, L., N. Flomenbaum, N. Lewin, M. A. Howland, R. Hoffman, and L. Nelson. *Goldfrank's Toxicologic Emergencies.* 7th ed. New York: McGraw-Hill Professional, 2002.

Turvey, B. E., and J. Savino. *Rape Investigation Handbook.* Amsterdam and Boston: Elsevier Academic, 2004.

Daubert Ruling

The 1993 U.S. Supreme Court ruling in *Daubert v. Merrell Dow Pharmaceuticals, Inc.* is a landmark decision on the admissibility of scientific evidence. Before *Daubert,* admissibility of scientific evidence was determined by the *Frye* ruling and its various state derivatives (see **Frye Rule**) and, for federal courts, Rule 702 of the Federal Rules of Evidence. In brief, the *Frye* test for admissibility is whether or not the technique at issue is one that has "general acceptance in the field," while Rule 702 assesses the potential usefulness of the proffered evidence. *Daubert* moved the focus to the scientific reliability of the testing and set out four basic principles to help determine it.

The four *Daubert* criteria are: (1) whether the methods upon which the testimony is based are centered upon a testable hypothesis; (2) the known or potential rate of error associated with the method; (3) whether the method has been subject to peer review; and (4) whether there is any remaining uncertainty, then the Supreme Court advocates a return to *Frye* by considering whether the method is generally accepted in the relevant scientific community.

The criteria will be considered briefly in turn.

Testable hypothesis. There can be no better way of determining the admissibility of scientific evidence than by asking whether the scientific method has been used in the investigation or in formulating the opinion expressed. It is the scientific method that distinguishes science from other activities that also require care and knowledge in their conduct. The scientific method demands that the matter can be expressed as a hypothesis and that the validity of the hypothesis can be tested. Let us take Newton's law of gravity as an example. We could express the hypothesis as "the huge mass of the Earth exerts an attractive force on objects around it." We can test the hypothesis by seeing what happens to an object exposed to the hypothetical attraction. As far as we can see, it will behave as hypothesized—the apple falls from the tree to the ground every time and does not remain suspended in midair or take off into outer space.

The scientific method is a little more complex than this but we need only concern ourselves with two features in regard to applying *Daubert*. One is that the hypothesis must predict something. The other is that testing the hypothesis consists of conducting tests designed to disprove it, and that the only thing we can do with certainty is to disprove the hypothesis—no amount of testing can prove it. The laws of nature are just hypotheses that have considerable value in regard to the things that they predict and that have withstood the many tests of time. The focus on disproving the hypothesis is called *falsifiability* and is associated with the scientific philosopher Karl Popper.

To pass this first of the *Daubert* ruling criteria, to be admissible the matter proposed as scientific evidence must satisfy the following:

1. Can it be expressed as a hypothesis that predicts an outcome?
2. Is the hypothesis capable of falsification—can it be tested in a way that will show that the predicted outcome has not been satisfied?
3. Have such tests been conducted and have they failed to falsify the hypothesis?

For example, DNA testing of semen from a vaginal swab meets this *Daubert* criterion. We can formulate the hypothesis that the DNA came from the suspect. The outcome of the hypothesis is that the DNA types in the semen will be the same as those from the suspect. We can falsify this by conducting the DNA typing—if any types are found that differ from those of the suspect then the semen is not his.

The ability of the testing used to detect a falsification of the outcome predicted by the hypothesis is a vital part of the scientific method, as is the ability to repeat the testing itself. Tests, or a combination of testing circumstances, that are not likely to find a falsification if one exists or that cannot be repeated on other occasions or by other observers do not present a sufficient challenge to the hypothesis for its subject matter to be regarded as "scientific." This is why the second and third criteria of the *Daubert* ruling, which require that the testing is reliable in itself (error rates) and has been subjected to the scrutiny of peers (published in the scientific literature), are important.

Error rates. The *Daubert* opinion included the following statement:

. . . in the case of a particular scientific technique, the court ordinarily should consider the known or potential rate of error, see, e.g., *United States v. Smith,* 869 F. 2d 348, 353–354 (CA7 1989) (surveying studies of the error rate of spectrographic voice identification technique), and the existence and maintenance of standards controlling the technique's operation. See *United States v.*

Williams, 583 F. 2d 1194, 1198 (CA2 1978) (noting professional organization's standard governing spectrographic analysis), cert. denied, 439 U.S. 1117 (1979).

Both sides, law and science, seem to have difficulty with this part of the opinion. There are two parts to the requirement: (1) providing the court with information that will allow it to consider error rates and (2) demonstrating the existence and maintenance of standards controlling the technique.

Error rate is better termed *uncertainty of measurement.* There is a mass of published data on uncertainty of measurement. The results of every test contain an element of uncertainty. For example, performing ten replicate analyses on a blood sample with a true alcohol concentration of 0.100 percent will not produce an answer of 0.100 each time. Comparing the average test result with the true value and measuring the spread of results around the mean can tell us much about the reliability of the test. The degree to which the mean of the ten tests approaches the true value is a measure of the accuracy of the test. The spread of results tells us the repeatability of the test. The spread is usually expressed by means of a statistical parameter, the standard deviation. An alternate expression of standard deviation used in analytical chemistry is the coefficient of variation (CV), which is the standard deviation expressed as a percent of the mean. Calculation of accuracy and CV is a normal part of validation of a quantitative (that is, one that measures the amount of material present in the sample) assay.

However, it is more difficult to give an objective measure of the potential error rate in a qualitative (that is, one that identified the presence of a target substance) assay. There will sometimes be information on the incidence of false positive and false negative results, but these can be situational and depend on the exact circumstances of the testing. Factors affecting confidence in the result of a qualitative test include the condition of the sample (perhaps degraded by heat, light, or moisture), the presence of similar materials (such as metabolites of a drug), the specificity of the analytical technique employed (for example, gas chromatography combined with mass spectrometry is much more specific that gas chromatography alone), and the amount of the target material in the sample (the more the better).

Finally, there may be instances of testing or data manipulation where there is no available data on uncertainty of measurement. In these cases, it may be possible to derive an "uncertainty budget" by considering the likely sources contributing to the overall uncertainty of measurement and estimating the magnitude of each.

The second element of this part of the opinion, the existence and maintenance of standards controlling the technique's operation, is much easier to deal with. Every test can be validated using controls and standards, and the court would be entirely justified in refusing to accept results from uncontrolled testing.

Peer review. Peer review is an integral part of science. Whether it concerns data on the discovery of a fundamental subatomic particle or one of many refinements of a measuring technique, the acceptance or rejection of the work hinges on peer review. The experimenter must be able to present the findings to peers for their scrutiny, normally by publication in a refereed journal. Confirmation or refutation of the findings will follow as others attempt to replicate the data. Findings that stand the scrutiny of peers and are of sufficient importance or utility become an accepted part of the field of science in which they are grounded.

There is a particular peer review process of value in determining the acceptability of an analytical technique, namely the development and publication of consensus "standard methods." The American Society for the Testing of Materials (ASTM) is the main body active in consensus method development in the United States. ASTM works through volunteer committees, including one for forensic testing (E30). The FBI and DEA support scientific working groups in various areas that also contribute to development of consensus standards in forensic science; for example the Scientific Working Group on DNA Analysis Methods (SWGDAM) has developed quality assurance standards for DNA testing.

General acceptance. The *Daubert* Court also recognized the role of general acceptance as an indicator of reliability, just as the *Frye* court had some seventy years earlier. In essence, general acceptance says that there is a degree of intrinsic reliability in a technique that has passed the test of time.

As determined, the *Daubert* ruling applies to scientific testimony and not to expert testimony in general. However, the more recent ruling of *Kumho Tire v. Carmichael* (1999) extends the principles to technical investigations.

See also Admissibility of Scientific Evidence; *Frye* Rule; *Kumho Tire* Ruling
References
Daubert v. Merrell Dow Pharmaceuticals, Inc., 509 U.S. 579 (1993).
James, S. H., and J. J. Nordby. *Forensic Science: An Introduction to Scientific and Investigative Techniques.* Boca Raton, FL: Taylor and Francis, 2005.
Kumho Tire Co. v. Carmichael, 526 U.S. 137 (1999).
Saferstein, R. *Criminalistics.* 8th ed. Upper Saddle River, NJ: Pearson Education, 2004.

Death

Death, in the context of most forensic circumstances, is simply diagnosed as cessation of the heartbeat and breathing. If a doctor or paramedical detects no heart sounds during a five-minute period then it is reasonable to pronounce life extinct. However, there is a broader and more complex set of factors surrounding the definition of death. It is known that some circumstances, such as severe hypothermia, or administration of paralyzing drugs, can mimic death. They can induce a state where no heart sound is detected by the human ear or sign of breathing detected by the classical feather or mirror, yet the individual is not dead. Patients on life-support systems add to the complexity.

There is a general consensus that the key indicator of death is cessation of brain activity.

According to the report of the Ad Hoc Committee in the *Journal of the American Medical Association,* four criteria must be met to declare death. These became known as the Harvard Criteria and are: (1) a total unawareness of externally applied stimuli; (2) no movements or breathing during a period of at least one hour in which the patient is continuously observed by physicians; (3) no reflexes, such as blinking, eye movement, and stretch-of-tendon reflexes; and (4) a flat electroencephalogram. The Harvard Criteria have proven to be reliable indicators of brain death, and physicians have generally reached a consensus about continuing to apply them.

Notwithstanding that, someone is not legally dead unless a qualified person (most jurisdictions specify a registered medical practitioner) certifies life extinct due to a specified reason, such as natural causes, accident, or homicide. There are very few cases indeed where it is necessary to call on the Harvard Criteria.

One problem that is unique to the forensic specialist is assigning a time of death. This is certainly not an exact science, but there is a wealth of experiential data that can give a reasonably reliable estimate in most circumstances. The data fall into four main categories: physical characteristics, temperature, biochemical measurements, and putrefaction.

Physical Characteristics

The main physical characteristics used are hypostasis and rigor mortis. When the heart stops, blood circulation also stops. There is no blood pressure and some of the structural barriers that in life keep blood in the veins and arteries are impaired. As a result, gravity acts to pull fluids to the lowest point of the body—this is hypostasis. The observable consequence is coloration as blood accumulates in the capillaries in the skin. To begin with there are small scattered patches, but these merge with time. The color starts as pink but changes to dark pink and then to blue as oxygen is removed. Carbon monoxide or cyanide poisoning each produce a bright cherry pink.

In the earlier stages, the lividity is not fixed: Finger pressure will cause blanching and turning the body will produce a new low point for the hypostatic shift in blood. However, as the interval increases, the blood loses its fluidity and these changes in response to pressure and rotation will not be seen. The time frame of hypostatic changes is:

- First patches seen at up to thirty minutes
- Coalescence seen from then to about four hours
- Shifting in rotation seen at about two to twenty-four hours, lessening with passage of time
- Disappearance on pressure seen at ten to twenty-four hours, the extent of disappearance lessening with passage of time

Rigor mortis is the stiffening of the body as biochemical changes cause muscles to become rigid. Rigor typically is first seen in the face, at about six hours postmortem, moving down the body to the arms (about nine hours) and legs (about twelve hours). Full rigor will last about twelve hours and disappear over the next twelve hours. However, rigor is one of the least certain of indicators and is affected by things like ambient temperature and body temperature.

Temperature

Biochemical and physiological events maintain the temperature of a healthy adult at about 37 degrees Celsius (98.6 degrees Fahrenheit). These homeostatic mechanisms are lost on death and the body obeys the laws of physics and cools to ambient temperature.

Like all the other indices of time since death, body temperature is an approximation influenced by many factors; for example, the ambient temperature, the environment (dry or wet), whether there is a significant wind, extent and type of clothing, body weight, and body position. It is of most value in the early period after death when indeed it is the only indicator of any use. In general, a clothed body in a temperate climate, not in water, and in calm conditions, will cool at about 1.5 degrees Celsius (2.7 degrees Fahrenheit) per hour for the first six hours and somewhat less thereafter.

Chemical Measurements

Many attempts have been made to correlate postmortem intervals with chemical changes. These, too, fail to be of any significant precision. Potassium levels in the vitreous humor (the clear fluid found inside the eye) are the most studied and most useful. The underlying principle is that metabolic processes in the living body maintain the chemical content of the fluid, including a low concentration of potassium. As cells die, the high levels of potassium inside them diffuse throughout the body and so levels rise in fluids with a low concentration of potassium in life. This is what happens to the potassium in vitreous humor. It takes about five to six days for the potassium in vitreous humor to equalize with that in blood. The increase in the early period occurs at a fairly constant rate.

Putrefaction

The organs of the body decompose after death, as microorganisms invade the tissues. Intestinal and chest organs are affected first, followed by brain and muscles. In temperate climates, with no special extraneous factors, putrefaction begins to be noticeable about forty-eight hours after death, and spreads over the following weeks. At two weeks, the abdomen is distended and organs are disrupted by gas. By three weeks, organs are disrupted and there is substantial disfigurement. At four weeks, there is slimy liquefaction of the whole body.

Overall Estimate of Postmortem Interval

None of these methods are intrinsically precise, but an experienced forensic specialist working in conditions that are not abnormal can give a reasonable estimate of the postmortem interval. The techniques have a sequence of applicability: Temperature is best in the early stages, then rigor and lividity, and finally putrefaction.

References
Byard, R., T. Corey, C. Henderson, and J. Payne-James. *Encyclopedia of Forensic and Legal Medicine.* London: Elsevier Academic, 2005.
Knight, B. *Simpson's Forensic Medicine.* London and New York: Oxford University Press, 1997.
Payne-James, J., A. Busuttil, and W. Smock. *Forensic Medicine: Clinical and Pathological Aspects.* London: Greenwich Medical Media, 2003.

Dental Records and Disaster Victim Identification (DVI)

Mass disasters, such as the Swissair Flight 111 accident, 1998, and the World Trade Center terrorist attack in the United States in 2001 have shown the power of DNA typing for identification of remains. However, dental identification is still one of the most rapid and reliable techniques available. The method is based on comparison of teeth in remains to dental records.

See also Odontology

References
Bowers, C. M. *Forensic Dental Evidence: An Investigator's Handbook.* Amsterdam and Boston: Academic, 2004.
Bowers, C. M., and G. Bell. *Manual of Forensic Odontology.* 3rd ed. Saratoga Springs, NY: American Society of Forensic Odontology, 1995.
Dorion, R. B. J. *Bitemark Evidence.* New York: Marcel Dekker, 2004.

Digital Evidence

Digital evidence is the term used to describe information of evidential value that has been stored or transmitted in digital form. It is popularly referred to as "computer forensics." There are many examples, from child pornography and Internet fraud to recovery of e-mail and business records in cases of environmental crime and illicit drug dealing.

The premise on which digital evidence is based is that examination of the computer-storage media may permit recovery of data. The recovered data can aid an investigation in the same way that written or printed documents can. The main difference is in the reliable recovery of the data. "Reliable" encompasses the process of recovery without corrupting or destroying the information, and also having checks in place to prove that there has been no alteration of the digital record.

Information recovery has been very successful. Mail messages and web-page files contain information that shows the routing of the mail. Erased data is most often not actually removed from the storage media, but rather the references to it in the computer file management system are deleted.

Reference
Casey, E. *Digital Evidence and Computer Crime.* 2nd ed. London and San Diego, CA: Academic, 2004.

Digital Imaging

Digital imaging is quite distinct from digital evidence. It is the capture of images in digital form, rather than as physical entities. The advantages are that the images can be stored, recalled, and manipulated readily from a computer workstation, unlike physical records. These benefits are also the problem with digital evidence. Proving the integrity of images demands detailed and stringent security procedures.

References
Blitzer, H. L., and J. Jacobia. *Forensic Digital Imaging and Photography.* San Diego, CA: Academic, 2001.
Russ, J. C. *Forensic Uses of Digital Imaging.* Boca Raton, FL: CRC, 2001.

Disaster Management

From a forensic science perspective, a *disaster* is a major incident that has caused or has the potential to cause death or serious injury to many people. There are three priority categories to be observed in the management of a disaster site: first is the safety of personnel (victims, bystanders, and responders); second is recovery of information that could indicate the cause of the disaster; and third is identification of the victims. Management of the scene has to deal with transportation of the injured or dead, and the calls from family and others affected by the incident as well. All are best served by implementation of a rational plan that clearly identifies available resources to be called on and the roles and responsibilities of the many responders.

Safety ranges from protection of everyone at the site to safety of an extended area in the case of biological, chemical, or explosive hazards. Good disaster-scene management often requires resolution of competing demands among or even within categories. For example, the first terrorist attack on New York City's World Trade Center, 1993, took place in freezing winter conditions. Water sprayed by fire department personnel to render the site safe from fire froze. The power to the site was out, and investigators were faced with finding their way around a dark subterranean garage with a hole through several floors and sheet ice underfoot.

The disaster site should be regarded as a crime scene until proven otherwise. The rules of crime-scene management apply: (1) control access to prevent loss and contamination; (2) record the site thoroughly and establish a chain of evidence system; and (3) identify, preserve, and collect materials that constitute or may contain evidence. Evidence items should be selected for their contribution to identifying what happened and any associative links with individuals.

Identification of bodies is a stressful but necessary part of all mass disaster investigations, whether accidental or criminal. Disaster victim identification (DVI) begins with inspection of the bodies and the immediate personal possessions. Physical appearance characteristics to be recorded include height and weight, sex, hair color, and distinguishing marks such as tattoos or surgical scars. Fingerprints should be taken. An inventory of personal items such as jewelry and clothing and the contents of pockets—wallets and perhaps even a driver's license or some other picture ID—should be made.

Technical procedures that can permit identity to be assigned include checks of fingerprints, DNA, and dental records. All need a sensitive liaison with the victim's family in order to obtain reference samples.

There is a significant stress in DVI, even for hardened forensic pathology and laboratory personnel. The numbers of victims and the contact with family personalize the activity in a way that forensic investigations of crime do not. Post-incident counseling should be available.

See also Mass Disaster Victim Identification
Reference
Byard, R., T. Corey, C. Henderson, and J. Payne-James. *Encyclopedia of Forensic and Legal Medicine.* London: Elsevier Academic, 2005.

Distance Determination (Firearms)

Determination of the distance from the victim at which a gun was fired can help in reconstruction of the shooting. A self-inflicted shot is made from very short range; one made during a struggle would also be short-range. The principle behind the range determination is that the debris from the firing of the ammunition disperses as it leaves the gun muzzle. Thus contact or near-contact shootings have a high density dispersed over a narrow radius and the density decreases and radius increases with distance. Contact wounds also show burning from the heat of the propellant blast.

The materials that make up the debris include lead from the bullet and residues of burnt propellant, typically nitrites from nitrocellulose in smokeless powder. The chemical tests used are the Griess test for nitrite and sodium rhodizonate for lead particles.

Testing is conducted by shooting the weapon at a target from various distances and comparing the radius of detectable residues on the clothing or body of the victim with those from the test fires. There is considerable variation and the results are to be taken as estimates and not as reliable determinations of firing distance.

See also Color Tests; Firearms

References

De Forest, P., R. E. Gaensslen, and H. C. Lee. *Forensic Science: An Introduction to Criminalistics.* New York: McGraw-Hill, 1983.

Saferstein, R. *Criminalistics.* 8th ed. Upper Saddle River, NJ: Pearson Education, 2004.

Schwoeble, A. J., and D. L. Exline. *Current Methods in Forensic Gunshot Residue Analysis.* Boca Raton, FL: CRC, 2000.

Warlow, T. *Firearms, the Law, and Forensic Ballistics.* 2nd ed. London and Bristol, PA: Taylor and Francis, 1996.

White, P. *Crime Scene to Court: The Essentials of Forensic Science.* Cambridge: Royal Society of Chemistry, 1998.

DNA Databases

The ability of DNA typing to get close to individualization of a sample means that databases can be used effectively (see **Databases**).

In the United States, the DNA database is maintained by the FBI and is titled CODIS, for Combined DNA Index System. Data is fed into the database at local and national levels. CODIS is proving to be a very powerful tool for public safety, with many cold hits being achieved even in the early stages of the system.

See also Combined DNA Index System (CODIS); Databases; DNA in Forensic Science; DNA Population Frequencies; Exclusion of Paternity; Phenotypes; Polymerase Chain Reaction (PCR); Population Genetics; Restriction Fragment Length Polymorphisms (RFLP)

References

Butler, J. *Forensic DNA Typing.* 2nd ed. Burlington, MA: Elsevier, 2005.

Federal Bureau of Investigation, Combined DNA Index System; http://www.fbi.gov/hq/lab/codis/index1.htm (Referenced July 2005).

DNA in Forensic Science

Methods to type stains of body tissues such as blood and semen were developed in the 1950s and refined in the 1960s and 1970s. However, the greatest advance in the identification of the origin of biological material came with the introduction of DNA typing in the mid-1980s. The immediate impact of DNA typing was in its ability to be more precise about the possible sources of biological material.

There are many ways to express how well a system used to type body fluids can differentiate between people. The simplest is to estimate the proportion of the population with the type or combination of types measured. Thus, a stain of group O is found in about half of the population. The typing systems used prior to the introduction of DNA testing would usually give frequencies of the range of 1 in 50 to 1 in 200 of the population. DNA immediately offered figures of 1 in several million of the population and, as systems have developed, now offers results of 1 in a billion or more.

As of this writing in 2005 it is barely fifty years since Crick and Watson elucidated the structure of DNA and hence how it was able to function as the body's carrier of genetic information. In that time, DNA and the genetic code have moved from the realm of Nobel Prize theory to basic school science. Almost every school student knows that the DNA molecule is a double helix, held together by the attraction between the constituent nucleotide bases adenine (A), cytosine (C), guanine (G), and thymine (T); the chemical structure of the bases is such that G pairs to C and A to T. The sequence of the bases provides the code for the structure of a protein. The entire sequence that produces a single protein is called a *gene,* and the genes are contained in chromosomes, found within the nucleus of cells.

However, most of the DNA in chromosomes is not in the genes and does not serve as a carrier of genetic information.

The DNA used in typing body fluids came from the discovery of variable number tandem repeats (VNTRs) in the noncoding DNA. These are segments in which a sequence of nucleotide bases is repeated end to end. The number of repeats varies between people, but because they are part of chromosomes, everyone inherits a sequence from each parent. There are many VNTR regions, each characterized by the number and the order of the nucleotide bases that make up the repeating unit.

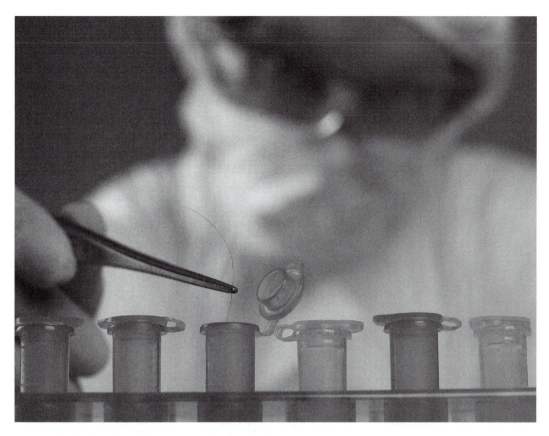

A scientist holds a hair for a DNA sample. (Andrew Brookes / Corbis)

The form of DNA typing introduced to forensic science in the late 1980s and early 1990s is called restriction fragment length polymorphism (RFLP). The name comes from the way in which the testing is performed. The complete DNA molecule is broken into hundreds of smaller units using a chemical called a restriction enzyme—the smaller units are the "restriction fragments." Restriction fragments made up mainly of a VNTR will have a size that varies according to the number of repeat units, provided that the size of the repeat unit is large enough. Hence the "length polymorphism."

The restriction fragments are then separated according to size by applying them to a slab of gel and applying an electric field, a technique named *electrophoresis*. Recognition of the fragment with the target VNTR is achieved by application of a *probe* of complementary nucleic acid to the VNTR repeat sequence. The probe binds more or less specifically to the sequence and can be located through visualization of an attached chemical or radioactive label. The size of the fragments determines how far they move in the electric field. A value is assigned to the fragment by comparison with a standard run at the same time.

Each VNTR region utilized gives an independent typing. Typing results from different regions, identified by specific probes, can be multiplied together. Using between four and eight probes gives types so rare that they are effectively unique, other than in identical siblings.

RFLP analysis takes some time to complete and needs fairly large samples to be successful—times of the order of two weeks and samples the size of a dime are needed for RFLP analysis. RFLP also requires undegraded DNA. For these reasons, in the early 1990s

RFLP analysis began to be replaced by polymerase chain reaction (PCR) techniques.

PCR is based on the way the body manufactures DNA. The double helix is separated and a particular area on one chain targeted by adding a primer. The primer is a nucleic acid sequence complementary to a stretch of DNA adjacent to the region of interest. The primer will thus bind to that site. A cocktail of chemicals, including DNA polymerase (the enzyme that synthesizes DNA in the nucleus) and the nucleotides A, C, G, and T, is added. The single chain acts as a template and the DNA molecule is rebuilt for the region of interest. The reaction is stopped, the chains are separated, and the process is repeated twenty to thirty times. In that way the amount of DNA containing the target sequence is multiplied about a million times, the quantity doubling each cycle.

Most PCR methods are variations on the theme of VNTR analysis. Unlike RFLP, the repeat units are short (four bases in the most common example of short tandem repeats [STRs]) and the overall size of the repeat segment of DNA is about one-tenth of that in RFLP analysis. For example, the TH01 marker, human tyrosine hydroxylase, has between five and eleven repeats of the sequence AATG.

PCR is extremely sensitive, the assay is rapid, and degraded DNA can be typed successfully. By combining results from several STR regions, population frequencies rare enough to approach individualization are possible. For example, there are thirteen standard core loci for STR data entry to the CODIS database, together with a sex-specific marker, and combined frequencies of 1 in 10 billion or rarer are regularly encountered.

The success of DNA typing in forensic science depends on the ability to reliably detect the DNA types in shed material. STR technology has been successfully applied to old case samples to exonerate convicted prisoners. It has given probative evidence from fingernail scrapings and from the mouth area of masks. It has even been used successfully in DNA recovered from a fingerprint where there were insufficient points for identification from the morphology of the print.

Future DNA developments include the typing of DNA from mitochondria and DNA specific to the Y chromosome. Mitochondrial DNA is not found in the nucleus, but is in structures in the cytoplasm. It is inherited solely from the mother, and is found in relatively high levels. It is of value in identification of old tissue or skeletal remains and of hairs. Y chromosome DNA is of value because of the problems that can arise in unique identification of the male fraction in typing vaginal samples from rape victims. The sensitivity of PCR analysis is such that low levels of female DNA carry through the separation process. Identifying DNA that is unique to the male material avoids this problem.

See also DNA Databases; DNA Population Frequencies; Exclusion of Paternity; Phenotypes; Polymerase Chain Reaction (PCR); Population Genetics; Restriction Fragment Length Polymorphisms (RFLP)

References

Butler, J. *Forensic DNA Typing.* 2nd ed. Burlington, MA: Elsevier, 2005.

De Forest, P., R. E. Gaensslen, and H. C. Lee. *Forensic Science: An Introduction to Criminalistics.* New York: McGraw-Hill, 1983.

National Research Council. *The Evaluation of Forensic DNA Evidence.* Washington, DC: Academy, 1996.

Saferstein, R. *Criminalistics.* 8th ed. Upper Saddle River, NJ: Pearson Education, 2004.

DNA Population Frequencies

The power of DNA as a tool in forensic science is its ability to give reliable assignment of the source of a sample. Source attribution depends on two things: the true rarity of the DNA characteristic typed and the ability to measure that rarity and express it in a meaningful way. (See also **Blood Grouping**.)

Recall that a genotype represents a pair of alleles, one inherited from each parent. A DNA profile is composed of the individual's gentotypes at several locations (loci). Estimation of the rarity of a blood type (DNA or any of the more traditional systems previously used) is seemingly straightforward. Blood collected from a number of people drawn at random from the population is analyzed and the profiles stored in a DNA population database and the frequency of the characteristic is measured. This group is called a sample of the population and is typically somewhere between 30 and 300 people. The frequency of occurrence of specific alleles in the genotypes can be counted. These frequencies can be used as inference tools for determining the prevalence of a particular allele or genotype or entire profile in the population as a whole. For example, if 16 percent of the sample has the allele 17 at the DNA location (locus) called D3S1179, it is assumed that 16 percent of the entire population has that same allele at that locus.

Because the loci examined in a forensic DNA profile are unrelated to one another, the allele frequencies can be multiplied to determine a frequency for the profile. Based on recommendations from the National Research Council, calculations of frequency are made for the genotype at each locus. These are then multiplied following the product rule. DNA population frequencies are also influenced by the composition of the reference or target population. A correction factor is incorporated in the final calculation to account for the possibility of population substructure or inbreeding. Calculating the rarity of a profile according to the NRC's recommendations and using population databases are relatively straightforward. Whenever a particular allele is observed in the sample population at a frequency of less than 0.01, a minimum allele frequency of $5/2n$ where n = number of individuals in the database is used. For isolated populations, such as islanders or Native Americans, databases may be constructed using samples from those populations. For larger ethnicity classifications, such as Caucasian Americans or black Americans of African descent, more broad generalizations or inferences can be made. Two other commonly used databases in forensic calculations are those for Southeastern and Southwestern Hispanics. Any reasonably common type will be detected and

there will be sufficient instances in the sample to give a reliable estimate of its frequency in the population as a whole. However, many of the DNA types are so uncommon that they may not be encountered in any of the blood from the samples. Even if we keep looking and increase the size of the sample, the frequency of occurrence will be so low that the estimate of the population frequency will not be as precise as we would wish. Note that there is no doubt that the characteristic is present in the population or that it is rare. The question is, exactly how rare is it?

When dealing with more than one typing characteristic (for example, type O blood that is also Rhesus positive) the frequency of occurrence of the combination is found by multiplying the frequencies of the individual components (the frequency of type O times the frequency of Rhesus positive in the illustration, or 50 percent of the population times 90 percent of the population or about 45 percent of the population overall). This means that the uncertainty in the frequencies of the individual DNA types are increased as we apply the multiplication rule. However, the frequency of occurrence of any combination of thirteen STR types is so rare that there is no alternative but to arrive at it using the multiplication rule.

DNA population frequencies are also influenced by the composition of the reference or target population.

Because scientists have no firsthand knowledge of the ethnicity of an individual involved in a crime, they often report a statistical value called a random match probability (RMP) for the entire population and/or the RMP for the profile in various ethnic groups. Differences do not often reach a single order of magnitude. Another commonly presented value is a likelihood ratio, which is essentially the reciprocal of the random match probability. In other words, the random match probability gives a value for the probability that any individual drawn at random from the population would have a particular profile. In the forensic context, a profile developed from a piece of evidence may be presumed to have been contributed by a particular perpetrator. A reference standard obtained from that perpetrator can then be collected and compared. If these profiles are the same, it is considered a match. For example, if a profile is obtained from a sample of blood taken from a broken window, and the profile obtained from a reference sample collected from John Johnson is a match, a random match probability is calculated to indicate the rarity of the profile. We know that John Johnson matched the blood on the window. How likely is it that anyone out there would have that same profile? Often, RMP is expressed in the 1 in X billion or trillion or even quadrillion range. If the RMP was 1 in 2,000,000,000,000, the statement would be that the probability of finding the particular profile in question in any given individual in the population would be one in 2 trillion. The alternative calculation or statement, the likelihood ratio, is a measurement based on the hypothesis that the two profiles in question (the one from the window and the one from John Johnson's standard) did in fact come from the same person, namely John Johnson, versus the likelihood that they came from different sources or that someone other than John Johnson deposited the blood on the window and had the same DNA profile. In the example above, the likelihood ratio would be 2,000,000,000,000, indicating that it is 2 trillion times more likely that John Johnson is the source of the blood on the window than someone else. Keep in mind that these values relate solely to the DNA profiles and are not related to guilt or innocence. It is not the job of the forensic DNA analyst to determine how John Johnson's blood was deposited on the window or if a crime was committed.

Although these values can be seemingly astronomical, the manner in which DNA profile frequencies are calculated is relatively conservative and thus tends to favor a defendant.

See also DNA Databases; DNA in Forensic Science; Exclusion of Paternity; Phenotypes; Polymerase Chain Reaction (PCR); Population Genetics; Restriction Fragment Length Polymorphisms (RFLP)

References
Butler, J. *Forensic DNA Typing.* 2nd ed. Burlington, MA: Elsevier, 2005.
National Research Council. *The Evaluation of Forensic DNA Evidence.* Washington, DC: Academy Press, 1996.

Document Examination

Since humans began using paper in the conduct of their affairs, there has been a need to establish the authenticity (or lack thereof) of the information contained in paper documents regardless of how it was applied (i.e., by hand or by machine). The early handwriting experts were teachers of penmanship, bankers, lithographers, engravers, or court clerks. They were self-taught and used simple equipment—magnifying lenses, rulers and, perhaps, monocular microscopes. The cases they examined were primarily for the purpose of detecting forgery, and the evidence was almost exclusively handwriting. Today, the professional document examiner must deal not only with the determination of the genuineness of handwriting but also with the identification of printing instruments and photocopiers, detection of erasures, additions or alterations to documents of all kinds, dating of documents, and the analysis of inks and papers, and must be familiar with all types of writing and reproduction implements. Indeed, the manner in which documents are produced today provides a much greater variety of challenges for the forensic document examiner than formerly. One thing they would emphasize that they do not do, however, is attempt to determine character or personality from handwriting. That they leave to so-called graphologists.

Prior to the 1890s, there is little information about the persons who were testifying in the courts about handwriting identification. In fact the only evidence that such testimony was being provided is contained in early court reports. The most publicized case (at least until 1934) in which handwriting identification played a major role was the notorious Dreyfus Affair in France, which began in 1894. A French spy in the German embassy in

Paris discovered a one-page handwritten document listing secret French army documents that had been turned over to the Germans. Captain Alfred Dreyfus, a French general staff officer, was accused of being the author and was tried for treason by court-martial. The main evidence against him was the testimony of Alphonse Bertillon, director of the Police Identification Service in Paris. As the father of the system for the identification of criminals that bore his name, Bertillon was held in high esteem in France and, although he had no training or true expertise in handwriting identification, his testimony that the incriminating document had been written by Dreyfus was sufficient to result in a conviction. Dreyfus was sentenced to life imprisonment at the infamous Devils Island. Anti-Semitism was rampant in the French army at the time, and Dreyfus was a Jew. His defenders claimed that was the only reason for his conviction and, for the next twelve years, France went through turmoil over the affair.

In 1896 another French officer, Ferdinand Esterhazy, was accused of having written the damning document, but he was acquitted at his court-martial in a matter of minutes and promptly fled to England. This prompted one of the most famous open letters of all time—Emile Zola's *J' Accuse*—written to the president of France, asserting that the army had framed Dreyfus. In 1898 it was discovered that much of the evidence used against Dreyfus had been forged by another army officer, and the high court of appeals ordered a new court-martial. There was worldwide indignation when Dreyfus was again convicted, the military court apparently being unable to admit error. Nevertheless, the president of the Republic issued a pardon, and in 1906 the court of appeals fully exonerated Dreyfus. He was reinstated into the army with the rank of major and awarded the Legion of Honor. His innocence was confirmed in 1931 with the release of the private papers of the German officer who had obtained the original incriminating document. These revealed that it had indeed been Esterhazy who had been the

source. Bertillon was embarrassed by his flagrant error, and the affair cast a dark cloud over the remainder of his career.

In the United States during the 1890s, handwriting identification had become sufficiently well-known that two New York experts published books on the subject; William E. Hagan published *Disputed Handwriting* and Persifor Fraser *A Manual for the Study of Documents.* Within a few years, Daniel T. Ames published *Ames on Forgery* in 1900. Although these authors were prominent in this emerging discipline, it was dominated for over half a century by Albert S. Osborne, also from New York.

Osborne, originally a teacher of penmanship in Rochester, New York, began performing handwriting identifications around 1887 and remained active until his death in 1946. He had published several papers on document examination, including one on typewriting identification in 1901, but it was the first edition of his book *Questioned Documents* in 1910 that cemented his position as the leader of the profession. This work totally overshadowed the earlier works and, with a revised version in 1929, formed the cornerstone for much of what document examiners do even today. In 1922 Osborne published a second book, *The Problem of Proof,* in which he discussed courtroom procedure, primarily from the point of view of the document examiner. In 1937, following another revision of *Questioned Documents,* he also published *The Mind of the Juror* and, in 1944 with his son Albert D. Osborne, *Questioned Document Problems.* His professional stature was such that John H. Wigmore (1863–1943), dean of law at Northwestern University in Chicago, wrote the introduction to several of Osborne's books. In the 1910 book, Wigmore wrote that with this book Osborne had established "a new profession." Another prominent jurist, Roscoe Pound (1870–1964), dean of the Harvard Law School, wrote about Osborne in the 1944 book: "It is not too much to say that he has created the profession of examiner of questioned documents in this country and has turned what had been largely a matter of superficial guess work or plausible advocacy into a matter of scientific investigation and demonstration." Osborne was awarded a doctor of science degree by Colby College, a small liberal arts college in Waterville, Maine, and, in 1942, became the first president of the American Society of Questioned Document Examiners, a formalization of an informal group of "friendly document examiners" whom Osborne had been inviting to meet together since 1913.

Other prominent workers in the field during this early period included John F. Tyrrell of Milwaukee, Wisconsin, who later published a technique for the decipherment of charred documents; Elbridge W. Stein of Pittsburgh, who published a paper on the use of ultraviolet light to detect forgery in 1913; J. Fordyce Wood in Chicago; Clark Sellars and John L. Harris in Los Angeles; Herbert J. Walter in Winnipeg; and Rafael Fernandez Ruenes in Cuba. All were examiners in private practice, as it was not until 1921 that the U.S. Treasury Department established the first government position for a document examiner, Bert C. Farrar, followed by Wilmer Souder at the National Bureau of Standards. When the Scientific Crime Detection Laboratory of Chicago was established in 1930, the staff document examiner was Katherine Applegate Keeler, one of the first women document examiners and wife of Leonarde Keeler, the prominent practitioner of polygraphy, who was also on staff. In Europe, Wilson R. Harrison of England published his massive work *Suspect Documents, Their Scientific Examination* in 1958, and the great Edmond Locard of Lyon in France published *Les Faux en Écriture et Leur Expertise* in 1959.

After the Dreyfus Affair at the turn of the century, the next case to direct the attention of the public to handwriting identification was what was described then as "The Trial of the Century" in the small community of Flemington, New Jersey. On March 1, 1932, the infant son of Charles and Anne Morrow Lindbergh was abducted from the family home in Hopewell, New Jersey, and,

although a $50,000 ransom was paid, the child's battered body was not found until May 12. Bruno Richard Hauptmann, a German-born carpenter, was subsequently arrested and tried for the kidnap/murder. Fourteen ransom notes were examined by eight document examiners for the state, including Osborne, Stein, Sellars, Tyrrel, Walter, and Souder, all of whom concluded that Hauptmann was the writer. After his conviction, Hauptmann was quoted in the *New York World Telegram* as complaining that "Dot handwriting is the worstest thing against me." He was executed on April 3, 1936.

Following World War II, major changes began to develop in the work of the questioned document examiner as a result of advances in the technology of writing and printing instruments. In 1945 the Reynolds ball-point pen was introduced (it sold for $16 and the advertising claimed that one could write under water with it, if one ever needed to). By the mid-1950s, the price had fallen to about $1 and the ball-point pen had almost completely replaced the fountain pen, just as the fountain pen had replaced the dip pen in the thirties. The characteristics of the ball-point pen had a direct effect on handwriting identification because the examiner had to learn to differentiate between faults from forgery and those due to defects in the ball point. Around 1960 the fiber-tipped pen started to become popular and brought its own challenges to identification work. In 1979 the PaperMate Company introduced an erasable ball-point pen, presenting an additional concern to the document examiner.

Typewriter technology also began to change with the introduction of the electric typewriter in the 1930s, proportional spacing typewriters in the early fifties, the single-element typeball machine in 1961, and the correcting "lift off" typewriter ribbon in 1973. The ability to change balls easily between machines and correct errors quickly presented another tool for the forger and challenge to the examiner. The print-wheel typing unit that became common in word processing and computer systems was introduced in 1972, followed not long after by the dot-matrix and laser printers that are so common today. All of the more recent developments in "type-writing" reveal much fewer of the conspicuous defects that the document examiner relies on to identify the product of a particular machine.

As the office photocopier became common, it introduced a whole new set of challenges to the document examiner, including the identification of the machine on which a particular copy had been produced. Initially, the major problem for the examiner was how to make a meaningful examination of the poor quality of the copies received. The thermographic process that used infrared energy to develop the image was introduced in 1950 but was quickly replaced by the much superior electrostatic plain-paper reproduction process in the late fifties.

The tools available to the document examiner have also expanded. Electronic ultraviolet and infrared viewers and chromatography for the differentiation of inks, electrostatic detection apparatus for revelation of pressure patterns and indented writing, specialized photographic procedures, and computerized digital imaging equipment have all been added to the examiner's arsenal. Questioned document examination has become a well-accepted part not only of criminal and civil litigation proceedings but also in dealing with problems of personnel, security, and commercial affairs. It is a distinct discipline within the forensic sciences.

Collection of Document Evidence

Excessive handling of questioned documents should be avoided. When collected, they should be placed in a clean envelope of proper size, and identification should be made on the container and not on the document. If it is absolutely essential to mark the document, it should be done on one corner on the back.

Standard or reference examples are called exemplars or knowns, and the suspected material is called the questioned material.

Because of the natural variation in writing, collection of exemplars is important. Requested handwriting exemplars must protect against attempts by the writer to disguise the normal hand. Some factors to consider are that they should:

- Be taken without distraction to the writer
- Be free from any knowledge on the part of the writer as to the questioned material (content, spelling)
- Use writing instruments similar to those used in the questioned material
- Should contain the same words and phrases as those in the questioned material (i.e., be a good representation of it)
- Should be repeated (to avoid successful disguise by the writer)
- Should be a mixture of written material (e.g., not solely signatures)

If the exemplars are preexisting writing, they must be contemporaneous with the questioned writing. There must be enough examples or text to cover the range of normal variations.

Writing Instruments

Ball-point pens leave a single rounded line. Worn ball-point pens accumulate ink on the side of the ball housing, and when the direction of movement is reversed, the ink is picked up by the ball and deposited as a smudge on the line. Wear and manufacturing imperfections can leave class or individualizing characteristics in the form of striations in the ink. These can be used to tell the direction of the stroke.

Fountain pens have more or less rounded nibs. As pressure is increased there is a separation of the nib, producing two marks that can be readily detected.

Fiber-tip pens often given a thicker line compared with ball-point pens. The ink fluid is uniform and constant, unless the ink supply is running low or the tip of the pen has been exposed to the air for too long a time. They usually do not leave indentations like other pens. Individualization of ball-point and fiber-tip pens is very difficult.

Pencil markings can easily be determined by their appearance. The lead of a pencil is a mixture of graphite and clay, the proportion determining the hardness of the pencil.

Paper

Most paper is made from wood pulp. Other nonwood fibers may be introduced to impart special characteristics. A bond paper, for example, contains 25 to 50 percent cotton fiber. Paper made from mechanically produced pulp is known as a ground wood. Newsprint is an example. Wood pulp used in making writing papers is generally treated with sodium sulfite. The product is known as sulfite paper.

The characteristics of paper used in its identification are color, size, shape, inclusions, pattern design, ruling and framing, watermarks, thickness, weight, bleaching, dying or other special processes, surface appearance, and fluorescence.

Inks

India ink. This is the oldest form of ink and is also known as Chinese ink. It is a fine suspension of carbon black in water and gum or in a solution of shellac and borax. It gives the most permanent of all ink colors.

Log wood ink. An extract of wood chips of the log wood tree, its chief color ingredients are hematoxylin and potassium chromate. It is not used today.

Iron gallotannate ink. This is made from tannic acid, gallic acid, ferrous sulfate, and an aniline based dye. This has been used for a long time and still is in many commercial inks.

Dye inks. These are another chief color source in inks. They are organic dyes and can be identified by thin layer chromatography.

The most generally applicable technique for the chemical analysis of ink is thin layer chromatography. Ultraviolet (UV) illumination will reveal any fluorescent components and infrared luminescence may be useful. UV and luminescence can be observed in instruments such as the videospectral comparator, which uses light sources, filters, and TV tubes.

Handwriting Comparison

Handwriting comparison accounts for more than 90 percent of document examination. The gross features of writing are class characteristics. They include features such as relative dimensions of letters, capitalization, and punctuation. Individual characteristics are less conspicuous and include features such as the formation of loops of letters such as *l* and *k* and whether letters such as *m* and *n* have rounded or pointed tops.

Careful observation of letter constructions, for example, by photographic enlargements, is the key tool.

Alterations, Erasures, Obliterations, and Indented Writing

Microscopic examination for damage to the paper fibers can reveal attempts to alter or erase. Alterations can be detected from ink composition (nondestructive testing by light and filter choices; destructive chromatographic testing by thin layer chromatography [TLC] or high performance liquid chromatography [HPLC] of ink dyes). Intersecting lines can also reveal which writing was first. Opaque obliterations can be read through by choice of light and filters, or from the reverse (clarify paper with oil). Oxidation erasures leave chemical changes to the paper (visible or UV light). Indentations resulting from writing on a page originally on top of the one being examined can be revealed using oblique lighting or a method known as electrostatic deposition analysis (ESDA). This instrument uses changes in surface charge on the paper to detect writings through several pages.

See also Charred Documents; Hitler Diaries

References

De Forest, P., R. E. Gaensslen, and H. C. Lee. *Forensic Science: An Introduction to Criminalistics.* New York: McGraw-Hill, 1983.

Hilton, O. *Scientific Examination of Questioned Documents.* Boca Raton, FL: CRC, 1993.

James, S. H., and J. J. Nordby. *Forensic Science: An Introduction to Scientific and Investigative Techniques.* Boca Raton, FL: Taylor and Francis, 2005.

Saferstein, R. *Criminalistics.* 8th ed. Upper Saddle River, NJ: Pearson Education, 2004.

Vastrick, T. W. *Forensic Document Examination Techniques.* Institute of Internal Auditors Research Foundation, 2004.

White, P. *Crime Scene to Court: The Essentials of Forensic Science.* Cambridge: Royal Society of Chemistry, 1998.

Driving while Intoxicated/Driving under the Influence

See *Alcohol*

Drowning

Drowning is death due to immersion in a fluid. The fluid fills the lungs, prevents oxygen from entering the blood, and so causes death by hypoxia. Immersion gives rise to a sequence of responses: breath holding, then involuntary inhalation of fluid and gasping for air under water. The amount of water inhaled is variable but can be large enough for it to pass through the alveolar membrane, which is the fine tissue separating the air sacs in the lungs from the capillary blood, and through which oxygen is taken into the blood and carbon dioxide removed from it.

Demonstrating that a body recovered from water died by drowning, rather than was dead before entering the water, depends on autopsy findings. In life the lungs are lined with a surfactant and drowning washes this away. The autopsy will probably reveal frothing at the mouth or in the airways, but this is not a definitive diagnostic feature.

Water contains tiny silica-covered plant bodies called diatoms. There is some evidence that diatoms enter the body in inhaled water and are carried through the lungs and to other organs in the blood before circulation ceases. The "diatom test" consists of examination of tissues from sites such as the lungs, liver, and kidneys. The presence of significant numbers in microscopic samples is used to support the diagnosis of drowning. However, this is a

somewhat controversial test and it is probably best not to base a conclusion of drowning solely on the presence of diatoms.

References

Byard, R., T. Corey, C. Henderson, and J. Payne-James. *Encyclopedia of Forensic and Legal Medicine.* London: Elsevier Academic, 2005.

Knight, B. *Simpson's Forensic Medicine.* London and New York: Oxford University Press, 1997.

Payne-James, J., A. Busuttil, and W. Smock. *Forensic Medicine: Clinical and Pathological Aspects.* London: Greenwich Medical Media, 2003.

Drug Enforcement Administration (DEA)

The mission of the Drug Enforcement Administration (DEA) is to enforce the controlled substances laws and regulations of the United States and bring to the criminal and civil justice systems of the United States, or any other competent jurisdiction, those organizations and principal members of organizations involved in the growing, manufacture, or distribution of controlled substances appearing in or destined for illicit traffic in the United States; and to recommend and support nonenforcement programs aimed at reducing the availability of illicit controlled substances on the domestic and international markets.

Laboratory support is one of the critical functions provided to DEA special agents and other law enforcement officers and officials. This support covers a variety of forensic disciplines and functions including the analysis of drugs, field assistance at clandestine laboratory seizures, and crime-scene investigations by forensic chemists. Specialists perform latent fingerprint identification and photographic development; evaluate digital evidence such as computers, diskettes, electronic organizers, and cameras; and develop, monitor, and process hazardous-waste cleanups and disposals. This support also includes the presentation of expert testimony that is essential for the successful prosecution and conviction of drug traffickers.

Additionally, the laboratory system provides support for intelligence activities through the

Heroin, Cocaine and Methamphetamine Signature Programs to determine the origin of the controlled substance and to highlight foreign drug distribution patterns. Intelligence activities are also supported through the Domestic Monitor Program, which helps monitor domestic drug distribution patterns and price/purity data at the retail level.

Reference
U.S. Drug Enforcement Administration; http://www.usdoj.gov/dea/ (Referenced July 2005).

Drugfire

Firearm evidence can be significant in crimes ranging from murder, attempted murder, suicide, and assault and rape cases to drug-related crimes. Using firearm evidence to its full potential can answer a number of significant questions, such as the nature of the weapon used, the condition of the weapon, the firing distance from the target, which direction the fire came from, what type of weapon fired a specific bullet, and who fired the weapon. Fundamental to firearms forensic examinations is the fact that no two firearms, even the same make or model, ever produce the same marks on fired bullets and cartridge cases. When a gun is fired, the firing leaves its own "fingerprint" in the form of unique microscopic grooves and striations on the bullet and its casings. The integrity of these characteristics changes little with time, allowing the potential identification of firearms recovered months or years after an incident. Significant developments in firearms identification technology have occurred over recent years. Drugfire is an automated computer technology system developed to assist the FBI and state and local law enforcement agencies in making links between firearms-related evidence. The system is used to digitize and compare evidence through image analysis, allowing comparisons to be made between images taken from cases within a single jurisdiction, or via networking, with images anywhere in the world.

See also Databases; National Integrated Ballistic Information Network (NIBIN)

Drugs

Drugs are encountered in forensic science laboratories in two main areas. These are the testing of suspected controlled substances and in toxicology, which is the testing of blood and other body tissues for the presence of drugs that may have contributed to death or altered behavior.

Testing suspected controlled substances is by far the most commonly encountered forensic examination. It is the greatest workload in a typical crime laboratory and can account for as much as three-quarters of the total laboratory resources. It is generally routine work but demands the very highest standards of security.

In contrast, many laboratories do not even have a toxicology section. However, the testing itself and the interpretation of the results can be among the most interesting of the types of work conducted in a crime laboratory.

The drug chemistry section of a forensic laboratory tests suspected controlled substances using color tests, and chromatographic and spectroscopic techniques. These tests are also used in the toxicology section, but with immunoassay being used as the screening technique.

See also Amphetamines; Cannabis; Cocaine; Color Tests; Crack Cocaine; Ecstasy; Gamma Hydroxybutyrate (GHB); Gas Chromatography; Hashish; Heroin; High Performance Liquid Chromatography (HPLC); Infrared Spectroscopy; Lysergic Acid Diethylamide (LSD); Mass Spectrometry; Methadone; Methamphetamine; Morphine; Narcotics; Opium; Phencyclidine, Phenylcyclhexyl, or Piperidine (PCP); Psilocybin; Ultraviolet (UV) and Visible Spectrometry

References
Christian, D. R. *Forensic Investigation of Clandestine Laboratories.* Boca Raton, FL: CRC, 2004.
De Forest, P., R. E. Gaensslen, and H. C. Lee. *Forensic Science: An Introduction to Criminalistics.* New York: McGraw-Hill, 1983.
Drummer, O. H. *The Forensic Pharmacology of Drugs of Abuse.* New York: Oxford University Press, 2001.
James, S. H., and J. J. Nordby. *Forensic Science: An Introduction to Scientific and Investigative Techniques.* Boca Raton, FL: Taylor and Francis, 2005.
Levine, B. *Principles of Forensic Toxicology.* Washington, DC: American Association for Clinical Chemistry, 1999.

Office of National Drug Control Policy; http://www.whitehousedrugpolicy.gov/ (Referenced July 2005).

Saferstein, R. *Criminalistics.* 8th ed. Upper Saddle River, NJ: Pearson Education, 2004.

U.S. Drug Enforcement Administration; http://www.usdoj.gov/dea/ (Referenced July 2005).

White, P. *Crime Scene to Court: The Essentials of Forensic Science.* Cambridge: Royal Society of Chemistry, 1998.

Duffy, John

This case presents the first British murder case in which psychological profiling played a major role.

Following a series of violent rape and murder cases, "Operation Hart" was implemented to find the offender. Included in the database of this operation was a profile of John Duffy, arrested for related but minor offenses. A psychological profile was sought, which predicted the offender would live in the Kilburn area of northwest London, be married and childless, have a history of violence, be unhappily married, and probably have two close male friends. Running this profile with the Operation Hart database resulted in a match for John Duffy, who, until the profile was run together with the database was a minor and relatively insignificant suspect. Duffy was later arrested and sentenced to serve several life sentences.

Reference
Evans, C. *The Casebook of Forensic Detection: How Science Solved 100 of the World's Most Baffling Crimes.* New York: Wiley, 1998.

Dyadic Death

Homicide followed by the suicide of the perpetrator is termed dyadic death. It is not common, and the most usual setting is within the family, including parents murdering children.

References
Byard, R., T. Corey, C. Henderson, and J. Payne-James. *Encyclopedia of Forensic and Legal Medicine.* London: Elsevier Academic, 2005.

Knight, B. *Simpson's Forensic Medicine.* London and New York: Oxford University Press, 1997.

Payne-James, J., A. Busuttil, and W. Smock. *Forensic Medicine: Clinical and Pathological Aspects.* London. Greenwich Medical Media, 2003.

Dynamite

The Swedish chemist Alfred Nobel invented dynamite in 1867. He found that nitroglycerine could be made safe to handle by mixing it with an inert substance, kieselguhr. The nitroglycerine did not lose its explosive force, but required a detonator to set it off.

See also Explosions and Explosives
References
James, S. H., and J. J. Nordby. *Forensic Science: An Introduction to Scientific and Investigative Techniques.* Boca Raton, FL: Taylor and Francis, 2005.

Saferstein, R. *Criminalistics.* 8th ed. Upper Saddle River, NJ: Pearson Education, 2004.

White, P. *Crime Scene to Court: The Essentials of Forensic Science.* Cambridge: Royal Society of Chemistry, 1998.

E

Ecstasy

Ecstasy, or XTC, is the street name for methylenedioxymethylamphetamine (MDMA) a Schedule I hallucinogenic amine and designer drug. MDMA was first synthesized in 1914 by the Merck company for use as an appetite suppressant, but never actually made it to the pharmaceutical market. During the 1970s and 1980s MDMA was used as a chemical aid to psychotherapy in marriage counseling in the United States, but was never officially approved for that use. MDMA is composed of chemical variants of amphetamine or methamphetamine and a hallucinogenic compound, which is most often mescaline. The drug is usually administered orally, in tablet or capsule form, resulting in effects that last for four to six hours. MDMA exerts its pharmacological effects by stimulating the central nervous system and recently gained popularity as a "rave" or "club" drug. The drug is reputed to produce feelings of well-being in the user and creates a high that stops fatigue, thus providing the ability to "rave all night." Some users have reported the stamina to party for two or three days continuously, and because MDMA has suppressive effects on eating and drinking, users often suffer from extreme dehydration and exhaustion. Reports indicate that the drug is not usually used in conjunction with alcohol because alcohol is reputed to diminish its pharmacological effects. MDMA also produces perceptual distortions that promote an intensification of feelings and emotions, the overwhelming desire to communicate, euphoria, and empathy.

Adverse effects associated with ecstasy use include nausea, hallucinations, chills, sweating, hyperthermia, and tremors. There may also be aftereffects of anxiety, paranoia, and depression. Studies on the long-term effects of MDMA use indicate damage to serotonergic neuronal pathways that may manifest as psychiatric and neuropsychotic disorders such as depression, anxiety, and paranoia. Deaths associated with ecstasy use have been related to hyperthermia, dehydration, and internal bleeding.

Ecstasy is synthesized in clandestine laboratories mainly in the Netherlands and Belgium, in tablet, powder, and capsule form. Most of the drug encountered in the United States comes from overseas laboratories, although a limited number of MDMA laboratories operate in the United States. Other suggested sources for U.S. drug distribution groups are foreign organized crime syndicates. Drugs are smuggled to the United States in large shipments via express mail, couriers, and freight shipments from large European cities.

See also Controlled Substances; Drugs
References
De Forest, P., R. E. Gaensslen, and H. C. Lee. *Forensic Science: An Introduction to Criminalistics.* New York: McGraw-Hill, 1983.
Drummer, O. H. *The Forensic Pharmacology of Drugs of Abuse.* New York: Oxford University Press, 2001.
James, S. H., and J. J. Nordby. *Forensic Science: An Introduction to Scientific and Investigative Techniques.* Boca Raton, FL: Taylor and Francis, 2005.
Levine, B. *Principles of Forensic Toxicology.* Washington, DC: American Association for Clinical Chemistry, 1999.
Office of National Drug Control Policy, Drug Policy Information Clearing House; http://www.whitehousedrugpolicy.gov/publications/factsht/mdma/index.html (Referenced July 2005).
Saferstein, R. *Criminalistics.* 8th ed. Upper Saddle River, NJ: Pearson Education, 2004.
U.S. Drug Enforcement Administration, Drug Descriptions, MDMA; http://www.usdoj.gov/dea/concern/mdma/mdma.html (Referenced July 2005)
White, P. *Crime Scene to Court: The Essentials of Forensic Science.* Cambridge: Royal Society of Chemistry, 1998.

Ejaculate

Human ejaculate consists of about 2 to 4 ml of fluid, containing 70 to 150 million spermatozoa. More than half of the ejaculate comes from the prostate and about a third from the seminal vesicles, with traces from other sources. The first part of the emission consists mainly of spermatozoa and prostatic secretions, with seminal fluid making up the bulk of the later parts.

The chemicals used in screening tests for semen—acid phosphatase and P30 (or prostate specific antigen)—are therefore carried in the earlier fraction of the ejaculate.

Postcoital samples are obtained by swabs from the female genital tract. Drainage of ejaculate can result in stains on underclothing and bedding. Obviously, all postcoital samples will contain mixtures of male and female secretions, and therefore contain genetic markers from both.

See also Color Tests; Rape; Semen Identification; Sexual Offenses

References
Butler, J. *Forensic DNA Typing.* 2nd ed. Burlington, MA: Elsevier, 2005.
De Forest, P., R. E. Gaensslen, and H. C. Lee. *Forensic Science: An Introduction to Criminalistics.* New York: McGraw-Hill, 1983.
James, S. H., and J. J. Nordby. *Forensic Science: An Introduction to Scientific and Investigative Techniques.* Boca Raton, FL: Taylor and Francis, 2005.
Saferstein, R. *Criminalistics.* 8th ed. Upper Saddle River, NJ: Pearson Education, 2004.
White, P. *Crime Scene to Court: The Essentials of Forensic Science.* Cambridge: Royal Society of Chemistry, 1998.

Electrical Injuries

Electric shock can be lethal due to induction of cardiac arrest. High-voltage shock, such as results from a lightning strike, will also cause burn injuries. Factors affecting the response to contact with an electricity source are, principally, the current, voltage, and duration of contact.

A current of 5 mA (milliamperes) is painful and one of over 50 mA can be fatal. The normal domestic supply of 120 volts is sufficient to cause death, particularly if the time of contact is of the order of 10 seconds or longer. The contact site often, but not always, shows a burn mark.

Injury from a lightning strike is quite different. The discharge is intensely powerful. It can cause burn and physical shock injuries. About half of the people hit by lightning strikes die as a result.

See also Burns
References
Byard, R., T. Corey, C. Henderson, and J. Payne-James. *Encyclopedia of Forensic and Legal Medicine.* London: Elsevier Academic, 2005.
Knight, B. *Simpson's Forensic Medicine.* London and New York: Oxford University Press, 1997.
Payne-James, J., A. Busuttil, and W. Smock. *Forensic Medicine: Clinical and Pathological Aspects.* London: Greenwich Medical Media, 2003.

Entomology

Forensic entomology is the study of the biology of insects to obtain evidence in legal

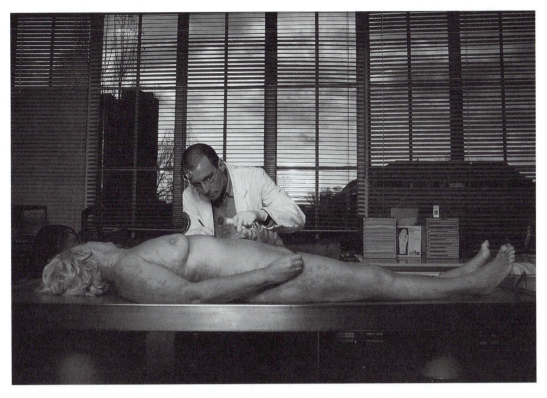

A postmortem is carried out by German forensic entomologist Mark Benecke who has worked worldwide on forensic entomology, the study of insects to solve crime. Dusseldorf University Clinic, Germany. (Volker Steger / Photo Researchers, Inc.)

investigations. Typically, the application is one of estimation of time of death from the history of fly larval development.

Blowflies will settle on a body and lay their eggs in it. The hatching of the eggs and the cycle of development of larvae to pupa to adult fly is well documented for many species and the time required to reach each stage of the cycle is known. Examination of the larvae will therefore allow a minimum time since death to be estimated.

Environmental temperature influences the rate of larval development, with the rate of development being faster at higher temperatures. Temperature also affects egg laying, as flies will not lay eggs if it is too cold.

See also Time of Death
References
Byrd, J. H., and J. L. Castner. *Forensic Entomology: The Utility of Arthropods in Legal Investigations.* Boca Raton, FL: CRC, 2000.

Greenberg, B., and J. C. Kunich. *Entomology and the Law: Flies as Forensic Indicators.* Cambridge and New York: Cambridge University Press, 2002.

Erased Writing

Documents may be altered after the original preparation in an attempt to hide the original meaning or contents, or perhaps as a means to commit a document forgery. Erasure of individual words, phrases, letters, or numbers is one of the most common ways in which a document can be altered. Erasures are always mechanical or chemical in nature, and often accompany alterations or obliterations. Over the years, an array of different tools have been used to try to mechanically erase writing, some of which include India rubber erasers, sandpaper, and razorblades or knives that may be used to scratch the writing from the surface of the document. Due to the damage caused in the process, mechanical erasures are very difficult to restore. Regardless of which method is used, the process of erasure damages and disturbs the fibers in the upper layers of the paper. Fiber damage can usually be seen quite easily using a microscope to examine the suspect area using direct or oblique lighting. This technique can generally show where an erasure has been made, but it will not necessarily expose the original writing or letters that had been present. In most cases, so much of the paper has been removed during the erasure process that it is not possible to see the original writings. Chemicals have also been used to erase writing. The chemical applications are usually strong oxidizing agents that conceal the ink; the oxidation reaction that occurs produces a colorless reaction product that usually can't be seen with the naked eye. The chemical erasure doesn't actually remove the ink, but changes the properties of the colored substance. However, microscopic examination may show areas of discoloration on the paper where the treatment was carried out, or there may be obvious staining on the document. In some situations, examination of the document using ultraviolet or infrared lighting will reveal any chemically treated area on the paper. Infrared luminescence is another technique that has been used successfully to disclose erased writing by highlighting invisible residues of the original ink left embedded in the paper.

Fuming the questioned document with a variety of chemicals may also reveal chemical erasures on a document. Chemicals such as ammonium sulfide and thiocyanic acid will react with lead residues left after the erasure of most inks. Iodine vapor, which is absorbed by paper, may also reveal areas where the paper fibers have been disturbed or damaged.

Although erasure by overwriting or crossing out occurs infrequently, it does occur. When the erasure is made using the same ink used in an original writing, it is virtually impossible to recover the original writing. However, if a different ink is used, the inks may react differently in the presence of specialized lighting

sources such as ultraviolet or infrared, revealing the erased writing.

See also Document Examination

References

De Forest, P., R. E. Gaensslen, and H. C. Lee. *Forensic Science: An Introduction to Criminalistics.* New York: McGraw-Hill, 1983.

Hilton, O. *Scientific Examination of Questioned Documents.* Boca Raton, FL: CRC, 1993.

James, S. H., and J. J. Nordby. *Forensic Science: An Introduction to Scientific and Investigative Techniques.* Boca Raton, FL: Taylor and Francis, 2005.

Saferstein, R. *Criminalistics.* 8th ed. Upper Saddle River, NJ: Pearson Education, 2004.

Vastrick, T. W. *Forensic Document Examination Techniques.* Institute of Internal Auditors Research Foundation, 2004.

White, P. *Crime Scene to Court: The Essentials of Forensic Science.* Cambridge: Royal Society of Chemistry, 1998.

Ethanol, Ethyl Alcohol

See *Alcohol.*

Ethics

Ethics means a set of guiding beliefs, standards, or ideals that define a group. They are situational and are codified; that is, they vary from place to place and exist as a formal or informal code for the organization. By comparison, *morals* are generally agreed-upon behaviors.

It is the duty of a forensic expert to be *ethical* in his or her approach to the selection, performance, and interpretation of all test results and analyses. It is his or her duty to conduct sufficient, appropriate tests to allow rational, truthful, and scientifically sound conclusions to be made, while conducting the best science at all times. Examination and interpretation of evidence must always be conducted without bias. Scientists or investigators employed by government agencies have a duty to approach, analyze, and report information as it presents, regardless of the potential outcome of a case. It's common for an expert to be called to testify by the prosecution, but the jury may inadvertently consider him or her as a prosecution witness. Many scientists have fallen prey to the prosecutorial way of thinking and should be aware of peer pressure from supervisors, attorneys, and law enforcement agents to change report notes or evidence descriptions, because the way the evidence is described, or an outcome is stated, biases the results toward the defense. An expert witness must know and stay within the limitations of his or her expertise, reporting exactly what he or she observes. He or she must be aware of providing an opinion on topics outside his or her field, and must not embellish notes or make assumptions because it fits in with the prosecution's case. When testifying, an expert must answer truthfully and to the best of his or her ability. An expert should never attempt to guess at answers or cover up when he or she does not know an answer. If mistakes are made, honesty carries a lot of weight, even if it is embarrassing.

The code of conduct of the American Board of Criminialistics contains some excellent examples of ethical guidelines for forensic scientists, for example:

- Treat all information from an agency or client with the confidentiality required.
- Ensure that appropriate standards and controls to conduct examinations and analyses are utilized.
- Ensure that techniques and methods that are known to be inaccurate and/or unreliable are not utilized.
- Ensure that a full and complete disclosure of the findings is made to the submitting agency.
- Ensure that work notes on all items, examinations, results, and findings are made at the time that they are done and appropriately preserved.
- Render opinions and conclusions strictly in accordance with the evidence in the case (hypothetical or real) and only to the extent justified by that evidence.

- Testify in a clear, straightforward manner and refuse to extend oneself beyond one's field of competence, phrasing one's testimony in such a manner so that the results are not misinterpreted.
- Do not exaggerate, embellish, or otherwise misrepresent qualifications when testifying.
- Maintain an attitude of independence and impartiality to ensure an unbiased analysis of the evidence.

See also American Board of Criminalistics (ABC)
Reference
American Board of Criminalistics; http://www.criminalistics.com/ (Referenced July 2005).

Exclusion of Paternity

Every child inherits one allele (DNA code for a gene) from the mother and one from the father. If we take it that the mother is known, then questions of paternity can be answered by testing blood from the mother, putative father, and child. If the child has an allele not found in either parent, then the putative father cannot be the true biological father. This is sometimes referred to as first-order exclusion. Also, if the putative father is homozygous (both his copies of the gene are the same) for an allele and that allele is not found in the child, then, again, he cannot be the true biological father. This is second-order exclusion.

However, in practice it is not quite as cut and dried. Genes mutate and there is always a slight chance that an apparent exclusion is due to mutation. For that reason, most laboratories will not attest to nonpaternity on the basis of a single exclusionary allele in the child. Laboratories are even more conservative in calling nonpaternity from second-order exclusions, as the father may not have been homozygous but rather may have a silent or recessive allele.

There is also the chance that no exclusion is revealed in testing but the match in alleles is a coincidence. Various statistical techniques have been evolved to deal with such circumstances. The paternity index (PI) is the most widely accepted of these. The underlying principle is that if there is no exclusion found on testing, then either the man is indeed the biological father or someone else is and the results match by chance. The paternity index is the ratio of these probabilities. For one allele, the ratio is the probability of the father passing on the allele divided by the frequency of the allele in the population. The value of the numerator of the ratio is 0.5 where the father is heterozygous and 1.0 where the father is homozygous for the allele in question, unless the mother shares both alleles with the child. In that case the numerators are 0.25 and 1.0. The total PI for more than one allele is obtained by multiplying all the individual PI values together.

See also Parentage Testing
Reference
Butler, J. *Forensic DNA Typing.* 2nd ed. Burlington, MA: Elsevier, 2005.

Exhumation

Forensic investigation of unnatural death sometimes requires that exhumations of remains be carried out. It may be the disinterment of a legally buried body, the examination of a homicide victim buried to conceal the crime, or the examination of one or more victims of war crimes.

All cases present problems, in that careful investigation to preserve all relevant information is critical. It is important that the forensic investigation team is aware of its goals before beginning the exhumation: For example, if time of death is an issue, then the collection of maggots and other infestations is important. If cause of death is an issue, then collection of body tissues and control material from the surrounding soil is important. If it is a war crime site, then the nature of the body— arms tied behind the back, for example—and wounds are important.

See also Cause of Death; Entomology; Time of Death

References
Byard, R. T. Corey, C. Henderson, and J. Payne-James. *Encyclopedia of Forensic and Legal Medicine.* London: Elsevier Academic, 2005.
Knight, B. *Simpson's Forensic Medicine.* London and New York: Oxford University Press, 1997.
Payne-James, J., A. Busuttil, and W. Smock. *Forensic Medicine: Clinical and Pathological Aspects.* London: Greenwich Medical Media, 2003.

Expert Testimony

The general rule governing admissibility of evidence is that witnesses may only testify to what they have personally observed or experienced. Witnesses are not allowed to testify with respect to their opinions except in certain circumstances. One of these exceptions is the expert opinion. An expert witness is considered to possess a particular knowledge gained through education or training in the course of a trade or profession and is permitted to give evidence whenever it would assist the jury in resolving questions outside the realm of experience of the average person. Expert evidence does not have to be scientific.

The forensic expert has a duty to evaluate scientific evidence and provide opinions on the significance of scientific findings regardless of whether the information provided is indicative of guilt or innocence. At all times, an expert witness must remain objective and be an advocate of his or her opinion and the truth. An expert should be able to defend the techniques and conclusions of any scientific analysis that he or she conducted pertaining to the case in question, and must be willing to discuss the shortcomings or limitations of a scientific technique, even though it may decrease the significance of a result, as well as any advantages.

The foundation for pretrial preparation, settlement negotiations, and courtroom testimony are the complete written crime-scene or laboratory reports. These reports and their conclusions may result in the expert witness not being called to testify, lead to settlement, and actually prevent trial.

For testimony to be of value to a case, it must impact the jury. An expert's testimony should be coherent, professional, and useful. The more expertise and educational and experiential background the witness can prove, the more impact the testimony will have on the jury. It is imperative that a witness is able to explain the intricacies of scientific data and conclusions clearly and logically to a judge and jury composed of nonscientists. An expert should never present biased testimony to help the case of the prosecution or defense. It is the jury who is responsible for determining whether experts' opinions are accepted or ignored during deliberations.

Testimony usually begins with qualification or establishing the ability and competence of the witness to testify in the area. Questions are presented to address training, education, and experience. For most forensic scientists, the prosecuting counsel presents these questions. If the information received is considered to be satisfactory, the court will declare the witness to be an expert in a specified field. Direct examination gives the expert an opportunity to demonstrate his or her professionalism, demeanor, and communication skills to the jury.

Direct examination is followed by cross-examination by the opposing attorney. During cross-examination, opposing counsel will try to uncover weaknesses in the background, knowledge, and experience of the witness. At times during the expert testimony, counsel may raise an objection to some aspect of the expert testimony, usually founded on hearsay or the improper foundation for an expressed opinion. Objections can be made to a question asked or to the answer the witness provides.

See also Admissibility of Scientific Evidence; *Daubert* Ruling; *Frye* Rule

References

Daubert on the Web; http://www.daubertontheweb.com/ (Referenced July 2005).

Daubert v. Merrill Dow Pharmaceuticals, Inc., 509 U.S. 579 (1993).

Frye v. United States, 293 F. 1013 (D.C. Cir. 1923).

Kumho Tire Co. v. Carmichael, 526 U.S. 137 (1999).

Explosions and Explosives

Terror incidents occurring in the United States that involve explosive and incendiary devices historically have been attributed to isolated individuals rather than organized terrorist groups. Because such occurrences are relatively uncommon, regular crime laboratories in the United States are rarely involved in the investigation of explosion scenes or explosive evidence. Investigation of simple homemade explosive and incendiary devices, including pipe bombs, may be conducted locally, but the investigation of major accidental or terrorist incidents generally uses the resources of the Bureau of Alcohol, Tobacco, and Firearms (BATF), the Federal Bureau of Investigation (FBI), or other major state facilities. However, in other countries where terrorist activities using such devices are more commonplace, the evidence is routinely investigated by appropriately equipped regular crime laboratories.

The chemical and physical reactions that occur in an explosion involve the processes of combustion. Combustion is an oxidation reaction in which oxygen combines with other substances to produce new products. Oxidation reactions always result in the release of large amounts of energy. An explosion is the result of combustion processes that produce a very large volume of gas almost instantaneously. The expanding gas produces the physical shock wave associated with explosions. By confining the explosive charge in a container, additional damage is caused, as the initial confinement of the generated gases provides sufficient energy to break the container into fragments, producing flying shrapnel. The fragments of the container or shrapnel cause damage and serious injury to anyone in the surrounding area of the exploding bomb. However, it is the very fast shock wave from the explosion that is responsible for most of the damage to property.

Explosives can be classified according to the speed at which the gases expand. Low explosives are extremely rapid-burning reactions that result in a pressure wave that travels at speeds up to 1,000 meters per second (about 3,000 feet per second). The burning rate for low explosives is called the speed of deflagration. Low explosives are more often used as propellants, but can become highly destructive if contained. Commercially available low explosives include black and smokeless powders, but almost any fuel and oxidizing agent combination can be made into an explosive. The ingredients required to make low explosives are readily available and can be purchased from gun stores and chemical supply houses. Potassium chlorate and sugar are well-known examples.

Pipe bombs often contain low explosives. The pipe provides the containment, and a safety fuse—often made of black powder in a fabric or plastic casing—is used to carry a flame to the explosive charge. With appropriate lengths of safety fuse, the fuse burns at a rate slow enough to allow an individual time to leave the site of the upcoming explosion. A pipe bomb typically contains black powder in a pipe with ends closed by threaded caps. Detonating the explosives from inside the pipe results in a huge volume of gas that expands almost instantaneously into the surrounding air when the pipe bursts, with quite devastating effects. Bomb disposal processes, whereby a bomb is exploded from the outside, results in a more harmless phase change, as there is no compression release.

High explosives have a faster pressure wave, moving at more than 1,000 meters (3,000 feet) per second. The subsonic pressure wave occurs inside the explosive charge. The burning rate for high explosives is called the speed of detonation. The instant detonation and rapid shock wave produce a shattering and shearing effect on the physical surroundings. Some high explosives, such as lead azide, are sensitive to shock and are used as primers. Others, such as TNT, will not burn unless detonated.

High explosives can be divided into two groups based on their sensitivity to heat and shock. The group that is highly sensitive to heat, friction, and shock and will detonate violently under normal conditions is called initiating explosives, and they are used as primers to detonate other explosives through a chain reaction. These explosives are so sensitive that they are rarely included in the main charges of homemade bombs but usually form the main ingredient in blasting caps. This group of explosives includes lead azide, lead styphnate, and mercury fulminate.

The second group of high explosives is quite insensitive to heat, shock, and friction, burning instead of detonating when ignited in small amounts under normal conditions. These noninitiating explosives are commonly used for commercial and military blasting and include compounds such as dynamite, trinitrotoluene (TNT), pentaerythritol tetranitrate (PETN), and cyclotrimethylenetrinitramine (RDX).

Quite often high explosives used for commercial or terrorism purposes are based on ammonium nitrate and fuel mixtures. The explosive called ANFO is based on ammonium nitrate soaked in fuel oil.

Dynamite is an explosive that is used when a quick-shattering effect is needed. Dynamite was developed by Alfred Nobel when he discovered that mixing nitroglycerin with an inert diatomaceous earth called kieselguhr resulted in a more stable product with equal explosive power. The pulp dynamite first developed by Nobel became the precursor to the straight dynamite series, the gradations of which were are based on the percentage of nitroglycerin present. Industrially available nitroglycerin-based dynamite has been superseded over recent years by ammonium nitrate–based explosives, such as water gels, emulsions, and ANFO.

Many terrorist organizations outside the United States have easy accessibility to a range of common military explosives for use in homemade bombs. One of the most common is RDX—more commonly known as C-4—that is encountered in a plastic, pliable doughlike form.

In general, four things are required to make a bomb: (1) an energy source; (2) a material that will undergo a rapid exothermic reaction, producing a large volume of gas when ignited by the energy source; (3) a means of containment so that the energy produced by the gas expansion and combustion processes is concentrated, causing as much destruction as possible when it finally dissipates; and (4) a method of initiation so the bomb explodes where and when intended.

Most homemade bombs delivered or hidden in packages, suitcases, or other containers are initiated with an electrical blasting cap wired to a battery. Many different switching mechanisms have been designed for setting off these devices, but the most common are clock and mercury switches.

There are many tests available for detection of explosives and their residues. They range from simple color tests (for example chlorate turns diphenylamine reagent blue, but does not give a color with Griess reagent) to expensive and complex field mass spectrometer tests.

For investigation purposes, explosions can be classified as diffuse or concentrated, regardless of the nature of the chemicals involved. Diffuse explosions are usually caused by mixed gases and dusts such as grain and coal dusts and are usually accidental in nature. Concentrated explosions are usually caused by the detonation of high or low explosives and are incendiary and deliberate in nature.

Once the explosion scene is carefully studied, and wherever possible photographed, the explosion is classified as diffuse or concentrated. If a diffuse explosion has occurred, investigators will try to trace the origin of the explosion and the nature of the explosive material. This can be often inferred from the type of the premises and the nature of the activities associated with it.

When a concentrated explosion is indicated, thorough searches must be conducted of the entire site in an attempt to recover any and all parts of the incendiary device, including the detonating mechanism, wires, switches, clockworks, and any artifacts that may not be normal to the environment.

See also Ammonium Nitrate–Based Explosives; Arson and Explosives Incidents System (AEXIS); Black Powder; Bombs; Dynamite; Nitrocellulose; Nitroglycerin; World Trade Center Bombing

References

De Forest, P., R. E. Gaensslen, and H. C. Lee. *Forensic Science: An Introduction to Criminalistics.* New York: McGraw-Hill, 1983.

James, S. H., and J. J. Nordby. *Forensic Science: An Introduction to Scientific and Investigative Techniques.* Boca Raton, FL: Taylor and Francis, 2005.

Saferstein, R. *Criminalistics.* 8th ed. Upper Saddle River, NJ: Pearson Education, 2004.

White, P. *Crime Scene to Court: The Essentials of Forensic Science.* Cambridge: Royal Society of Chemistry, 1998.

Eyewitness Testimony

Forensic science evidence has been under considerable public and legal scrutiny during the last few decades. It is important that the value of scientific investigations is not lost in concerns about some instances of tests not conducted properly. Sometimes forensic science lives up to its name of "the silent witness," for example in a murder or when the perpetrator was disguised or hidden.

The alternative to scientific evidence is eyewitness testimony. Despite the routine use of interrogation, examination-in-chief and cross-examination in the investigative and trial processes, the fact is that eyewitness testimony can be extremely unreliable. Mind and memory do not function as a camera. They are perturbed by the circumstances surrounding the event being recalled and the manner in which the recall is elicited. The most obvious example is in the prohibitions in most jurisdictions regarding the conduct of lineup identifications. These include avoiding unconscious bias by conducting the lineup after showing the witness photographs and by avoiding clothing clues.

F

Falls

Falls that require forensic investigation vary from falling out of bed to a lethal descent from a height of many feet. They present two forensic challenges: Could the injuries on the body have been caused by the fall alone or do they conceal others, and how did the fall contribute to death? A fall can easily result in death, for example, if there are head injuries and if the injured is aged or infirm. The fall need not be from a great height. The most difficult cases to work through are those, for example, in abuse of the very young or very old, in which there may have been repeated injury by intent or neglect, but in which the sufferer is quite likely to have had several accidental falls. There is little that an investigation can do to provide a clear-cut differentiation between accident and abuse. Frequently the best that can be done is to consider fairly and honestly the postulate presented as the cause and reply as to whether it could have resulted in the injuries observed.

Postmortem investigation will usually answer the question of whether injuries from the fall were fatal: Gross physical damage to brain tissue, intracranial hemorrhage, frank cardiovascular injury, and/or spinal damage will usually suffice, for example. Sometimes the question is asked about the height of the fall. Although there is an intuitive assuredness that the greater the height, the greater the damage, it has not proved possible to develop a reliable relationship between observed injuries and the height of the fall. Variables include the physical condition of the victim (especially age), the course of the fall (whether there was any contact to dissipate energy before the final collision), and the nature of the impacted surface.

See also Autopsy
References
Byard, R., T. Corey, C. Henderson, and J. Payne-James. *Encyclopedia of Forensic and Legal Medicine.* London: Elsevier Academic, 2005.
Knight, B. *Simpson's Forensic Medicine.* London and New York: Oxford University Press, 1997.
Payne-James, J. A. Busuttil, and W. Smock. *Forensic Medicine: Clinical and Pathological Aspects.* London: Greenwich Medical Media, 2003.

Feces

Feces are food residues passed out of the body after completion of travel through the digestive system. A laboratory may be asked to examine materials for feces in cases of anal rape, or in vandalism acts. Feces have a characteristic odor mainly due to skatole. Urobilinogen is a bile pigment excreted in feces, which may be detected using its fluorescent reaction to Edelman's reagent.

Reference
Butler, J. *Forensic DNA Typing.* 2nd ed. Burlington, MA: Elsevier, 2005.

Fibers

Textile fibers provide associative evidence in many crime situations. For instance, they can be easily transferred back and forth from clothing, furnishings, floor coverings, and vehicles. The forensic examination of fibers relies on the measurement of the characteristics of the fiber type and associated dye materials. The color, diameter, and other physical features can be used for comparison and identification purposes. Textile fibers can be divided into two categories: natural or synthetic.

Fiber examination used to be one of the centerpieces of crime laboratory investigation, as seen, for example, in the Wayne Williams case. Williams murdered twenty-seven young black boys in Atlanta in the period from October 1979 to May 1981. The first significant link came from fibers found on one of the bodies that could have come from the trunk of Williams's car. However, as time has gone on, fiber evidence has become regarded as very much second-rate to DNA testing in regard to reliability and evidential value. This is not a true reflection on properly conducted fiber testing and this section of the text therefore attempts to provide a good base for understanding what is involved in fiber evidence.

Natural Fibers

Natural fibers are produced or manufactured from plant or animal materials. Animal fibers are made from the hairs of an animal and include wool (sheep), mohair (goats), furs (mink), and angora (rabbits). They are protein in structure, and a number of distinct morphological features can be used in their identification. The main morphological features of animal hairs and wools are the root, medulla, cortex, and cuticle.

Silk is another natural fiber composed of the protein fibroin, produced by the caterpillar *Bombyx mori,* but with morphological features that are quite different from animal hairs. The classic features of silk include crossover marks and a triangular cross section with rounded corners. The remarkable properties of silk are its tensile strength, toughness, luxurious appearance, and a large capacity for absorbing dyes. Because silk is a smooth-surfaced continuous filament fiber, fiber transfer during contact is minimal or unlikely. The best appreciation of form and structure requires viewing actual images in full color—this can be achieved by going to the following website: http://www.fbi.gov/hq/lab/handbook/examhair.htm.

The most commonly encountered natural fiber is cotton, but the widespread use of white cotton limits its forensic value. Cotton fibers also have very characteristic microscopic features, and appear flat and spiraled or twisted under the microscope. Cotton is a seed fiber of the cotton plant shrub. These fibers can be dyed with a wide range of dyes, and because of their convoluted structure transfer easily during contact.

Other commonly encountered plant fibers include linen and hemp. Linen fibers, made from flax, are twice as strong as cotton and are used in both heavy and lightweight fabrics. Hemp fibers are three times stronger than cotton; very resistant to mildew, microbes, and rotting; and take up dyes very well. Plant fibers may be encountered as technical fibers, such as those found in cords, ropes, sacks, and mats, or as individual plant cells found in some fabrics and papers. Microscopic examination of technical fibers may reveal epidermal tissue, cellular crystals, and cross-sectional features. Chemical tests for the presence of lignin can also be used to positively identify plant materials. During the examination process, technical fibers can be macerated, fabrics teased apart, and paper can be repulped for the examination of individual plant cells. Characteristics such as the relative thickness of cell walls and lumen; cell length; and the presence, type, and distribution of dislocations are important in sample identification and comparison. The direction of twist of the cellulose in the plant cell wall can also be of use.

Synthetic Fibers

The majority of modern fabrics are made from synthetic fibers. There are hundreds of trade names and many generic classes of synthetic fibers, but some of the more commonly encountered fiber types include acetate, acrylic, aramid, nylon, polyester, and rayon.

The first synthetic or man-made fibers, including rayon and acetate fibers, were made from regenerated cellulose. The first truly synthetic fiber was nylon. These synthetic fibers are polymers that are treated to produce a fine filament. Often the extrusion processes used to make a fiber filament leaves characteristic striations on the fibers that are usually visible using microscopy and can be used to characterize them.

Other physical characteristic features include surface pitting due to the addition of delusterant to remove the sheen associated with synthetic fibers, variations in fiber diameter, and differences in cross-sectional shape. When viewed longitudinally by a microscope, the apparent cross-sectional shape of fibers can often be determined by slowly focusing through the fiber, a technique called optical sectioning. The presence or absence, size, shape, distribution, and relative abundance of delusterant particles can also be used for comparison purposes.

Dyes

Textile fibers are usually dyed to impart color to the fabrics made from them. Colors can be the product of the application of several dyes, and fabrics with the same bulk color may have dyes with different chemical compositions. Fiber color may be uniform along the length of a fiber, or it can vary.

Characterization of fibers by color begins with visual observation under the microscope. A degree of objectivity can be obtained by using a microspectrophotometer to record the visible spectrum of the dyed fiber, but greater discrimination of fiber dyes is obtained by extracting the dye using chemical solvents and performing thin layer chromatography (TLC) or high performance liquid chromatography (HPLC) analysis. However, these techniques unfortunately result in sample destruction. Thin layer chromatography can be used in some cases to detect and compare dye components. This is an inexpensive, simple technique that can be used to complement the use of visible spectroscopy in comparisons of fiber dyes. TLC should be considered for fiber comparisons only when it is not possible to discriminate between the fibers of interest using other techniques, such as comparison microscopy and microspectrophotometry.

Analysis of Fiber Evidence

In the preparation of fiber evidence in the laboratory, evidentiary items are visually inspected, and fine forceps are used to remove fibers of interest. Simple magnifiers and stereomicroscopes with a variety of illumination techniques are useful in the recovery process. Once a visual examination has been completed, tape lifting or scraping may be conducted to recover loose fibers. Tape lifts are placed on clear plastic sheets or glass microscope slides, which eases the search and removal of individual fibers of interest. Tape lifts or scrapings are usually examined using a stereomicroscope and fibers of interest are isolated for further analysis. Fibers on tape lifts can be easily removed for further examination using fine forceps, microscopic tools, and a solvent such as xylene.

During the recovery and examination of fiber evidence, care must be exercised to ensure that contamination does not occur and can be minimized by examining questioned and known items in separate areas or rooms, by thoroughly cleaning examination tools, and by changing lab coats before examining items that are to be compared. Contamination is most likely with fibers that have high shedability. Shedability is the ease with which fibers are released from a fabric by rubbing or gentle contact. Generally wools, furs, and loosely woven and piled fabrics have high shedability, transferring a large number of fibers on a single contact. Fibers with low shedability include tightly woven, smooth synthetic fabrics. Shedability tests are easily conducted by applying, and rapidly removing, a strip of sticky tape to a garment of interest, and examining the resultant fiber transfer by eye or stereomicroscope.

Microscopic examination is the quickest, most accurate, and least destructive means of determining the microscopic characteristics and polymer type of textile fibers. Side-by-side microscopic comparison is usually the most discriminating means of determining whether two or more fibers are consistent with originating from the same source.

Fibers are first examined with a stereomicroscope, noting physical features such as crimp, length, color, relative diameter, luster, apparent cross-section, damage, and any adhering debris. At this point, fibers can be tentatively classified into broad groups such as synthetic, natural, or inorganic.

If all of the compared characteristics are the same under the stereomicroscope, the fibers are then examined by comparison microscopy. The physical characteristics of the fibers are then compared visually side by side under a comparison microscope, using the same lighting and magnification, to determine if they are the same in the known and questioned samples.

Many synthetic fibers display a feature known as birefringence, which can be observed using a polarizing light microscope. This characteristic arises from the alignment of the polymer molecules in the filament during extrusion. The regular arrangement of atoms imparts crystallinity to the structure of the finished fiber, subsequently producing a number of optical effects that can be used in fiber characterization, identification, and comparison.

Light passing through a synthetic fiber emerges polarized, perpendicular and parallel, to the axis along the fiber. Each polarized plane shows a characteristic refractive index depending on the polymer type. Plane polarized light passing through a birefringent fiber is always retarded in one plane relative to the rays passing through the perpendicular plane. For a birefringent fiber, the sign of elongation is positive ($+$) if the light passing through the length is retarded in relation to the light passing across the fiber, and negative ($-$) if the light passing across the fiber is retarded in relation to the light passing along the length of the fiber. Generally, all common manufactured fibers with a birefringence higher than 0.010 have a positive sign of elongation.

The retardation of the slower ray can be estimated by observing the interference color displayed at a point where the thickness of the fiber is known, and by comparing it to a chart—the Michel-Lévy chart. For a full-color presentation on the Michel-Lévy chart go to http://www.microscopyu.com/articles/polarized/michel-levy.html.

To see what a nylon fiber looks like (a) under normal illumination, (b) under polarized light with crossed polars, and (c) viewed with a wedge compensator in place for estimation of retardation, go to http://www.microscopyu.com/articles/polarized/polarizedintro.html. The differences are quite dramatic and show how compelling microscopic identification of fibers can be.

A polarized light microscope equipped with a hot stage can be used to observe the effect of heat on thermoplastic fibers. Using slightly uncrossed polars, it is possible to observe droplet formation, contraction, softening, charring, and melting of fibers over a range of temperatures.

Many synthetic fibers exhibit fluorescent properties that can also be used in their identification and comparison. Fluorescence is the emission of light of a certain wavelength by an object when excited by light of a shorter wavelength and higher energy. Fluorescence may result from the chemical composition of fibers or from dyes and other additives. Examination and comparison can be carried out using different combinations of excitation and barrier filters and noting the color and intensity or absence of fluorescence emission at each excitation wavelength.

Solubility testing is a destructive method that can provide supplemental information to nondestructive methods of fiber comparison and identification. Possible reactions of fibers to solvents include partial and complete solubility, swelling, shrinking, gelling, and color change.

The chemical compositions of synthetic fibers are routinely determined using infrared analyses. Infrared spectra of fibers can be obtained using an infrared spectrometer with an IR microscope attachment. Fiber polymer identification is made by comparison of the fiber spectrum with reference spectra.

Mineral Fibers

In certain crime scenarios an investigator may also encounter mineral fibers, including glass fibers and asbestos. Asbestos fibers are often found alone or mixed with other components in building materials and insulation products. Asbestos minerals can be easily identified by their optical properties using polarized light microscopy. Scanning electron microscopy with energy dispersive spectrometry can also be used. Nonmicroscopic techniques for asbestos identification include x-ray diffraction and infrared spectroscopy.

Glass fibers, often found in building materials and insulation products, can be classified into three groups: (1) fiberglass (continuous and noncontinuous), (2) mineral wool (rock wool and slag wool), and (3) refractory ceramic fibers (glass ceramic fibers). Light microscopy can be used to determine the refractive index for the classification and comparison of glass fibers, and solubility tests can also provide important identifying information. An ultraviolet fluorescent binder resin may also be present on some glass wool products. Scanning electron microscopy with energy dispersive spectrometry can also provide elemental composition that can be used for comparison of glass fibers.

See also Comparison Microscope; Gas Chromatography; High Performance Liquid Chromatography (HPLC); Mass Spectrometry; Microscope; Microspectrophotometry

References

De Forest, P., R. E. Gaensslen, and H. C. Lee. *Forensic Science: An Introduction to Criminalistics.* New York: McGraw-Hill, 1983.

James, S. H., and J. J. Nordby. *Forensic Science: An Introduction to Scientific and Investigative Techniques.* Boca Raton, FL: Taylor and Francis, 2005.

Microscopy U, Michael-Levy Birefringence Chart; http://www.microscopyu.com/articles/polarized/ michel-levy.html (Referenced July 2005).

Microscopy U, Polarized Light Microscopy; http://www.microscopyu.com/articles/polarized/ polarizedintro.html (Referenced July 2005).

Robertson, J., and M. Grieve. *Forensic Examination of Fibers.* 2nd ed. London and Philadelphia, PA: Taylor and Francis, 1999.

Saferstein, R. *Criminalistics.* 8th ed. Upper Saddle River, NJ: Pearson Education, 2004.

White, P. *Crime Scene to Court: The Essentials of Forensic Science.* Cambridge: Royal Society of Chemistry, 1998.

Field Sobriety Tests

Field sobriety tests are preliminary tests conducted by law enforcement officers to enable them to assess the level of a person's physical impairment due to alcohol or drug use, and to determine whether a breath or blood test is required or justified for evidential purposes. Up until the mid-1970s, there was no consistency in the field sobriety tests used by police departments around the United States. Because there were many tests being used with no standardization, the National Highway Traffic Safety Administration (NHTSA) initiated studies to identify the best tests for enforcement use and standardize the way they were administered and scored. From the many different tests in use at the time, researchers identified three that were considered sufficiently accurate, practical, and reliable for assessment of sobriety: the walk-and-turn, one-leg-stand, and horizontal-gaze nystagmus tests. Further studies conducted in 1981 led to the development of a standardized set of administration and scoring values to promote consistency in the application of these tests. These three tests are now known as the standardized field sobriety test battery and are the basis of an NHTSA training program for police officers. Although the three-test battery is currently used in all states, their use is not mandatory, and many other field sobriety tests still remain in use.

The standard field sobriety test (SFST) battery is made up of a series of psychophysical tests and a preliminary breath test, if the breath test is available or authorized for use in that jurisdiction. Portable handheld roadside breath-test devices can be used, but any results obtained are considered to be preliminary and nonevidential. These breath-test results can be used only to establish the probable cause for requiring a person to submit to a more thorough breath or blood test.

Horizontal gaze nystagmus is an involuntary jerking of the eye as it moves to the side. Under normal physiological conditions, nystagmus occurs when the eyes are rotated at a high peripheral angle. A person is usually unaware of the nystagmus and is unable to stop or control it. During the test, the test subject is asked visually to follow a penlight or a pen as it is moved horizontally in front of his or her field of view. The more intoxicated the person is, the less distance the eye is able to track an object before jerking or nystagmus occurs. It is considered that the higher the blood-alcohol level or impairment, the smaller the angle of movement to the side before the eye jerks. Phencyclidine, barbiturates, and other depressant drugs can also cause nystagmus, and the combination of such drugs with alcohol is thought to decrease the angle at which nystagmus occurs even further.

The walk-and-turn and one-leg-stand tests are called divided attention tasks. These exercises test the suspect's ability to understand and carry out two or more simple instructions at one time. The ability to complete divided attention tasks is significantly affected by increasing blood-alcohol concentrations. The walk-and-turn test requires the suspect to maintain his or her balance while he or she walks heel to toe for nine steps in a straight line, to turn around on the line, and then repeat the process, all while listening and responding to the test instructions.

The one-leg-stand requires that the suspect maintain his or her balance while standing with heels together during the time the test instructions are given. The suspect then has to lift one foot off the ground and main-tain balance while counting out loud for thirty seconds.

If the test subject is a disabled driver who physically cannot perform the standard field sobriety tests, other tests may be supplemented. In these cases tests may include counting out loud, reciting the alphabet, or finger dexterity tests.

See also Alcohol
References
Levine, B. *Principles of Forensic Toxicology.* Washington, DC: American Association for Clinical Chemistry, 1999.
Saferstein, R. *Criminalistics.* 8th ed. Upper Saddle River, NJ: Pearson Education, 2004.

Fingernail Examination

Useful evidence can be developed from material found under the fingernails in some circumstances. For example, if a victim scratched the assailant, then the DNA of the assailant may be found in the scrapings from the victim's nails. If a sexual assault included digital penetration then material from the victim may be found in nail scrapings from the suspect. This is one of the areas in which the sensitivity and discriminating capability of DNA testing has proved so valuable.

Fingernails have longitudinal ridges. These have been an attractive source of potential evidence for a long time. The hypothesis is that the ridges are produced randomly, and so nail clippings can be associated with an individual with the same certainty as fracture matches or ammunition to weapon associations from individualization characteristics.

See also DNA in Forensic Science
References
Payne-James, J., A. Busuttil, and W. Smock. *Forensic Medicine: Clinical and Pathological Aspects.* London: Greenwich Medical Media, 2003
Saferstein, R. *Criminalistics.* 8th ed. Upper Saddle River, NJ: Pearson Education, 2004.

Fingerprints

Fingerprinting is the most well established of the forensic sciences used for identification of individuals. The first paper on its application was published in *Nature* in 1880 by a physician, Dr. Henry Faulds. Over 100 years of application and acceptance were shaken in the United States in 2002 when Judge Louis Pollak ruled in the Pennsylvania Supreme Court that experts could not testify that fingerprints matched because the procedure did not meet the standards required by the U.S. Supreme Court in *Daubert v. Merrill Dow Pharmaceuticals, Inc.* (1993). The judge later modified his view, but the decision has had major repercussions on the image of fingerprinting as the most reliable of identification sciences.

The uniqueness of fingerprint patterns and the ability to visualize even faint (latent) residues deposited on surfaces result from the way that skin and its friction ridges are formed. Skin is composed of two layers known as the outer skin, or epidermis, which is constantly worn off and replaced, and the inner skin, or dermis. The dermis layer contains several anatomical structures, but of interest to fingerprint formation are the sweat glands that exude perspiration via the sweat ducts and the sebaceous glands that secrete oily substances. The epidermis is made up of a stratified layer of cells, produced by the papillae of the basal layer, which is an undulating layer of cells that divide constantly. As the cells divide, they pass upwards through the layers of the skin until they are shed from the surface as dead cells.

The skin found on the palms of the hands and the soles of the feet is called friction-ridge skin. The basal layer has more pronounced undulations that produce the patterning called ridges and furrows. The undulations of the ridges and furrows produce fingerprints, palm prints, and sole prints. The ridge patterns are formed by the number and distribution of the dermal ridges or papillae. The basal layer contains many sweat glands, and the sweat will leave a negative of the patterning—the latent print. Fingerprint impressions are the marks left when a finger is placed on a surface.

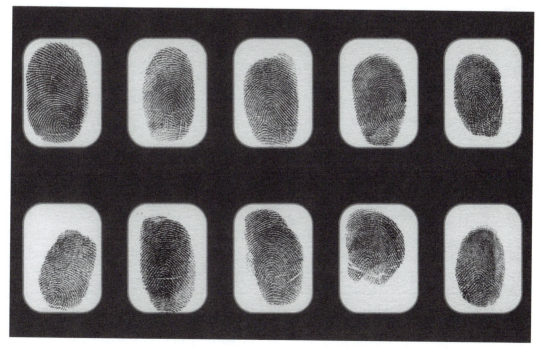

A set of fingerprints for both hands. (iStockphoto.com)

Fingerprints are formed from around the thirteenth week of gestation. Their pattern never changes during a person's life. In adults, the skin on the ball of the finger is approximately one- to two-twenty-fifths of an inch thick. In men, a fingerprint ridge is approximately one-fiftieth of an inch wide, but slightly less in women and even smaller in children. Men usually have coarser fingerprint ridge patterns than women. Friction ridges may be temporarily or permanently destroyed due to some kind of trauma or injury. If the ridges are permanently destroyed, by extreme injury to the epidermis, extending into the dermal ridges and dermal papillae, scarring will result, altering the original pattern. If temporary damage occurs, through trauma to the upper layers of the epidermis, the ridges will eventually return to their original pattern. Historically, attempts that have been made to remove or obliterate fingerprints by deliberate mutilation or plastic surgery have all failed.

There are three types of glands producing secretions that may produce fingerprints on a surface. These are the eccrine, apocrine, and sebaceous glands. Ecrine glands are particularly prevalent on the palms of the hands and the sides of the feet. These secretions are mainly water, with a small percentage of inorganic and organic components. The sebaceous glands are associated with hair follicles and are not found on the soles of the feet or the palms of the hands. However, secretions from these glands can be transferred to the hands by touching skin that contains sebaceous glands, such as the face. The sebum functions as a lubricant and contains a high percentage of glycerin, free fatty acids, wax esters and squalene, and a small percentage of sterol esters, steroids, and hydrocarbons. Due to the anatomical distribution of apocrine glands, apocrine secretions contribute little to fingerprint deposits.

It is accepted that it was Henry Faulds, living in Japan at the same time, who was first to recognize and publish the idea that latent prints from a crime scene could provide evidential value.

Fingerprint Classification

There are four basic principles of fingerprints, similar to the basic requirements for an effective forensic science blood-typing system:

- They are an individual characteristic.
- They remain unchanged throughout a person's lifetime.
- They do not alter when deposited on a surface.
- An accurate representation of the print can be obtained from the surface on which it was deposited.

The ten to fifteen years following Faulds's paper in *Nature* saw considerable advances in the documentation of systems that would allow fingerprints to be classified and so retrieved from a database. Juan Vucetich, an Argentinean police officer, established his own fingerprint classification system in 1891. The following year, Francis Galton published "Fingerprints," which was the first manuscript on the role of fingerprints in identification and crime investigation. A few years later, Edward Henry developed a fingerprint classification scheme and cataloged sets of prints that could be used for identification purposes. Henry's classification was adopted in Europe and North America in the early 1900s. Fingerprints are classified in three ways:

- By the shapes and contours of ridges in individual patterns
- By noting the finger positions of the pattern types
- By relative size, determined by counting the ridges in loops and by tracing the ridges in whorls

Traditional classifications are based first on the general pattern of the print, which can be in the form of an arch, a loop, or a whorl.

Arches

Arches are the least common but simplest fingerprint patterns. The pattern is formed when ridges running across the fingerprint rise in the center to form an arch. The flow of the ridge is continuous from one side of the print to the other. In a plain arch the curve is gentle, giving a wave-like appearance. The curve may be more acute and rise sharply, giving the appearance of a tent—hence the description of tented arch. About 5 percent of prints are classified as arches.

Loops

Loops are the most common pattern, accounting for about 65 percent of all prints. The flow of the ridge becomes reversed and the ridge exits on the same side that it entered. Each loop must be bounded by two diverging ridges, called type lines. There are two other features of a loop: the core, which is the visible center of the loop, and the delta, which is a point or ridge end nearest the place where the type lines diverge. All loops have at least one delta. Ridges entering and exiting from the little finger side of the hands result in an ulnar loop; the ridges slant toward the little finger. If ridges enter and exit from the thumb side the pattern is a radial loop; the ridges slant toward the thumb.

Whorls

The remaining classification category is that of whorls. Whorls have type lines and at least two deltas. There are four subtypes of whorl; two are characterized by an easily recognizable loop formation. They are the plain and the central pocket loop whorls, each of which has at least one ridge that makes a complete perimeter to the whorl. The positioning of the deltas is used to differentiate between plain and central loop whorls. An imaginary line between the two deltas will cross through the enclosed ridges in a plain loop but not in a central pocket loop. A double loop whorl is found when two loops are combined into one print. Any other print that meets the definition of *whorl* (type lines and at least two deltas) is classified as an accidental whorl.

Accidental whorls contain more than one pattern, but not the plain arch, and often have more than the two deltas. It can be a combination of a loop and a whorl, a loop and a central pocket loop, or any combination of two different loop and whorl type patterns.

Ten Print Records

The original Henry fingerprint classification system converted ridge patterns for each finger into letters and numbers. However, the system was too cumbersome to deal with more than about 100,000 sets of prints. The basic system used in the United States today is a modification of the Henry system. There are a number of subsets in the classification, but the primary set is a record of the presence or absence of whorls on each finger. The record pairs right index and thumb, right ring and middle, left thumb and right little finger, left middle and index, and left little and left ring fingers to give five ratios. If no whorl is present, a value of zero is recorded for that finger. If a whorl is present, the value depends on which finger. If it is in the first finger set it has a value of sixteen, falling to eight, then four, then two, then one as we go from (R index/R thumb) to (L little/L ring). The values are added to give a numeric ratio that classifies the patterns of the set of prints from the ten fingers of the individual. These reference classification sets are termed *ten print records*.

To be used for comparison purposes, fingerprint impressions taken from a person need to be absolutely clear and visible. A rolled impression can be obtained to provide the entire friction surface of a finger or thumb, from the tip to one-fourth inch below the first joint. These impressions are made by rolling the finger or thumb from one nail edge to the other and can be used to provide an investigator with sufficient ridge characteristics for a correct classification. Plain impressions can be used to verify the order of the rolled impressions and to show characteristics that may be distorted in rolled prints.

Palm prints are sometimes found on evidence or at a crime scene, where the whole hand can leave a distinctive impression. Major case prints are a complete set of prints made of all parts of the hand including the fingertips, palm, sides of the fingers, and sides of the palm, and can be very useful in forgery cases. Major case prints are always obtained from corpses associated with an investigation, and are used to identify or eliminate latent print evidence and in the identification of the deceased. Prints can be obtained before or after the onset of rigor mortis, or after the start of decomposition. The hardest prints to obtain are those from a body that is already undergoing decomposition, especially when the hands are badly charred or decomposed. Problem prints and impressions can also come from hands that are too dry or wet from perspiration, and deformities and scarring can also be a problem when obtaining impressions.

Fingerprint Detection

Fingerprint impressions are the marks left by a finger that has been placed on a surface. They can be faint or sometimes even invisible to the naked eye, hence the adjective *latent*. They also represent only a part of the print from the whole finger. There are a number of different categories of impressions that can be encountered, and the parameters that affect the quality of those impressions and the nature of the surface will have some bearing on the fingerprinting techniques required to recover those prints.

The first step in fingerprint detection at a crime scene is the examination of all surfaces and objects. Location of fingerprints can be by naked-eye observation (perhaps aided by a magnifier), by physical enhancement such as a laser or alternative light source illumination, or by chemical treatment (including traditional powdering processes). Once located, all prints must be recorded and items collected and preserved for laboratory examination.

The print is deposited as direct residues from eccrine sweat glands, or from secondary residues from sebaceous or apocrine glands picked up on the fingers by touching and then left in the fingerprint. Other sources include

prints left in blood or imprints from greasy materials such as lubricants that might have been used in a rape.

The main factor arising from the nature of the touched surface is whether or not it is porous. Porous surfaces include paper, unfinished wood, and cardboard. Prints on these are usually preserved well because print residue can soak into the porous surface. Nonporous evidence, like the prints that are found on plastic, glass, metal, and foil, is considered more fragile because the print residue usually just lies on the surface and is not absorbed. Even the most careful handling can wipe away or destroy a latent print on a nonporous surface. However, treatment of porous surfaces may react with the surface material and damage it or obscure the latent image.

The most common process for processing latent prints is by powdering or chemical treatments.

Powders

Most powders come in kits in a variety of colors that can be applied with brushes or other application instruments. The crime-scene technician uses the powder that provides the best contrast with the background for photography purposes, but black and white powders are used most. Specialized powders are also available for processing prints on brightly colored or textured surfaces. Powders are applied to the surface with brushes that are made from a variety of materials including camel hair, squirrel hair, and fiberglass. The brush type chosen for dusting is generally the preference of the investigator. The investigator's experience and comfort with brushing and the brush type are a large factor in obtaining useable or the best prints. The powder is applied to the surface using a light touch, with more powder being added as necessary. The brush is used with gentle, light strokes to avoid wiping away print residues with too much pressure.

Regular, nonmagnetic powder is the most commonly applied powder for processing prints on windows, countertops, television sets, metal surfaces, painted doors, mirrors, and broken glass. Magnetic powders are used as part of more specialized processing and are good for the development of latent prints left on shiny magazine covers, coated surfaces, and plastic materials, like food storage containers or storage and sandwich bags. Fluorescent powders for the appropriate surface usually include the use of an ultraviolet light source while applying the powders.

Chemical Processing

Chemical processing of fingerprints is usually conducted in the laboratory. Latent prints on paper products are traditionally developed with chemicals. Paper provides a porous surface that absorbs skin secretions; latent prints tend not to rub off paper as they would a nonporous surface. The quality of latent prints on paper is determined by the amount of finger contact and the amount of pressure applied. These prints can be developed by exposure to chemicals that react with minerals and organic matter in the skin secretions left on the print residue.

Chemical processes include ninhydrin, which can be used to develop latent prints on porous surfaces (paper, cardboard, and wood). Ninhydrin reacts with amino acids commonly found in latent print residues, forming a purple compound. The prints appear reddish-purple to brownish-purple in color. A frequently used analog of ninhydrin 1,8-diazafluoren—9-one (DFO), when applied to paper, develops 2.5 times more latent prints than ninhydrin alone.

Sticky-side powder is a chemical process used for the processing of latent prints on the sticky side of adhesive tapes and labels.

Iodine fuming is a chemical process that can be applied to surfaces that are impractical for dusting with fingerprint powder or may have residue that will damage the dusting brush. On surfaces that are greasy, the iodine fumes will be absorbed at different rates by different fats or oily residues, but the latent impressions may still be visible. This technique can usually reveal high-quality latent prints on most paper and hard, smooth surfaces.

Fuming with cyanoacrylate esters, such as superglue and similar adhesive products, can also be useful in the development of latent prints. Known as the superglue technique, it is now used on many surfaces that at one time were considered unsuitable for the recovery of latent fingerprints. The superglue reacts with amino acids, fatty acids, and proteins in the latent print and the moisture in the air, producing a visible, sticky white material along the ridges of the fingerprint.

Dye Stains

Dye stains like Sudan black are also used in detecting latent prints. Sudan black is a dye that stains the fatty components of sebaceous gland secretions and is used for developing latent prints on smooth or rough nonporous surfaces contaminated with greasy, oily, and sticky substances. The dye works well on glass, metal, or plastic surfaces and is good for candles or waxed-paper milk cartons. Sudan black can also be used to enhance glue-developed latent prints.

Ardrox is a fluorescent dye stain also used to enhance cyanoacrylate-developed latent prints. Once applied, the dye can be illuminated with an ultraviolet lamp, causing the latent prints to brightly fluoresce. This technique allows the visualization and subsequent photography of weakly developed latent prints that can't be seen under normal lighting conditions.

Amido black, naphthalene blue-black, or naphthalene black 12B is a protein dye that is sensitive to proteins including those found in blood. The application of this dye will stain protein residues in a blood-contaminated latent print blue-black in color. This dye does not stain the normal protein constituents found in latent print residues.

Crystal violet, also known as gentian violet, is another protein dye that stains the fatty components of sweat a deep purple color. This protein dye can also be used to enhance the appearance of bloody fingerprints. Crystal violet dye can also be used on many types of adhesive tapes, as normal powders stick to the entire sticky side of the tape and not just the latent prints.

Mechanical Methods

A number of mechanical methods are also routinely employed in fingerprint detection. When a secreted material is left on a surface, the properties of that surface are modified. By applying an electrostatic field using electrostatic detection apparatus (ESDA) these modifications can be made apparent. For this technique to be successful, the prints have to be very fresh to obtain significant results.

Another mechanical method is x-ray. The latent prints are revealed by dusting with lead powder and then x-rayed. This rather difficult technique, which is rarely used, has many advantages for use on difficult surfaces, such as multicolored paper, fabric, wood, rubber, leather, and human skin.

Vacuum metal deposition is a mechanical recovery method that works on the principle that fingerprint contamination hinders the deposition of a metallic film following evaporation under vacuum. This extremely sensitive technique for fingerprint detection on a variety of surfaces can be used in conjunction with cyanoacrylate fuming.

Fingerprint Impressions

Fingerprint impressions in blood are generally processed using specialized light techniques. Weak fingerprints can be visualized and photographically enhanced using excitation wavelengths at 400 nm. With older blood, the protein hemoglobin absorbs light at 400 nm. If the surface reflects light at 400 nm then the prints are visible as dark ridges against a light background. If the bloodstain is on a shiny surface, then the natural luminescence of blood can be used to enhance the print. Blood exhibits a weak photoluminescence with excitation at 360 nm. Chemical processing should only be carried out on blood if visual techniques fail. Leuchomalachite green can be used to stain fingerprints in blood that can then be

photographed. However, the most conventional treatment for blood prints is amido black. DFO can also be used for processing blood prints on porous paper surfaces. Once treated with DFO, stains can be visualized and photographed under ultraviolet illumination.

Depending on the crime scene, it may be necessary to develop latent prints on skin. Skin has a number of qualities that distinguish it from other specimens examined for latent prints. Live skin is a dynamic surface; the skin tissue constantly renews itself, shedding old cells that might contain the imprint of an assailant's grip. Skin is a pliable medium that allows movement, allowing the possible distortion of fingerprints. Latent prints left on a skin surface can be washed away with water excreted through perspiration or with general hygiene. In the situation of homicide victims, the skin may have been exposed to harsh conditions, including mutilation, contamination with body fluids, weather effects, and decomposition. Additionally, during crime-scene processing, several people may handle the body, which may partially or fully destroy existing prints, as well as add new prints to the skin. There are a number of recommended procedures for processing a corpse for latent prints. These include the application of fuming techniques and encompass the complications caused by refrigeration and temperature effects on the outcome of processing.

Identification from Fingerprints

It should be obvious that identification of someone from a recovered print from a scene does not use the features of the Henry classification system. The latent is from one finger only and is probably a partial record of the complete fingerprint. Identity by fingerprints is established by comparing the type and position of certain ridge characteristics. Where they appear in the same order, in the same relationship to each other, and in sufficient number, then it is established that they were made by the same digit.

Identification from scene prints uses detail in the ridges as its basis. These minutiae are of three basic types: endings, bifurcations, and dots. Endings of short ridges are sometimes called spurs, and sometimes short ridges join longer ones in a crossover. Traditional manual matching compares individual latent prints to candidate reference specimens on the basis of the nature and arrangement of minutiae.

The concept of point matching has stirred considerable controversy in recent years. Different jurisdictions traditionally required different numbers of matching points in order to call a match. For example, continental Europe accepted eight to twelve points, the United Kingdom required sixteen, and the United States accepted twelve. However, an expert panel convened by the International Association for Identification reported in 1973 that there was no valid basis in requiring a minimum number of friction ridge characteristics to provide a positive identification.

That view is now generally accepted and identifications are made on a sequence of increasingly specific observations, referred to as Level 1, Level 2, and Level 3 characteristics. Level 1 deals with the general ridge flow and pattern configuration. It is not sufficient for individualization, but can be used for exclusion and for preliminary sorting of data. Level 2 detail includes the traditional minutiae and permits individualization. Level 3 detail includes all dimensional attributes of a ridge, such as ridge-path deviation, width, shape, pores, edge contour, incipient ridges, breaks, creases, scars, and other permanent details.

Many examiners in the United States now follow the so-called ACE-V method for identification: *a*nalysis, *c*omparison, *e*valuation, and *v*erification. Analysis consists of the objective, qualitative, and quantitative assessment of Level 1, 2, and 3 details to determine their proportion, interrelationship, and value to individualization. Comparison is the objective examination of the attributes observed during analysis in order to determine agreement or discrepancies between two friction-ridge impressions. Evaluation is the cyclical procedure of comparison between two friction-ridge impressions to effect a decision; that is,

made by the same friction skin, not made by the same friction skin, or insufficient detail to form a conclusive decision. Verification is the independent analysis, comparison, and evaluation by a second qualified examiner of the friction-ridge impressions. Evaluation is a subjective step and that is why the verification process is so important.

Automated fingerprint identification systems (AFISes) use computer algorithms to locate the presence of ridge characteristics and their special locations. A map of the detail and its direction is created and stored. Comparisons are based on comparisons of the maps.

Fingerprint Age Determination
There is no reliable way to determine the age of a latent fingerprint. The best approximation will be from corroborated history of the object on which the print was deposited. For example, a print on a window that was cleaned fourteen days earlier must be less than fourteen days old.

Fingerprint Forgeries
Because fingerprints are created by the papillae in the dermis, surface damage to the finger will not destroy the intrinsic fingerprint pattern. Thus, attempts to alter or destroy fingerprints will not work unless the finger is severely damaged. Probably the best-known example is that of the gangster John Dillinger, who tried to destroy his fingerprints using acid. However, prints collected from his body at the morgue after his death by shooting matched those taken at an arrest before the acid treatment.

Because latent prints are a physical deposit, it is theoretically possible to lift a print from one solid object and transfer it to another. The transfer could also be made from a facsimile of the original finger. The case involving the Mickelberg brothers illustrates some of the issues surrounding fingerprint forgery.

A well-planned sting operation netted gold bullion worth $650,000 Australian from the Perth (Australia) mint in 1982. In brief, two men posing as dealers made some legitimate small purchases, followed by the larger one for $650,000 in gold bullion. In this instance, their check bounced. West Australian police eventually arrested and charged three brothers, Peter, Ray, and Brian Mickelberg. The brothers were tried and convicted in 1983, largely on the basis of confessions and a partial print recovered from the check that was a match to Ray. The Mickelbergs protested their innocence, claiming to have been beaten by police, that the confessions were fabricated, and that the fingerprint was a plant. Police had raided their apartment and found a bucket of silicone rubber casts of the hand and fingers of Ray. The brothers claimed that the police had then used one of the finger casts to plant the partial print by coating it with sebaceous secretions from the nose and rolling it over the check. None of the appeals succeeded, including a lengthy inquiry that involved many independent fingerprint experts. Issues raised at the hearing included whether or not the latent image was of the shape expected from a soft finger pad or the harder rubber cast and whether or not the images of the pores were negatives. The arguments were that the impression from a real finger is always somewhat flattened with indistinct boundaries and so would be different from one from the relatively rigid casting. Further, the pores on a true print are the source of the sweat that reacts with chemical treatment. In contrast, rubbing a cast of a finger over the nose and rolling a false print has the body fluid on the ridge, not the pore. The hearing was not able to resolve these questions.

After twenty years the affair resurfaced when one of the original investigators confessed that the officer in charge had indeed beaten the accused and that the confessions were fabrications. However, he had no comment to make on the fingerprint, and it is still not known whether it was a plant or a real impression.

See also Automated Fingerprint Identification System (AFIS); Integrated Automated Fingerprint Identification System (IAFIS); Latent Prints

References

De Forest, P., R. E. Gaensslen, and H. C. Lee. *Forensic Science: An Introduction to Criminalistics.* New York: McGraw-Hill, 1983.

James, S. H., and J. J. Nordby. *Forensic Science: An Introduction to Scientific and Investigative Techniques.* Boca Raton, FL: Taylor and Francis, 2005.

Komarinski, P. *Automated Fingerprint Identification Systems (AFIS).* Amsterdam and Boston: Elsevier, 2005.

Lee, H. C., and R. E. Gaensslen. *Advances in Fingerprint Technology.* 2nd ed. Boca Raton, FL: CRC, 2001.

Lee, H. C., T. Palmbach, and M. T. Miller. *Henry Lee's Crime Scene Handbook.* London and San Diego, CA: Academic, 2001.

Saferstein, R. *Criminalistics.* 8th ed. Upper Saddle River, NJ: Pearson Education, 2004.

White, P. *Crime Scene to Court: The Essentials of Forensic Science.* Cambridge: Royal Society of Chemistry, 1998.

Firearms

Firearms examination in the laboratory deals with the weapons and their ammunition.

Ammunition

A live round of ammunition is properly referred to as a *cartridge*. The *bullet* is the projectile that is fired through the barrel of the firearm by ejection from the cartridge case when the firing mechanism detonates the primer that in turn ignites the propellant. Bullets are generally made of lead, but may be encased partially or entirely in a jacket of copper, aluminum, or steel. Bullets are also found with a coating of copper or brass, which reduces fouling of the bore and allows the projectile to travel through the barrel more easily. A variety of bullet styles exists, such as round-nose, wadcutter (for target shooting), semiwadcutter, full metal jacket, soft-point, hollow-point, and semijacketed hollow-point. The harder surfaces of metal-jacketed bullets make them more suitable for use in semiautomatic weapons, as they are less likely to jam when feeding out of the magazine and into the chamber.

The cartridge case is usually made of brass. It bears a stamp (the headstamp) on the base with information on the caliber and manufacturer. The case holds all the components of the cartridge: the primer, the powder or propellant, and the bullet, which is held in place by the crimp. The end of the case has a rim, which holds the ammunition round in place in the chamber and which is the part that is caught by the mechanical ejector. The primer is usually a mixture of chemicals that may include lead styphnate, barium nitrate, and mercury and antimony sulfide. These are toxic chemicals, and it has been estimated that about 80 percent of the lead in the air in a firing range originated in the ammunition. The powder or propellant in modern ammunition is smokeless powder, which is made up largely of nitrocellulose and burns very rapidly on ignition. Before smokeless powder, the low-explosive black powder was used. Shotgun cartridge cases are essentially similar, except that they have sides of paper and the shot is contained within the cartridge.

Rimfire cartridges were common in many older cartridges but are now exclusively 0.22 caliber. They are not reloadable. Centerfire cartridges have the primer located in the center of the cartridge case. Almost all calibers other than 0.22 are centerfire. The primers can be removed after firing and replaced, making this type of cartridge reloadable.

Bullets are traditionally made from lead. It is malleable and heavy and so responds to the rifling of the barrel and transfers high energy on impact. Safety concerns about lead exposure are resulting in alternative materials being used to make bullets. Iron powder, tungsten, copper, and steel have all been used.

Firearm Components

The basic components of a firearm are its barrel, firing chamber, stock, and action.

The *barrel* of a firearm is simply a metal tube created by a series of drilling, reaming, and smoothing operations performed on a piece of steel rod. Barrels may be rifled or smooth. The *firing chamber* holds the cartridge in the gun and must be capable of containing the very high pressures that result from the ignition of the propellant, which forces the bullet out through the barrel. Access to the chamber is achieved through the *breech*. The *stock* is the means by which the shooter holds and controls the firearm.

The mechanism for firing, loading, and unloading is known as the *action*.

Barrel Rifling

A rifled barrel consists of a series of spiral grooves cut into the inner surface of the barrel, known as the *bore*. The uncut areas between the grooves are known as the *lands*. The lands cut into the bullet and cause it to rotate on its longitudinal axis as it passes along the barrel. This rotation gives the bullet stability in flight, prevents it from tumbling end over end, and gives the weapon greater accuracy. The tools used to produce the rifling leave unique, individual microscopic imperfections on the lands and grooves. These imperfections change each time the tools are used, and so each rifled barrel has its own unique set of characteristic microdefects that it in turn imparts on the bullet as it passes through the barrel.

There are four main ways to create rifling of a gun barrel. These are called hook, broach, button, and hammer. The hook and broach methods are based on pulling cutters through the tube that will become the rifled barrel. In the hook method, a cutter is pulled from end to end repeatedly until the desired depth is achieved. In the broach method, a series of progressively larger cutters form rifling with a single pass. The button method pushes or pulls a carbide plug with a rifled cross-section through the bore. In hammer-forged rifling, a mandrel with a rifling configuration is forced into the bore by external hammering. One pass of the tool creates the rifling.

Firearms Evidence

Marks on bullets and cartridge cases are compared by microscopy. In a traditional setup, a comparison microscope is essential for side-by-side viewing of tests and evidence. Oblique, reflected lighting is used to highlight individual characteristics. Today, most firearms examiners use digital images from the National Integrated Ballistics Information Network (NIBIN) database systems to prepare high-resolution magnified images that are then the bases of their comparisons. These images also provide an excellent record of the work done.

Every man-made object (not just bullets) shows generic characteristics arising from its manufacturing process. These are known as *class characteristics*. Even when no suspect weapon is available, information that can aid investigators may be obtained from a detailed examination of a fired bullet's class characteristics. These characteristics serve to reduce the population of firearms that may have been involved in a shooting incident. Determining the bullet's caliber, the number of lands and grooves, the direction of twist, and land and groove width dimensions can provide valuable information about the type of weapon that was used. However, it is rare to be able to state with absolute certainty the exact make or model of a weapon based only on class characteristics of a bullet fired from it.

The ammunition manufacturer can sometimes be determined from a bullet's weight, shape, and composition.

Individual characteristics are the result of tool marks impressed onto the bullet as it travels through the barrel, and of tool marks impressed on the cartridge case by the firing and ejector mechanisms. These markings result from microscopic imperfections transferred from the tools used to create the barrel or firing mechanisms. These markings are unique to each gun. Just as with class characteristics, any manufactured physical object (not just bullets) will show individual characteristics that may be used for individualization. They may also arise from random wearing patterns. These characteristics can be altered by extensive use, cleaning, or abuse. The areas of ammunition bearing identifying marks are the primer area, the rim, the side of the case, and the bullet.

When a gun is fired, the firing pin is pressed into the soft metal of the primer, leaving a negative impression. Ignition of the propellant forces the base of the cartridge case backward against the breech face, leaving another set of impressed markings on the base of the primer and/or cartridge case.

Pressure from the expanding gases from the burning gunpowder forces the sides of the cartridge case into tight-fitting contact with the inside surfaces of the chamber. Markings from irregularities in the chamber can be scratched into the sides of the cartridge case when it is extracted from the chamber.

The loading and ejector systems leave characteristic markings on the casing. Markings caused by pressure, scratching, and scraping on the sides of the cartridge case by the magazine or loading mechanism may also bear individual characteristics.

The clawlike extractor, which grips the base of the cartridge case and pulls it out of the chamber, may also leave striated tool marks. In the same manner, the ejector may leave its mark when it strikes the base of the cartridge case, throwing it out of the weapon.

In addition to determining the type of weapon that may have fired a bullet, an examiner can also check to see if a bullet bears sufficient individual characteristics for comparison, either with other bullets from the same or different crime scene, or with test bullets fired from a specific weapon.

Test firing must be done in a manner that prevents damage to the test bullets. Traps containing cotton waste used to be used, but most laboratories now conduct test firing into water in a recovery tank. It is essential that the same type of ammunition be used for test firing. Differences in ammunition can cause differences in the individual characteristics between the evidence and tests. Test specimens are first compared with each other to determine if the weapon produces identical markings each time it is fired. Finally, one of the tests is chosen and compared with the evidence. Occasionally, all tests may have to be compared individually with the evidence, due to poor reproducibility.

Cartridge Cases

Cartridge cases left at the scene of a shooting can contain evidence that allows them to be associated with the particular gun that fired the ammunition. The evidence consists of physical marks left by the various mechanical actions involved in inserting the ammunition, detonating the primer, removing the cartridge case, and ejecting it.

Shotgun Pellets and Wadding

Shotgun pellets are usually lead spheres, although pellets used for hunting waterfowl must be made of materials other than lead, such as steel, bismuth, or tungsten. Pellets are surrounded by wadding and are classified by number sizes, with the larger the number, the smaller the pellet. Birdshot pellets are smaller than buckshot pellets, ranging in size from 0.05 to 0.17 inch in diameter. Buckshot pellets range from 0.24 to 0.36 inch in diameter. In addition to pellets, elongated, hollow lead rifled slugs may be used. Pellets can be sized by weight and diameter if they are not badly deformed.

Wadding helps to cushion and protect the shot from the hot gases produced by the burning gunpowder, as well as to keep the shot together as it exits the barrel. On close-range shots, the wadding may enter the wound along with the shot. Wadding may be made of paper, felt, plastic, or plastic granules. The wadding usually helps to identify the ammunition manufacturer. The type of wadding used is often characteristic of a particular ammunition manufacturer. Plastic wadding may be marked with identifying markings, particularly when fired from sawed-off barrels with rough edges, or when the wadding scrapes against adjustable chokes or front sights that protrude into the barrel.

Firearms Identification and Testing

The field of firearms identification is sometimes referred to as forensic ballistics or simply ballistics. Strictly speaking, *ballistics* is the physics of the trajectory of a projectile. Firearms as a discipline within forensic science is the examination of guns and their ammunition for a variety of reasons. Most work in firearms examinations is carried out to determine if a bullet, cartridge case, or other ammunition component was fired by a particular firearm. The tool marks created during the firing and ejection of ammunition provide the basis of the testing.

Firearms testing also includes:

- Checking a weapon for proper function. These tests have safety and investigative implications.
- Identifying weapons by class characteristics. This is helpful to investigators so they know what type of weapon they are looking for.
- Testing for gunshot residue. Residues may provide information on whether a gun was fired and, if so, whether or not it was recent.
- Distance determination. A muzzle-to-target determination estimates the distance that the muzzle of the firearm was from the target when the weapon discharged.
- Serial number examination. Obliterated serial numbers are treated to attempt to restore the original number.

Firearm Types

There are thousands of types of firearms in existence. They can be classified either as long guns or shoulder weapons (such as rifles and shotguns), or as handguns, which are designed to be held and fired with one hand (such as revolvers, pistols, derringers, and single-shot pistols).

Rifles and Shotguns

A *rifle* has a rifled barrel, with a bore diameter between 0.17 and 0.50 inches, and fires single projectiles. A *shotgun* has no rifling and a larger bore (except for the 0.410), and fires multiple projectiles.

Single-shot weapons are usually break-open types and must be loaded and reloaded by hand after each shot.

Repeating weapons have a manually operated loading and unloading mechanism. The ammunition is fed from a magazine. This group includes guns with bolt, lever, and pump action.

In *semiautomatic* rifles and shotguns the forces generated by the burning of the propellant operate the action. In a gas-operated weapon, the combustion gases are siphoned out of the barrel and push a piston rearward to operate the action. In recoil-operated systems, the action is operated due to rearward pressure against a spring-loaded breech. The ammunition is magazine fed. Fully automatic weapons are similar to semiautomatic in how the action operates, but the major difference is that the weapon will continue to fire as long as the trigger is depressed and ammunition remains in the magazine.

Hand Guns

A *revolver* is a repeating weapon with a revolving cylinder that contains between five and nine firing chambers. With each shot, the cylinder turns and aligns with the barrel, locking in place before the weapon is discharged again. The cylinders may be fixed within the gun or may be detachable. Revolvers with fixed cylinders are loaded by inserting ammunition through a loading gate, and spent cases are ejected through the same gate. Other revolvers have a detachable cylinder, held in place by a locking device.

The *top-break revolver* has a latch at the top of the frame that allows the hinged barrel to be tipped downward, exposing the rear of the cylinder. This action usually causes a star-shaped extractor at the rear of the cylinder to lift all the cartridges or cartridge cases partially out of their chambers so they can be manually removed. This type of design was popular in the late 1800s and early 1900s.

In a *swing-out revolver,* the cylinder rod is attached to a hinge-like device known as a crane, which is unlocked by sliding a cylinder release located on the side of the frame. An ejector rod is manually activated to expel all the cylinder contents at one time.

In a *single-action revolver,* the cylinder rotation and alignment takes place when the hammer is pulled back and cocked. Because the shooter does part of the work, the force required to pull the trigger is substantially reduced, usually falling in the three to five-pound range.

In the *double-action* mode, pulling the trigger rearward causes the cylinder to rotate and align itself with the barrel, as well as to cock and release the hammer. This requires more force to be applied to the trigger by the shooter, typically about twice that required in single-action guns.

Revolvers may operate with single action only, double action only, or both single and double action.

In a *semiautomatic pistol,* the forces generated by the burning gunpowder force the slide rearward against spring pressure. In a recoil-operated weapon, the breech and barrel move rearward locked together for a short time after discharge. In a blowback operation, the barrel is fixed, and only the breech moves rearward. Cartridges are loaded into a spring-loaded box device known as the *magazine* and inserted into the grip of the weapon. As the slide is forced back by the firing of the weapon, the empty cartridge case is extracted from the chamber and ejected from the weapon, and a new cartridge is stripped from the magazine and fed into the chamber as the slide moves forward under spring tension. These weapons typically have higher magazine capacities (typically six to fifteen rounds) and can be either single action only, double action only, or both single and double action.

Derringers have been around for years and consist of multiple barrels—usually two but sometimes four—cut into the same piece of steel. The barrels are very short, typically not much longer than the cartridge itself, and are designed solely for personal defense. Because of their size, they are difficult to fire. They usually fire only in the single-action mode and have very small grips. Because of the short barrel, they have poor accuracy.

Single-shot pistols must be manually loaded and unloaded one at a time, usually by means of a break-open breech. They have a single-action trigger mechanism and are largely used for target shooting.

Firearms Discharge Residue Testing

Discharge residues consist of unburned propellant and the gaseous products of the discharge reaction. The gaseous-phase products are emitted from the muzzle and breech of firearms upon firing. The metals from the primer—such as antimony, lead, and barium—condense into tiny particles with a characteristic shape when seen in the scanning electron microscope, as well as having a distinctive elemental composition.

Firearms discharge residue (FDR) or gunshot residue (GSR) tests are all based on identification of these materials. Most are based on identification of the metals, as the combination of antimony, barium, and lead is not common in the environment. Early tests used techniques such as atomic absorption spectrophotometry to measure the elements in bulk samples obtained by washing the hands of a suspect with weak acid solutions. Controls consisted of washings from the nonfiring hand. The technique is not very sensitive, and for a time neutron activation analysis was proposed as a better alternate. It is certainly much more sensitive but requires specialized facilities (to produce the neutron source and detect the radioactive decay) to conduct the testing. The currently accepted method of choice is scanning electron microscopy in conjunction with x-ray spectrometry (SEM-EDX). This offers the required sensitivity but couples the chemical analysis with the well-nigh unique morphology of the discharge residue particles in a single test. The SEM uses an electron beam to visualize the particles, but at the same time, the beam results in emission of x-rays that have energy spectra characteristic of the elements in the target. An analyzer unit can be coupled to the SEM to detect and record the spectrum of the emitted radiation. Modern instruments also have the capability to conduct the particle searches automatically. Samples are collected by dabbing surfaces with adhesive tape fixed to the stub that is inserted directly into the SEM apparatus.

The major problem with gunshot residue is that it can easily be washed or rubbed off, meaning a negative result does not eliminate the possibility that an individual may have discharged a weapon. Deceased subjects may retain residues longer, and their hands may be covered with paper bags when being transferred to a morgue to protect these residues before sampling. The best way to ensure positive results is sampling as soon as possible after a shooting. Investigators should have an ample supply of kits available at all times. Most 0.22 rimfire ammunition (other than Federal brand) does not contain barium and antimony.

Distance Determination—Muzzle to Target

When a firearm is discharged, unburned and partially burned particles of gunpowder, as well as gunpowder and primer combustion residues, are propelled out of the barrel along with the bullet by the tremendous pressures developed by the burning gunpowder. Some of these residues also escape from the opening breech in semiautomatic weapons and from the gap between the cylinder and barrel on revolvers. Some of these residues can be deposited on the hands of the shooter.

When a firearm is discharged within close proximity to the target, gunpowder and primer residues can also be deposited on the target. The appearance of a residue pattern can be helpful in estimating the distance that the muzzle end of the barrel was from the target when the weapon was fired. If the weapon was perpendicular to the target, the residue pattern is usually circular around the bullet entry hole. The residue typically consists of unburned and partially burned gunpowder particles and residues from the combustion of the gunpowder and primer. The diameter and density of the residue pattern can be reproduced by making a series of test firings using the same firearm and ammunition from various distances into a material similar to the evidence.

The type of firearm, barrel length, and type of ammunition are all factors that affect the size and density of residue patterns. For this reason, all cartridges recovered in the weapon and any partial boxes of ammunition associated with the scene of the shooting or in the possession of the subject should be submitted for testing.

Residue patterns can be compared by a visual examination and by utilizing a stereomicroscope. In addition to residues, observations about tearing and singeing of cloth should be noted, as this is indicative of a contact shot.

Chemical tests, such as the Griess test for nitrite residues and the sodium rhodizonate test for lead, can also be very beneficial, especially when dealing with colored garments. Infrared photography may also be used to visualize vaporous lead (soot) residues on heavily bloodstained clothing.

Firearms Function Testing

Functional tests are carried out on firearms held in evidence to answer the following basic questions:

- Does the weapon actually fire ammunition?
- Do the safety mechanisms operate properly?
- Will the weapon discharge without pulling the trigger?
- Has the weapon been modified (shortened, made to fire fully automatic, etc.)?

See also Ammunition; Bullets
References
De Forest, P., R. E. Gaensslen, and H. C. Lee. *Forensic Science: An Introduction to Criminalistics.* New York: McGraw-Hill, 1983.
Heard, B.J. *Handbook of Firearms and Ballistics: Examining and Interpreting Forensic Evidence.* New York: Wiley, 1997.
James, S. H., and J. J. Nordby. *Forensic Science: An Introduction to Scientific and Investigative Techniques.* Boca Raton, FL: Taylor and Francis, 2005.
Lee, H. C. T. Palmbach, M. T. Miller. *Henry Lee's Crime Scene Handbook.* London and San Diego, CA: Academic, 2001.
Saferstein, R. *Criminalistics.* 8th ed. Upper Saddle River, NJ: Pearson Education, 2004.
Warlow, T. *Firearms, the Law, and Forensic Ballistics.* 2nd ed. London and Bristol, PA: Taylor and Francis, 1996.
White, P. *Crime Scene to Court: The Essentials of Forensic Science.* Cambridge: Royal Society of Chemistry, 1998.

Fires
See *Arson; Accelerant Residues*

Footwear

Footwear evidence may be recovered from crime scenes in the form of an impression in soil, as a print on a piece of paper, or perhaps as an impression of a bloody footprint on a floor or carpet surface made by the footwear tread or sole. Like all other impression evidence, the primary goal is to preserve the impression or to reproduce it for examination in the laboratory. Several recovery and enhancement techniques can be applied to achieve this. The first step is usually to photograph the print using a measurement scale to show all the details of the impression. Various side-lighting techniques may be utilized to highlight many ridge details, which can then be photographed and compared with print details from a suspect shoe. For shoe marks that are impressed in soil, the best recovery method is photography and casting of the print. A number different chemicals are available that can be used to enhance and develop footwear impressions made in blood, snow, and underwater.

The number of class and individual characteristics associated with a print determines the evidentiary value of such an impression. Characteristics that are associated with size, design, and shape can be used to illustrate that a particular shoe could have made an impression, but cannot be used to exclude other shoes with the same class characteristics as the source of the print. Individual characteristics that are present as a result of wear, cuts, gouges, or damage are much more significant in determining the source of the print. With sufficient individual characteristics providing several points of comparison, it may be possible to say that both the test impressions and the evidence originated from only one source.

An automated shoeprint identification system has been developed in England that may help the forensic scientist in making shoe-print comparisons. The Shoeprint Image Capture and Retrieval (SICAR) system uses specialized software and multiple databases to search known and unknown footwear files for comparison against footwear specimens. Impressions can be compared to a reference database to find out the type of shoe that made a specific imprint. Such impressions can also be searched in suspect and crime databases to determine if that shoeprint matches a shoeprint of someone who has been in custody or matches shoeprints found at other crime scenes.

See also Casts; Individual Characteristics; Shoe Examination

References
Bodziak, W. J. *Footwear Impression Evidence.* New York: Elsevier, 1991.
De Forest, P., R. E. Gaensslen, and H. C. Lee. *Forensic Science: An Introduction to Criminalistics.* New York: McGraw-Hill, 1983.
James, S. H., and J. J. Nordby. *Forensic Science: An Introduction to Scientific and Investigative Techniques.* Boca Raton, FL: Taylor and Francis, 2005.
Lee, H. C., T. Palmbach, and M. T. Miller. *Henry Lee's Crime Scene Handbook.* London and San Diego, CA: Academic, 2001.
Saferstein, R. *Criminalistics.* 8th ed. Upper Saddle River, NJ: Pearson Education, 2004.
White, P. *Crime Scene to Court: The Essentials of Forensic Science.* Cambridge: Royal Society of Chemistry, 1998.

Forensic Quality Services

Forensic Quality Services (FQS) is a not-for-profit corporation established by the National Forensic Science Technology Center in 2003. It is the third in the series of companies that provide quality systems support to the forensic sciences and that can trace their origins to the American Society of Crime Laboratory Directors. The most significant contribution of FQS to quality assurance in forensic science is from its accreditation programs. Provided under the title "FQS-I," these were the first forensic accreditation programs in the United States to comply with the ISO 17025 international standards for testing laboratories, the first to be offered without conditions to all forensic science laboratories, and the first to be recognized by the National Cooperation for Laboratory Accreditation (see www.forquality.org for information on FQS, and www.nacla.net for information on NACLA).

Forensic Science

Forensic science can be defined as science as it pertains to the law. That simple definition encompasses many different scientific or technical procedures and at least three basic principles. The basic principles describe the evidential use to which testing is applied. The testing may elicit *inceptive evidence,* in which the information shows that a crime has been committed. Measuring the amount of alcohol in the blood or identifying a white powder in someone's possession as cocaine are examples. Scientific testing can be used to address the identity or origin of something—a bloodstain, a fingerprint, or a paint chip recovered from a scene. If the properties of the material tested differ from those of the putative source, then a conclusion of nonidentity can be safely drawn. This is the principle of *testing for exclusion.* However, if no differences are found, the specific nature of the test results has to be considered. If it is a fingerprint, then a conclusion of identity can be made. For other evidence, the evidence may be best regarded as *corroborative.* DNA is a good example of the distinction: Some DNA testing results in a combination of types so rare as to be regarded by any reasonable measure as being unique and therefore identifying the origin of the body tissue. Others may be less compelling and leave the probability of a chance match in types as worthy of consideration.

Exclusionary and corroborative evidence are specific examples of the general class of *associative evidence*. Many of the examples of associative evidence are based on the principle enunciated by Edmund Locard in the early twentieth century, usually quoted as "every contact leaves a trace." However, the Locard principle is not a defining law for forensic science (and indeed there is no clear statement of it in Locard's writings).

When examining and interpreting associative evidence, the results must be given careful consideration. With this type of evidence, there is a link with the falsification approach to basic science (see **Daubert Ruling**). Testing may produce evidence to exclude an association. For example, DNA types in recovered sperm may differ from those of the suspect. If they do not exclude, then we are left with corroborative evidence that supports but does not prove the hypothesis of association.

The objective of forensic science is the identification, individualization, or classification of physical evidence. For some types of evidence, identification or individualization is only possible after a number of thorough chemical or scientific tests are conducted. The types of evidence that require such extensive testing to ensure accurate identification include bloodstains, body fluids, drugs, arson accelerants, and other chemicals. The identification of unknown substances or objects may be achieved by comparing their characteristics with those of known standards or references, well-established criteria, or database information. In the forensic examination of fibers and hairs, determinations, of fiber type, form, dye composition, elucidation of color, species, or anatomical origins use class characteristics for such identification. Class characteristics can place an item into a distinct class or group. However, the ultimate goal of forensic identification is individualization—to be able to say that the evidence in question originates from a specific single source, locus, scene, or person. In reality, few types of evidence can be

unequivocally individualized, such as fingerprint and DNA evidence. All other types of evidence, if appropriately identified, can be said to be consistent with originating from a particular source, site, or individual.

References
De Forest, P., R. E. Gaensslen, and H. C. Lee. *Forensic Science: An Introduction to Criminalistics.* New York: McGraw-Hill, 1983.
James, S. H., and J. J. Nordby. *Forensic Science: An Introduction to Scientific and Investigative Techniques.* Boca Raton, FL: Taylor and Francis, 2005.
Saferstein, R. *Criminalistics.* 8th ed. Upper Saddle River, NJ: Pearson Education, 2004.
White, P. *Crime Scene to Court: The Essentials of Forensic Science.* Cambridge: Royal Society of Chemistry, 1998.

Fracture Matching

Fracture lines in glass are formed when the glass is bent by an applied force. An impinging force that stretches the glass beyond its limits of elasticity will result in the glass fracturing and breaking apart. The fracturing process results in a pattern of fracture marks that are left on each fragment that, when examined closely, can provide information on the force and direction of the impacting force. This fracture information may be useful in the reconstruction of events that happened at a crime scene.

When a projectile penetrates a glass window, it leaves a characteristic fracture pattern: Cracks radiate outwards from the center of the site of penetration and encircle the site of penetration or impact. The radiating lines are called *radial fractures* and the circular lines are called *concentric fractures*. A high-velocity projectile, such as a bullet, will generally produce a round crater-shaped hole surrounded by an almost symmetrical pattern of radial and concentric cracks. The hole tends to be wider on the exit side of the glass and can be a means of determining the direction of impact. As a force pushes against one side of a pane of glass, the natural elasticity of the glass medium allows it to bend in the direction of the force. As the elasticity of the glass is exceeded, the glass begins to crack and the first fractures begin to form on the opposite side of the penetrating force and form into radial lines. The application of more force will place more tension on the front surface of the glass, resulting in the formation of the characteristic stress or fracture marks—arch-shaped striations that are present on the edge of each broken glass fragment. These marks run perpendicular or begin at right angles to one surface and curve to become nearly parallel with the opposite surface. The perpendicular edge always faces the surface on which the crack originated, so looking at the stress marks on a radial crack close to the point of impact, the perpendicular striations are always on the edge opposite the surface on which the force was applied (the side on which the first cracks begin to form). When examining concentric fractures, the stress-mark striations are always perpendicular to the edge of the glass on the side that was dealt the extreme force, or the side of impact.

The patterns of fracture and stress lines can be used to show a mechanical fit between two pieces of broken glass. Because the structure of glass is naturally amorphous, two glass objects will never break in the same way and will therefore produce different fracture-line patterns. The fracture lines and ridges are minutiae that can help in reconstructing glass fragments. Photography of groups of fracture ridges and lines can be used to demonstrate identifying points between two glass fragments. When it is possible to demonstrate a mechanical fit, the stress and fractures ridges of the two matching glass fragments will mesh or fit together. The two matching fragments can be viewed edge to edge using low-power microscopy, allowing the analyst to observe, compare, and verify the matching ridges and lines and to obtain a photographic record of the evidence.

The so-called three-R rule is a convenient means of remembering the directionality of glass fracture markings for radial and concentric cracks: *R*adial cracks form a *r*ight angle on the *r*everse side of the force.

See also Bulbs (Automobile, Examination of in Accidents); Glass

References
Curran, J. M., T. N. Hicks, and J. S. Buckleton. *Forensic Interpretation of Glass Evidence.* Boca Raton, FL: CRC, 2000.
De Forest, P., R. E. Gaensslen, and H. C. Lee. *Forensic Science: An Introduction to Criminalistics.* New York: McGraw-Hill, 1983.
James, S. H., and J. J. Nordby. *Forensic Science: An Introduction to Scientific and Investigative Techniques.* Boca Raton, FL: Taylor and Francis, 2005.
Saferstein, R. *Criminalistics.* 8th ed. Upper Saddle River, NJ: Pearson Education, 2004.
White, P. *Crime Scene to Court: The Essentials of Forensic Science.* Cambridge: Royal Society of Chemistry, 1998.

Frye **Rule**

The skills of a forensic scientist incorporate the application of scientific principles and techniques to the analysis of many different types of evidence that is recovered from a crime scene. In applying and conducting their scientific tests and methodologies forensic scientists must also be aware of the evidentiary demands and constraints imposed by the judicial system. Any analytical or scientific techniques, principles, or methods used in the investigation and examination of evidence not only must be scientifically sound, but must also satisfy the criteria of admissibility imposed by the judicial system. The guiding principles that were adopted as standard guidelines on the judicial admissibility of scientific evidence, in thirty-one states in the United States, were first presented in 1923, in the case of *Frye v. United States,* when the Court of Appeals for the District of Columbia rejected the scientific validity of the polygraph test:

Just when a scientific principle or discovery crosses the line between the experimental and demonstrable stages is difficult to define. Somewhere in this twilight zone the evidential force of the principle must be recognized, and while courts will go a long way in admitting expert testimony deduced from a well-recognized scientific principle or discovery, *the thing from which the deduction is made must be sufficiently established to have gained general acceptance in the particular field in which it belongs* (293 F. 1013, at 1014, emphasis added).

For any procedure, scientific technique, or principle to meet *Frye* standards, the courts must first determine if the procedures or methods in question are generally accepted by an appropriate segment of the scientific community. In the process of a *Frye* challenge, an advocate for the test presents a series of appropriate experts to the court, who testify that the questioned test is scientifically reliable, meets sound scientific principles, and is accepted by that scientific field or community. Over the years, there has been much discussion and debate on the *Frye* principle and its flexibility to cover the many new scientific issues and techniques that are now being accepted in the scientific community. In a significant ruling made in 1993, in the case of *Daubert v. Merrill Dow Pharmaceutical Inc.,* the U.S. Supreme Court concluded that the *Frye* standard should not be an absolute prerequisite to the admissibility of scientific evidence. According to the Court, Rule 702 of the Federal Rules of Evidence provides that trial judges are now assigned the responsibility for the admissibility and validity of scientific evidence presented in their courts.

See also Admissibility of Scientific Evidence; *Daubert* Ruling

References
Daubert v. Merrill Dow Pharmaceuticals, Inc., 509 U.S. 579 (1993).
Frye v. United States, 293 F. 1013 (D.C. Cir. 1923).

G

Gacy, John Wayne

This case demonstrates the difficulties faced with identification of human remains.

John Wayne Gacy was an active member of his community and well respected for his involvement in many community groups, despite some rumors of his homosexuality. Gacy was first married in 1964 and soon after was accused of sexually harassing his employees. He was convicted in 1968 of sodomizing a teenage boy, for which he was sentenced to serve ten years in state prison. After serving eighteen months, he was released and remarried after moving back to Chicago. He bought a house, and it later came to light that both his wife and neighbors were known to complain about the stench coming from the house. He started a contracting business, through which he employed several teenage boys (supposedly to save money). Soon, however, rumors of his homosexuality reappeared and again he was divorced. In 1978 a fifteen-year-old boy was reported missing after having left to meet with a contractor who had offered him a job—John Wayne Gacy. Gacy was questioned by police who, after researching his background and discovering his prior record, returned to question him further. Soon after, Gacy voluntarily showed police the crawl space beneath his home where he was eventually found to have buried around thirty bodies.

The bodies were all young teenage boys, and so this presented a problem to investigators tasked with identifying them. Many of the

John Wayne Gacy, 36, seen in this undated file photo. Gacy, a convicted sodomist who acted as a part-time clown for neighborhood children, was charged with the slaying of a suburban Des Plains youth and is suspected in the sex slayings of as many as 32 young men and boys. December 28, 1978, Chicago, Ill. (Bettmann / Corbis)

bodies were identified using dental records, x-rays, and fingerprints. However, due to the homosexual nature of the crime, many families did not come forward to assist in the identification of their missing family members. Even after using the technique of facial reconstruction and releasing the resulting images to the media, nine bodies remained unidentified. In 1994 John Wayne Gacy was executed by lethal injection.

See also Identification

References
Court TV's Crime Library, Criminal Minds and
    Methods; http://www.crimelibrary.com/
    serial/gacy/gacymain.htm (Referenced July
    2005).
Evans, C. *The Casebook of Forensic Detection: How Science
    Solved 100 of the World's Most Baffling Crimes.*
    New York: Wiley, 1998.

Gamma Hydroxybutyrate (GHB)

Gamma hydroxybutyrate (GHB) is also known on the street as "Liquid X," "Georgia Home Boy," "Goop," "Gamma-oh," "E-Z Lay," and "Grievous Bodily Harm." This drug is a powerful, rapidly acting central nervous system depressant and putative neurotransmitter that is actually produced in the body naturally in very small amounts, but its physiological function is largely unclear. GHB was first synthesized in the 1960s as a human anesthetic and was discontinued because of intolerable side effects. The drug was used as a sleeping aid and bodybuilding supplement during the 1980s and was abused as a recreational psychoactive substance in the 1990s, and as a result was designated as a scheduled substance in the United States in 2000. GHB is encountered on the street as an odorless, colorless liquid or in white-powder form. It is used most commonly in the form of a chemical salt, which is mixed with water. The drug is generally taken orally, quite often in combination with alcohol, and is abused for its euphoric effects, hallucinations, and sedation. At low doses the drug produces physiological effects similar to those experienced with alcohol. Reports also document GHB abuse by

bodybuilders for anabolic effects, as it has an alleged ability to release growth hormone and stimulate muscle growth. At one time GHB was considered to be a safe and "natural" food supplement, and was commercially available in health food stores. A number of adverse effects are associated with GHB use, including drowsiness, dizziness, nausea, unconsciousness, seizures, severe and potentially fatal respiratory depression, vomiting, discoordination, and coma. A major concern is the fact that the recreational dose range of GHB is extremely narrow. It has a high toxicity index, and a minor overdose can cause unconsciousness; large overdoses can be fatal. At higher overdose levels, GHB can produce both unconsciousness and vomiting, an extremely dangerous combination, leading to aspiration or inhalation of vomit and possible suffocation. A recreational dose of pure GHB powder is generally between one and three grams, but frequent users who have developed a tolerance may take as much as four or five grams in a single dose. The liquid form of GHB also prevalent on the streets comes in widely variable concentrations. The onset of action of GHB is usually between ten and twenty minutes, and the main effects of GHB last for one to one and one-half hours. Secondary effects of the substance, which are generally milder, may last for a further one to two hours. Recreational users may also consume a single dose of GHB slowly over the period of an evening rather than taking a full dose at one time. GHB is most commonly produced by combining gamma butyrlactone and sodium hydroxide. Two other chemicals often used as GHB counterparts are 1,4-butanediol and gamma butyrlactone.

Young adults and teenagers at nightclubs and parties are the most common users of illicit GHB. High school students, college students, and rave partyers often use it as a pleasure enhancer that depresses the central nervous system and causes intoxication. It can also be used as a sedative, both to reduce the effects of stimulants such as cocaine, methamphetamine, and ephedrine, or hallucinogens such as LSD and mescaline, as well as for the prevention of physical withdrawal symptoms. The development of physical dependence has also been demonstrated with consistent use of high doses of the drug. There also are numerous documented cases in which GHB has been used to incapacitate women with the purpose of committing sexual assaults. GHB is illegal to possess or sell in the United States. In March 2000 GHB was placed on Schedule I of the Controlled Substances Act.

See also Controlled Substances; Date Rape; Drugs
References

Drummer, O. H. *The Forensic Pharmacology of Drugs of Abuse.* New York: Oxford University Press, 2001.

Levine, B. *Principles of Forensic Toxicology.* Washington, DC: American Association for Clinical Chemistry, 1999.

Saferstein, R. *Criminalistics.* 8th ed. Upper Saddle River, NJ: Pearson Education, 2004.

Gas Chromatography

Chromatography refers to the technique used for the analysis and separation of chemical compounds or analytes. There are many different types of chromatographic procedures that are routinely used in the forensic analysis, separation, and quantitation of chemical evidence. The types of evidence routinely analyzed by chromatographic methods include illicit drug substances, fiber dyes, inks, petroleum products, solvents, and blood-alcohol samples. Thin layer chromatography is a separation method using a thin layer of an adsorbent material such as silica gel. Liquid and gas chromatography involve the separation and quantitation of analytes based on their chemical and physical interaction between a liquid solvent or gas and a stationary phase within a column.

With gas chromatography, a form of partition chromatography, chemical separation is achieved by the partitioning of the analytes

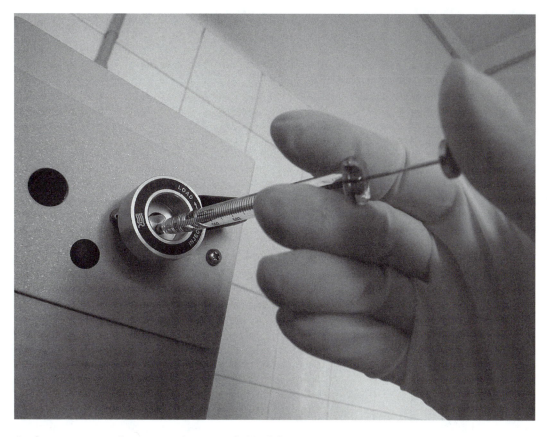

A technician injects samples into a gas chromatograph. (iStockphoto.com)

between an inert gas, called the mobile phase, and a liquid stationary phase held in place on a fine capillary column. Different compounds have different affinities for each of these two phases and partition between the two phases accordingly. At constant temperatures and gas-flow rates, different compounds will be retained by the stationary phase for different amounts of time. A compound that preferentially partitions into the stationary phase will be retained longer on the column than a compound that preferentially partitions more into the gaseous mobile phase.

One of the major drawbacks of gas chromatography is that it requires the analyte to be volatilized at an elevated temperature, a process that may break down heat-labile compounds. Many compounds may not be suitable for analysis by gas chromatography because of their lack of volatility, resulting in poor separation. By producing a chemical derivative of the drug of interest, using a suitable derivatizing agent, these problems can be circumvented. Several derivatizing reagents are commonly employed for this purpose and include silylating and acetylating agents.

An appropriate detector is chosen based on the requirements of the analysis, including sensitivity and specificity.

See also Drugs; Fibers; Mass Spectrometry
References
De Forest, P., R. E. Gaensslen, and H. C. Lee. *Forensic Science: An Introduction to Criminalistics.* New York: McGraw-Hill, 1983.
James, S. H., and J. J. Nordby. *Forensic Science: An Introduction to Scientific and Investigative Techniques.* Boca Raton, FL: Taylor and Francis, 2005.
Saferstein, R. *Criminalistics.* 8th ed. Upper Saddle River, NJ: Pearson Education, 2004.
Tebbett, I. *Gas Chromatography in Forensic Science.* New York: Horwood, 1992.
White, P. *Crime Scene to Court: The Essentials of Forensic Science.* Cambridge: Royal Society of Chemistry, 1998.

Gasoline

Crude oil is a mixture of aliphatic hydrocarbons, which can be characterized by the length of the carbon chain. Petroleum refining does this by separating the constituents according to their boiling points. The shorter-chain, more volatile hydrocarbons distill off at lower temperatures. Gasoline is the second fraction recovered after naphthas and consists of a mixture of chains of seven to eleven carbon atoms. Gasoline is volatile and flammable and, aside from its legitimate role as automobile engine fuel, is encountered as an accelerant in arson fires and as a constituent of Molotov cocktail incendiary devices. Traces of gasoline are readily detected and identified in the laboratory.

See also Accelerant Residues; Arson
References
Almirall, J., and K. Furton. *Analysis and Interpretation of Fire Scene Evidence.* Boca Raton, FL: CRC, 2004.
Nic Daeid, N. *Fire Investigation.* Boca Raton, FL: CRC, 2004.
Saferstein, R. *Criminalistics.* 8th ed. Upper Saddle River, NJ: Pearson Education, 2004.

Genetic Markers

Genetic markers used in forensic science have three essential properties related to their variability and constancy. First, they are stable in stains and other shed material—that is, they persist long enough to be measured in the laboratory and will not change from one form of the marker to another as they degrade. Second, they are stable in the body—that is, the characteristic will not alter with time and so can be compared reliably with the corresponding marker in a stain or shed material no matter when the reference blood sample is taken. Third, they show sufficient variability among individuals that they offer a degree of discrimination between people.

The first two properties are a matter of chance, depending on the chemical nature of the potential marker and the way in which it responds to environmental challenge outside of the body. The third is achieved if the marker of interest is the result of expression of a gene, because genes are stable within an individual. Mutations at a sufficiently early point in the human evolutionary tree can result in there being a number of alternate forms of any gene within the population. These alternate forms are called *alleles.*

When someone inherits the same allele from each parent, he or she is said to be *homozygous* for the inherited characteristic. When a person inherits a different allele from each parent, he or she is said to be *heterozygous.* The combination of alleles present is the *genotype* of the individual. The observed characteristic that results from the expression of the alleles is the *phenotype* of the person. There are some alleles that do not result in a measurable characteristic in the individual. These are said to be *recessive,* while those that do produce an observable characteristic are said to be *dominant.* Thus, in the ABO system, the A and B alleles are dominant, but the O allele is recessive. Two people who are type B can have a different genotype, either BB (homozygous) or BO (heterozygous), for the same phenotype of B.

See also ABO Blood Groups; Blood Grouping; DNA in Forensic Science

References

Butler, J. *Forensic DNA Typing.* 2nd ed. Burlington, MA: Elsevier, 2005.

Saferstein, R. *Criminalistics.* 8th ed. Upper Saddle River, NJ: Pearson Education, 2004.

Geology

The field of geology is one of the physical sciences and is defined as the science of the earth's history, composition, and structure, and their associated processes. Geology draws upon other scientific disciplines, such as chemistry, biology, physics, astronomy, and mathematics in its application. Geologists may play a role in forensic investigations of soil evidence. The forensic examination of soil can involve the analysis of naturally occurring rocks, minerals, vegetation, and animal matter. A geologist having knowledge of local geology may be able to help police by indicating the general area where soil may have been originally picked up and a crime committed.

Reference

Murray, R. C. *Evidence from the Earth: Forensic Geology and Criminal Investigation.* Missoula, MT: Mountain, 2004.

Glass

Glass was first made as early as 2000 BC and is an amorphous substance made of silica and borates or phosphates that have been melted and fused at extremely high temperatures. Glass also occurs naturally as a volcanic-based substance called obsidian. Glass is neither a solid nor a liquid, but exists in a vitreous, or glassy, state. Although its molecules are arranged in a noncrystalline manner, sufficient intermolecular cohesion still exists to impart mechanical rigidity. Glasses can come in a variety of forms and can be transparent, translucent, or opaque—the color varying with changes in chemical composition. Glass as a medium is a poor conductor of heat and electricity. Molten glass is plastic in nature and can be shaped using a range of techniques. At low temperatures glass is extremely brittle and will break and fracture easily. The basic ingredient of most glasses is silica, which is derived from sand, flint, or quartz. Silica can be melted at very high temperatures to form fused silica glass.

For most glasses, silica is combined with other raw materials in varying proportions. Carbonates of sodium or potassium lower the fusion temperature and viscosity of silica. Limestone or dolomite (calcium and magnesium carbonates) are added as stabilizers, and additional ingredients such as lead and borax are added to impart certain physical properties. Most manufactured glass is a soda-lime composition used to make bottles, tableware, lightbulbs, and window and plate glass. Fine-quality glass or glass crystal is made from potassium-silicate formulas that include lead oxide. Leaded glass is heavy and refracts light, making it suitable for lenses and prisms, as well as imitation jewels. Borosilicate glass contains borax as a major ingredient, along with silica and alkali. This type of glass is extremely durable, resistant to chemicals and high temperatures, and is often used for cooking utensils, laboratory glassware, and chemical processing equipment. The color of glass is usually the result of impurities that are present in the raw materials. Manganese may be added to clear or colorless glass to offset traces of iron that tend to impart green and brown tints. Glass can also be colored by dissolving in it metallic oxides, sulfides, or selenides, or by the addition of microscopic particles that become dispersed throughout the glass. Generally, most glass formulas include the addition of broken waste glass of a similar composition, called *cullet*. Cullet promotes melting and homogeneity of the new batch of glass, as does the addition of fining agents such as arsenic or antimony.

The melting temperature of a specific glass is determined by its chemical composition. Some glasses will melt around 500° C (900° F), while others won't melt until temperatures reach at least 1650° C (3180° F). The tensile strength of glass normally ranges from 280 to 560 kg per sq cm (4000–8000 lb per sq in), but can exceed 7000 kg per sq cm (100,000 lb per sq in) in glasses that have undergone specialized treatments. Specific gravity tends to range from 2 to 8, which is less than the specific gravity of aluminum but greater than that of steel.

There are many different types of glass. Window glass, which has been used since the first century AD, was first made by casting, or blowing hollow cylinders that were then slit and flattened into sheets. Today, window glass is made mechanically by drawing molten glass through a slotted block submerged on the surface of the molten glass pool, into an annealing furnace. Once the glass emerges after the annealing process, it is cut into sheets.

Because of the nature of the manufacturing process, ordinary drawn window glass is usually not uniformly thick. It is this variation in thickness that can result in distortions in the appearance of objects viewed through the glass. The use of ground and polished plate glass, or the use of float-glass manufacturing processes, can be used to correct the problem of variable thickness and, hence, distortion.

Plate glass is manufactured by rolling molten glass sheets between double rollers. After the sheet has been annealed, both sides are finished simultaneously. By using the float-glass process, flat surfaces are ensured on both sides by floating a continuous sheet of glass on a bath of molten tin at temperatures high enough to enable surface imperfections to be removed by the fluid flow of the glass against the tin.

Wire glass is a safety glass that doesn't shatter when it is struck. This glass is made by introducing wire mesh into the molten glass before it is passed between the rollers. Laminated glass is a safety glass used for vehicle windshields and is made by laminating a sheet of transparent polyvinyl butyral plastic between two sheets of thin plate glass. The plastic adheres tightly to the glass and holds the broken pieces in place when the glass is broken or smashed.

Bottles and glass containers are manufactured using an automated process that combines pressing and blowing to form the hollow body and the open end of the container. Lenses used in eyeglasses, microscopes, telescopes, and cameras are made from optical glass, which differs from other glass in the way in which it bends, or refracts, light. Optical glass is made from extremely pure materials.

Forensic Significance of Glass

Broken and shattered glass fragments that are produced in the perpetration of a crime can be used to link a suspect to that crime scene. Chips or fragments of broken glass from a window may become stuck in the perpetrator's shoes or clothing during a burglary, and particles of headlight or reflector glass found at hit-and-run scenes may be used to link a suspect car back to the locus and may be used to confirm the identity of the hit-and-run vehicle. Due to the prevalence of glass in today's environment, the evidentiary value of glass is maximized when it can be individualized to a single source. This only happens when the suspect and crime-scene glass can be assembled and physically fitted back together. This process involves piecing together the irregular edges of broken glass and matching any irregularities and striations found on the glass surfaces (see **Fracture Matching**).

Identifying Glass Particles

Glass particles can be identified by three physical characters: conchoidal fracture, amorphous structure, and isotropism. During the examination of glass and debris, a few simple tests can be applied to eliminate particles that might be confused with glass. Fragments of plastic can be separated and eliminated by testing for indentation with a needle; cubic crystals such as salt can be identified by shape, fracture, and water solubility; and mineral grains can be identified using polarized light. Under the polarized light microscope, isotropic particles remain dark, whereas anisotropic or birefringent particles show up colored or bright.

Refractive Index

In the forensic analysis of glass, one of the first measurements made is the refractive index of the glass fragments. Determination of refractive index is usually carried out using an immersion method and involves finding the temperature and wavelength at which a particular glass fragment, immersed in a liquid, has the same refractive index as that liquid. There are a number of different methods that can be applied to achieve this, including the Becke line method, dispersion staining, and phase microscopy. Most refractive index testing today is conducted using the unfortunately titled GRIM equipment (for *G*lass *R*efractive *I*ndex *M*easurement).

Density Determinations

Density gradient techniques can also be used to identify particles, including glass. A density gradient can be made by layering a number of miscible oils or liquids in a tube. The heaviest liquids are layered at the bottom, and liquids of lesser density are then sequentially added. Over time, the solution forms a uniform gradient and any particles added to the tube will settle at the level of equivalent density. There are a number of variations on this technique,

which include using a plummet and balance or a density meter.

Elemental Analysis

Elemental analysis of glass can also be conducted for forensic comparison purposes. Wet chemical analysis, neutron activation, atomic absorption, mass spectrometry, and emission spectrography are techniques that may be applicable to the analysis of silicate materials, including glass. The problems associated with these techniques result from the fact that particles of glass routinely recovered in forensic work are small, irregularly shaped, and often recovered in insufficient amounts for successful analyses. Elemental analysis for the determination of the presence or absence of detectable amounts of arsenic, cobalt, titanium, sulfur, lead, boron, zirconium, strontium, barium, iron, and chromium may be significant, as these elements may be deliberately added in small amounts to glass during the manufacturing process or be present in trace levels in the raw materials.

See also Bulbs; Fracture Matching; Refractive Index

References

Curran, J. M., T. N. Hicks, and J. S. Buckleton. *Forensic Interpretation of Glass Evidence*. Boca Raton, FL: CRC, 2000.

De Forest, P., R. E. Gaensslen, and H. C. Lee. *Forensic Science: An Introduction to Criminalistics*. New York: McGraw-Hill, 1983.

James, S. H., and J. J. Nordby. *Forensic Science: An Introduction to Scientific and Investigative Techniques*. Boca Raton, FL: Taylor and Francis, 2005.

Saferstein, R. *Criminalistics*. 8th ed. Upper Saddle River, NJ: Pearson Education, 2004.

White, P. *Crime Scene to Court: The Essentials of Forensic Science*. Cambridge: Royal Society of Chemistry, 1998.

Glue Sniffing

The recreational inhalation (also called "huffing" or "sniffing") of volatile hydrocarbons and anesthetics, including substances such as ether and nitrous oxide, first became a common activity in Europe, Great Britain, and North America in the 1800s. It was during the early 1900s that alcohol, ether, and chloroform became the most commonly inhaled substances, and during the 1940s and 1950s there was a trend toward gasoline sniffing. The 1960s heralded the onset of inhalant abuse of glues and acetone, which are still popular inhalants today.

Inhalant substances are gases or vapors that are inhaled to induce psychoactive, or mind-altering, effects. There are three main groups of inhalants: nitrous oxide, volatile nitrites, and petroleum distillates. Most inhalant abusers don't consider these kinds of substances to be harmful or deadly because most are found in commonly available household items such as cleaning fluids, modeling glues, nail-polish remover, hair sprays, paints, cooking sprays, and aerosol propellants that include freon. Toluene is one of the most common solvents that is inhaled recreationally, but other solvents such as naphtha, methyl ethyl ketones, gasoline, and trichloroethylene also produce comparable physiological effects.

Immediately after the user inhales the substance, he or she generally experiences feelings of exhilaration and euphoria, accompanied by slurred speech, impaired judgment, double vision, and, in some situations, drowsiness and stupor. These depressant effects then gradually wear off and the user returns slowly to a normal state.

Consistent inhalant abuse can significantly affect the day-to-day functions of a user. The first signs of regular abuse begin when the user starts to neglect his or her personal appearance and shows a lack of interest in school or work, with the loss of any professional or academic drive. These behaviors are commonly accompanied by the gradual withdrawal from family, friends, and social and athletic activities. Conversations with the user may become senseless and silly and often the user may exude a chemical odor on his or her breath or clothing. The user is also often in the possession of inhalant paraphernalia such as bags, discarded aerosol cans, or rags.

Among the most commonly abused inhalants, nitrous oxide is a light anesthetic gas used legally as a painkiller in medicine and dentistry. It may be present as a propellant in cans of whipping cream and is also used as a boosting agent to accelerate combustion in auto racing. Inhalation of nitrous oxide provides an intense, brief "high" that lasts for only a few minutes. Nitrous oxide does not completely block pain, but is used to relieve anxiety and create a dreamy or floating sensation for the patient.

Volatile nitrites are stimulant drugs with associated powerful cardiac effects. Amyl nitrite is prescribed for people suffering from angina pectoris and asthma and is usually dispensed in the form of a clear yellow liquid in single-dose glass ampoules. The ampoule is crushed between the fingers and the vapors of the drug inhaled. The pharmacological effects of the drug cause vasodilation and an increased heart rate. The effect of sudden peripheral vasodilation reduces blood flow and oxygen to the inner regions of the brain, resulting in the onset of sudden, intense weakness and dizziness that may last from thirty to sixty seconds.

As a means of circumventing current drug regulations, butyl nitrite products can often be sold in bars and specialty shops as room deodorizers called "Locker Room" or "Rush." The pharmacological effects of this drug are identical to those of amylnitrite and include brief euphoria, changes in blood pressure, flush, dizziness, and headaches. Again, the euphoric effects have a rapid onset and dissipate quickly. More experienced users try to maintain or prolong the euphoric effects by using increasing doses of the drug.

Petroleum distillates are gases or vapors of liquids made from petroleum. They are inhaled or abused for their powerful depressant, intoxicating, and mind-altering effects. Commonly abused sources of petroleum distillates include gasoline, lighter fluids, kerosene, toluene, naphtha, model airplane glue, typewriter correction fluids, magic markers, and aerosol gases used as propellants in cooking sprays, spray paints, and cleaning agents.

Although there is some debate on the issue, many experts believe that these substances are physiologically addictive. Addictive or not, inhalant abusers undoubtedly have a high risk of inducing liver, heart, and brain damage or even death from the chemicals they inhale. Inhalant abuse has also been associated with some other serious side effects, including central nervous system (CNS) effects, such as chronic tiredness; depression, irritability, and poor balance or lack of coordination; nervous system and brain damage; and nosebleeds, headaches, eye pain, anemia, muscle and joint pain, loss of muscle control, and sores around the nose and mouth.

In some states glue sniffing is considered a Class B misdemeanor. A person is considered guilty of glue sniffing if he or she inhales or ingests "model glue, or any substance containing toluene, acetone, benzene, N-butyl nitrite, or any aliphatic nitrite," for the purposes of becoming intoxicated. It is also considered a Class B misdemeanor to distribute nitrous oxide for the purpose of producing intoxication.

See also Controlled Substances; Drugs
References
Saferstein, R. *Criminalistics*. 8th ed. Upper Saddle River, NJ: Pearson Education, 2004.
Winger, G., F. G. Hofmann, and J. H. Woods. *A Handbook on Drug and Alcohol Abuse: The Biomedical Aspects*. 3rd ed. New York: Oxford University Press, 1992.

Gm and Km Typing

In the late 1970s and early to mid-1980s, just before DNA typing was discovered, attempts were made to increase the discriminating power of tests used to type blood and its stains by adding to the range of genetically determined protein markers uses. One such example that showed initial promise was the polymorphisms on the gamma and kappa chains of immunoglobulins. These are two of the protein structures that make up antibody molecules. They were found to have genetically determined variations in structure that could be identified by reactions with antibodies. Those on the gamma chains were named Gm and those on the kappa chains were named Km.

However, the reliability of the testing never achieved the standards required for routine application in forensic science, and no further research was conducted due to the introduction of successful DNA typing systems.

See also DNA in Forensic Science
Reference
Butler, J. *Forensic DNA Typing*. 2nd ed. Burlington, MA: Elsevier, 2005.

Gun Shot Residues

See *Firearms*

Guns

See *Firearms*

H

Haigh, John George

Police in London, England, became suspicious when John George Haigh, together with several rich elderly women, reported the disappearance of Olivia Durand-Deacon, another wealthy older woman. Initial checks revealed that Haigh had prior convictions for forgery, conspiracy to defraud, obtaining money by false pretences, and fraud. Further investigations led police to Haigh's place of business, where they found such items as containers for storage of acid, rubber boots, rubber gloves, and a gas mask. A dry-cleaning receipt was also found, which police were able to trace back to Olivia Durand-Deacon. When questioned by police, Haigh initially attempted to explain away the receipt, among other things, but eventually confessed to one detective what he had done. Haigh had dissolved Olivia Durand-Deacon in acid, but boasted to the detective that without evidence they could not charge him with anything. He wrote police a complete statement detailing that he had killed eight other people in the same way, saying that he had an uncontrollable desire to drink their blood. When the remaining "sludge" was examined, recovered were three gallstones, twenty-eight pounds of human fat, part of a left foot, eighteen human bone fragments, and a set of dentures. Haigh was sentenced to be hanged in the summer of 1949.

References
Court TV's Crime Library, Criminal Minds and Methods; http://www.crimelibrary.com/serial/haigh/ (Referenced July, 2005).
Owen, D. *Hidden Evidence: Forty True Crimes and How Forensic Science Helped Solve Them.* Willowdale, Ontario: Firefly, 2000.

Hair

Hairs are commonly encountered as physical evidence and can count as crucial evidence in any type of case. Traditionally, they have particular importance in cases of personal contact, including rape and homicide, as associative evidence. Human head hair and pubic hair are commonly encountered, but hairs from other body regions are encountered occasionally. Animal hairs as evidence can come from domesticated animals, pelts and furs used for clothing, and wild animals or game. With the advent of refined and more sensitive analytical methods, the presence of cellular materials adhering to the root and hair shaft can be a source of samples for DNA analysis, which can be further utilized as a means of personal identification. New developments in chemical analysis can now be exploited to illustrate patterns or timelines of drug use or poison exposure in hair samples obtained for toxicology studies.

Anatomy of Hair

Hair is an appendage of mammalian skin. On some animals the hair serves to keep the animal warm and to protect it from environmental assault. The hair shaft that protrudes above the skin surface is made of a strong structural protein called *keratin*—the same protein that makes up the nails and the outer layer of skin. There are three types of hair associated with humans. *Lanugo* hairs are formed in the embryo beginning at five months gestation, and are shed in the seventh to eighth month of gestation. Lanugo hairs, which are fine, soft, and nonpigmented, are similar to, but often thicker and longer than vellus hairs. *Vellus* hair, which replaces lanugo prior to birth, is colorless and very fine and is typically located on the forehead and balding scalp. *Terminal* hair is the type of hair we normally consider when talking about human hairs; these hairs, which replace some areas of lanugo and vellus hair after birth, are usually pigmented, medullated, and are coarser and thicker than vellus hair. These hairs are the same as the underhairs, wool, and fur of other mammals.

Each strand of hair consists of three layers constructed like that of a pencil in cross section. The first or inner layer is composed of a core of condensed cells and is called the *medulla*. It is usually present only in large, thick hairs. During the formation of the medulla, medullary cells collapse, resulting in the appearance of a cellular network with spaces and gaps that are filled with air. The medullary appearance can be distinctive between different species of mammals.

The second or middle layer is called the *cortex* and provides tensile strength, as well as color and texture. The cortex is made of spindle-shaped cells that form a core surrounding the medulla. Cortical cells may contain some nuclear remnants and pigment granules, but they are mainly filled with keratin microfibrils embedded in a matrix of sulfur-rich proteins. In some hairs small structures called *cortical fusi* may appear in the cortex. These structures, which may be confused with pigment granules, are actually spindle-shaped air inclusions. Their distinctive shape can be used to identify them from the rounded pigment granules. In colored hairs, the cortex also houses the pigment granules. These ellipsoidal granules contain the pigment melanin. Two types of melanin impart hair color; black-brown eumelanins give dark colors and red-yellow phaeomelanins give lighter colors. There are fewer pigment granules in graying hair and none in white hair.

The third or outer layer is called the *cuticle,* a thin layer of flattened, overlapping, colorless cells that protect the cortex. The cuticle cells slope outwards with their edges pointing toward the tip of the hair. The overlapping cuticle cells form a distinctive pattern, called the scale pattern, which can be observed microscopically. The pattern varies between species and can be used for species identification. The cuticle surface of a fully developed hair is very hard when it is dry and imparts protection to the cortex. Cuticles can become damaged as a result of environmental assault, combing, brushing, washing, and abrasion. Because the hair shaft is composed of dead material and is unable to repair itself, the majority of damage will be found closer to the tip than to the root area.

The root of the hair lies below the surface of the skin and is housed within a hair follicle. The estimated total number of hair follicles for an adult human is in excess of 5 million, with approximately 1 million of those follicles distributed on the head. At the base of each active hair follicle is a pear-shaped organ called the *dermal papilla.* In normal hair follicles the dermal papilla consists of a highly active group of cells responsible for the production of a hair. The dermal papillae are surrounded by cortical or matrix epidermal cells, which divide and proliferate rapidly during the growth phase of the follicle. As the cells proliferate and the follicle grows, the epidermal cells are pushed up the follicle to produce the hair shaft that protrudes above the skin surface. The epidermal cells undergo differentiation to produce the keratinized hair fiber. The natural growth process of the follicle causes the regular shedding of hairs. It is this periodic shedding that results in a person's hair being found on clothing and at crime scenes.

The part of the hair shaft that remains inside the follicle, below the skin surface, is called the *root*. The root is surrounded by an inner root sheath and an outer root sheath. The inner root sheath is also made of three layers (the cuticle, Huxley layer, and Henle layer) and stops at the level of the sebaceous gland, leaving only the hair cortex and the surrounding cuticle to protrude above the epidermis. The outer root sheath is distinct from other epidermal components of the hair follicle and is continuous with the epidermis. The outer root sheath is the site of attachment for the erector pili muscle. When the erector pili muscle contracts, the hairs are erected or stand on end, and goose bumps appear on the skin. A glassy membrane called the *basement membrane* separates the outer root sheath and the dermal sheath by providing a physical dividing line between epidermal and dermal cells. The basement membrane also acts as a physical barrier, which is vital to our immunological protection. Hair follicles grow in repeated cycles, and can be divided into three distinct phases. Each hair passes through the phases independent of the neighboring hairs.

Anagen (Growth Phase)

Approximately 85 percent of all hairs are in the growing, or anagen, phase at any given time. This active growth phase can vary from two to six years. Hair grows approximately 10 cm (4 inches) per year but any individual hair is unlikely to grow to lengths any greater than 1 meter (3 feet). During anagen, the follicle is at full size and maximally biochemically active.

Catagen (Transitional Phase)

At the end of the anagen phase, hairs enter into the catagen phase, which lasts one or two weeks. During the quiescent or regressive catagen phase, the hair follicle shrinks to about one-sixth of the normal length; the lower part is destroyed, and the dermal papilla breaks away. During this phase metabolic activity and hair growth begins to slow down.

Telogen (Resting Phase)

The telogen, or resting, phase follows the catagen phase and normally lasts from five to six weeks. During this time metabolism and hair growth are arrested; the hair does not grow but stays attached to the follicle, while the dermal papilla remains in a resting phase below. Approximately 10 to 15 percent of all hairs are in this phase at any given time.

At the end of the telogen phase, the hair follicle reenters the anagen phase. The dermal papilla and the base of the follicle join together again and a new hair begins to form. If the old hair has not already been shed, the new hair pushes the old one out, and the growth cycle starts all over again.

Forensic Examination and Analysis

The forensic investigator is often asked to associate recovered hairs with an individual. This is a difficult task that can't be done with any degree of certainty unless DNA analysis is possible. However, known and questioned hairs can be unequivocally excluded from having a common origin if they are found to be significantly different.

Collection

Hairs are most often collected by visual inspection, as they are large enough to be seen by the human eye, but special collection methods might be required to meet the needs of special cases. For example, the collection of pubic hairs from a sexual assault victim and the suspect (if in custody) requires visual inspection for loose hairs, hair combings, as well as plucked hairs for comparison purposes.

Control Samples

If known head hairs are submitted for comparison, then representative hair samples should be obtained from all over the scalp. Hairs should not be cut; rubbing, massaging, or combing the scalp will yield about thirty telogen hairs. If the suspect sample contains anagen hairs, then plucked hairs must be obtained for comparison. Each sample should be packaged and labeled separately.

Laboratory Examination

Microscopy is the main technique used in the forensic examination of hairs. The first step is to determine that the sample is really hair. This is carried out by a quick microscopic examination for the presence of cuticle scales, medulla, and pigment granules. The next step is to determine whether it is of human origin or from some other species. Determination of human origin is based on the relative size and appearance of the medulla, the appearance of the scales, and the diameter of the hair shaft. Most animal hairs have a smaller diameter and larger medullary index than human hairs. The medullary index is the ratio of the medulla to that that of the overall hair shaft, which for human hairs is about one-third, but one-half or greater for most other animals. The medulla of most animal hairs is more regular or geometrically patterned than human hair, which tends to have no regular structure. The tips of human hair may be blunt from cutting and styling processes, whereas animal hair tends to be more pointed. Cuticular scales that extend from the shaft of an animal hair tend to be more uniform in shape and size and are arranged in more regular patterns than those of human hairs.

Gross characteristics of hairs can also provide information on racial origin. However, overlap of features can also make racial identification impossible, especially in cases that involve mixed racial background where hairs may have a blend of characteristics or show predominant characteristics of a particular race. The characteristics that are most obvious in race determination include pigmentation, cuticle thickness, hair diameter, the cross-sectional shape of the hair, and the degree of wave or curl.

Common Racial Characteristics include:

- Negroid hairs tend to be deeply pigmented, with a coarse clustering of pigment granules. The hairs are usually tightly curled, due to having flattened elliptical cross-sections.

- Mongoloid hairs tend to be very straight with circular cross-sections. The hairs are deeply pigmented with thick cuticles.
- Caucasian hairs come with a broad range of colors, pigment densities, and shades. The cross-sectional shape, degree of curl, and range of variability of the hair diameter are partway between those of Negroid and Mongoloid.

Hairs from one area of the human body differ from hairs from other areas. Head hairs grow to the longest length of human hairs, but generally do not exceed one meter (3 feet) in length. Because human head hair is often cut at regular intervals, it is quite uncommon to find a head hair that naturally tapers at the tip. The tips of human hairs are usually abraded or cut. The length of scalp hair can vary enormously on an individual. Finer scalp hairs may not have a medulla, but when a medulla is present it can be fragmented or continuous. Medullae tend to be more prominent in coarser hairs.

Pubic hairs are generally shorter and coarser than head hair. They tend to be curlier, with flatter cross-sections, and the medulla usually continuous and very broad, which may impart the wiry characteristic of these body hairs. Pubic hairs tend to show fluctuations in diameter, and the tip may be somewhat tapered but abraded.

Eyebrow and eyelash hairs are usually short and curved with finely tapered tips. These hairs usually have a thinner diameter than pubic hairs and are usually some of the darkest hairs on the body.

Axillary hairs from the underarm are very similar to pubic hairs with respect to coarseness, medullation, twisting, and curling. These hairs tend to show less fluctuations in diameter than pubic hairs.

Morphological Comparison by Microscopy
Microscopic examination of hair usually includes the comparison of color, diameter range, degree of twist or curl, pigment granule distribution, and size or shape of granules. Because nearly all these factors can

vary widely throughout a sample or even in a single hair, the overall objectivity of a comparison tends to be quite limited. Even after a lengthy comparison is made, and the investigator is unable to determine two separate origins, the results are usually reported as "the two hairs could have had a common origin."

Genetic Markers

Some genetic markers are present in hair and can also be helpful in the comparison of samples. ABO blood grouping and enzymatic analysis is theoretically possible in hair samples that contain root sheaths. For this reason, plucked anagen samples are necessary when collecting test samples. However, this type of analysis has not been quite as successful in practice, partly because most evidentiary hairs are recovered as loose hairs in the telogen phase. Hairs in the anagen phase can be used for mitochondrial DNA analysis, which is addressed in the DNA entries of this book.

The presence of an external root sheath can also help determine the gender of a person. These cells can be removed and stained to determine the presence of Barr bodies—cellular structures associated with the nuclei of female cells. If Barr bodies are found in a high percentage of cells, the donor is most likely female. Again, this determination is not clear-cut because male hairs may show a small percentage of Barr bodies. Another method for gender determination involves staining the cells with the fluorescent dye quinacrine. This stain highlights the Y chromosome of freshly formed cortical cells in the hair root.

Elemental Analysis

A large number of chemical elements have been identified in human hair. Those present in the highest concentration tend to be the essential elements, which play a significant physiological role in the body as catalysts and biochemical cofactors. These elements include sodium, potassium, phosphorus, sulfur, and zinc. Other elements that are considered nonessential are thought to be ingested in some form as environmental contaminants and may include toxic elements such as arsenic, cadmium, mercury, and lead. Although there is significant variability in the elemental composition of hairs among individuals and on a single individual, studies have shown that there is little variability between the physiological elements. However, it is thought that significant variability in the presence of nonessential elements may allow for individualization of a hair sample. From a forensic standpoint, the deposition of chemical elements from the blood into a hair shaft that grows about 1 cm (0.4 inch) each month could provide a record of the elemental status of an individual over a determined time period.

However, the value of chemical profiling and elemental analysis of hair for forensic purposes is a highly debatable issue. Skepticism stems from the fact that elemental tests are traditionally clinical diagnostic tests. There are also a number of published limitations and problems associated with inconsistency in results from intra- and interlaboratory studies; difficulties in interpretation of results because of variation in factors such as age, gender, color, and anatomical region of hair growth; and problems associated with control and evaluation of external contamination and lack of correlation between elemental concentrations in hair and metabolically important or related tissues. Factors affecting elemental composition of hair are as follows:

Gender

The effects of gender on the elemental composition of hair are largely unclear. Some studies have suggested that there is no difference in elemental concentrations in hairs from either gender. Other studies have indicated that there were higher concentrations of trace elements in the hairs of female test subjects compared to those of male test subjects, which may be attributable to differences in organic content between male and female hair.

Age

From a forensic perspective there seem to be no clear trends in the elemental composition of hair that can be attributed to age.

Color

The relationship between color and elemental composition of hair is very complex. Studies have suggested that blonde hair tends to have higher levels of zinc than brown and black hair, but also has lower levels of other elements than darker hair. Differences were also found between brown, black, and red hair.

Racial Origin

The effect of racial origin on the elemental composition of hair is equally unclear, although it has been proposed that racial differences in hair type and color may be the reason for observed differences in elemental composition.

Although it has been established that there are indeed population differences in the elemental composition of hair, the credibility of elemental hair testing is undermined by the uncertainty of the source of the elements identified. Are the elements measured a result of endogenous elements supplied to the hair follicle by the abundant vascularization of the area, being deposited and incorporated into the growing hair? Or, is there an exogenous route of exposure due to environmental factors? For example, hair products can contain zinc, mercury, and selenium, including heavy metals in hair colorings and dyes. Also, it has been demonstrated that substances may enter and permeate the hair shaft via the edges of the cuticle, and that this may provide a route for entry of chemicals present in soil and dust residue that may settle on the hair. Several studies have shown longitudinal differences in elemental composition that may be related to environmental and not dietary exposure to certain elements in structurally altered hair. It is the uncertainty of whether an elemental profile is a result of both endogenous and exogenous elemental exposure that hinders the use of elemental profiling for forensic purposes.

Drug Analysis of Hair

Urine is the most commonly utilized biological specimen for drug testing. However, the acquisition of the sample is generally considered invasive and the sample itself is susceptible to adulteration and alteration. Because most drugs are eliminated in the urine within two or three days after administration or use, urine as a test matrix is considered to provide limited information on drug exposure.

Like trace elements, which have been shown to accumulate in hair over time, illicit and licit drug substances are also thought to pass into and be incorporated into the matrix of the growing hair. Although the actual mechanism involved in drug absorption and incorporation has not been fully determined, recent finding suggest that drugs may be incorporated into hair in a variety of ways. One proposed mechanism of drug incorporation suggests that drugs present in the circulating blood may passively diffuse from the highly vascularized hair follicle into the growing hair shaft. Another mechanism suggests that drugs may passively diffuse into the hair shaft via secretions from the sebaceous glands and from the external environment. Others have suggested drug uptake into hair is a result of metabolic processes and membrane transport systems in conjunction with the drug's affinity for the pigment melanin. Whatever the mechanism, hair can potentially provide a record of a person's drug history and may be a more appropriate matrix for the measurement of drug exposure. The use of hair as an analytical sample for the determination of the presence of drugs of abuse has several advantages compared to the use of other body fluids:

- The hair sample for testing can be obtained without using invasive techniques.
- Acquisition of a test hair sample doesn't require specialized facilities or training.
- The sample obtained for analysis is less prone to adulteration or alteration.
- The stability of drugs in hair has been demonstrated.
- Hair samples do not need specialized storage after being obtained for analysis.

- Because the growth rate of hair is approximately 1 cm (0.4 inch) per month, the sample may provide a history or long-term record of drug use, holding several weeks, or months, worth of information.

Like other types of forensic hair examination, a sample size of 50 to 100 hairs is recommended for analysis. Hair from the crown of the head is considered to be the best analytical sample; however, hairs from other head regions, as well as different body regions, have been successfully tested.

As with the elemental analysis of hair, the reliability of forensic hair testing is steeped in controversy. The main concern is that the interpretation of data obtained from analysis can be very difficult. This is based on the consideration that drugs sequestered in different body compartments may be selectively released to the tissues surrounding the hair follicle, while others are not readily released from tissue compartments. This would result in some drugs being easier to detect than others, which would in turn provide inconsistent histories of drug use and abuse. This validates the need for further elucidation of the mechanisms involved in the incorporation of drugs by hair.

The same factors that affect trace-element incorporation into hair—such as age, gender, hair color, and disease state—are also considered to affect drug incorporation. Concerns also lie with certain practicalities of hair analysis, such as the lack of standard reference materials for the validation of techniques, the lack of standardized techniques and procedures for the analysis of hair samples, and the fact that the relative abundance of drugs and their metabolites in hair is still unknown.

In the laboratory the first test applied to hair analysis, back in the 1970s, was immunoassay. Immunoassay methods are now available for heroin/morphine, phencyclidine, cocaine, marijuana, and benzodiazepines. Prior to assaying test hair samples, the samples must first be decontaminated by washing with ethanol and a phosphate buffer to remove surface materials. Drugs are then extracted from the hair matrix using solvent-based extraction methods that involve boiling in solvents for extended periods of time, acid-base extractions, or alkali digests of the hair sample. Other enzymatic digestions using the enzyme protease have also been suggested as appropriate extraction techniques.

Once a hair sample is decontaminated and the hair extracted, the same analytical techniques that are applicable to blood and urinalysis can be applied to hair analysis. Such techniques include immunoassay, thin layer chromatography, high performance liquid chromatography, gas chromatography, gas chromatography mass spectrometry, and Fourier transform infrared analysis. Fluorescent polarization immunoassay also is often used for drug-screening purposes.

See also Comparison Microscope; Genetic Markers; Microscope

References

De Forest, P., R. E. Gaensslen, and H. C. Lee. *Forensic Science: An Introduction to Criminalistics.* New York: McGraw-Hill, 1983.

James, S. H., and J. J. Nordby. *Forensic Science: An Introduction to Scientific and Investigative Techniques.* Boca Raton, FL: Taylor and Francis, 2005.

Robertson, J. *Forensic Examination of Human Hair.* London: Taylor and Francis, 1999.

Saferstein, R. *Criminalistics.* 8th ed. Upper Saddle River, NJ: Pearson Education, 2004.

White, P. *Crime Scene to Court: The Essentials of Forensic Science.* Cambridge: Royal Society of Chemistry, 1998.

Hallucinogens

Hallucinogenic compounds are a group of drugs that have no prescribed medical use and are abused or used recreationally to induce perceptual alterations. These drugs are more correctly termed *psychomimetics,* drugs that simulate natural psychoses. These drugs are also associated with physiological effects such as intense and variable emotional changes; ego distortions; and thought disruption accompanied by marked changes in normal thought processes, perceptions, and moods. Drugs in this group include lysergic acid diethylamide (LSD), which was first created in a laboratory setting in the 1940s. This drug is considered the prototypical hallucinogen because of its extensive use and its extensive study. Mescaline, found in the Lophophora cactus and used by North and Central American Indians in the form of cactus buttons, and psilocybin, found in a number of species of mushrooms and used as magic mushrooms, share some chemical and pharmacological features even though they are chemically distinct compounds. LSD is a semisynthetic drug related to the ergot alkaloids, whereas mescaline and psilocybin are phenylethylamine and indolethylamine derivates that are naturally occurring compounds. The three compounds also chemically resemble three main neurotransmitters: norepinephrine, serotonin, and dopamine.

Phencyclidine, also called PCP and "Angel Dust," is a central nervous system (CNS) depressant that causes hallucinogenic effects. It was first used as a dissociative anesthetic back in the 1950s. These drugs were utilized to make patients insensitive to pain by separating their body responses and functions from their minds without inducing unconsciousness. PCP was first used in the 1960s and had the reputation of being a "bad" drug. PCP became widely accepted in the 1970s and quickly became one of the most commonly used hallucinogens. Ketamine replaced PCP as an anesthetic for use in humans, but it also produces hallucinogenic effects.

Other hallucinogenic compounds include marijuana, which is a depressant drug, and so-called Ecstasy, which has pharmacological effects like amphetamine; both can cause hallucinogenic effects when used in very high doses.

Hallucinogens have powerful mind-altering effects and have the ability to change how the brain perceives time, everyday reality, and the surrounding environment. These drugs affect regions and structures in the brain responsible for coordination, thought processes, hearing, and sight, and consequently cause people who use them to hear voices, see images, and feel sensations that do not exist. Although the mechanism is unclear, some people who use these compounds develop chronic mental disorders. PCP is considered to be an addictive drug, but LSD, psilocybin, and mescaline are not.

Another group of hallucinogenic agents that promote the same physiological effects as LSD are the hallucinogenic amphetamines that have no approved medical use. These synthetic substances are chemically related to amphetamine and mescaline. Dimethyltryptamine (DMT) was first synthesized in 1931 and demonstrated to be hallucinogenic in 1956. This compound is present in many different plants and can be synthetically produced. It has been used in a few common hallucinogenic snuffs and intoxicating beverages, but is ineffective when taken orally unless combined with another drug that inhibits its metabolism. Generally, it is sniffed, smoked, or injected. A number of other hallucinogens have very similar structures and properties to those of DMT. Diethyltryptamine (DET) is a less potent analogue of DMT that produces the same pharmacological effects. Methyldimethoxymethyl-phenethylamine is also called STP ("serenity, tranquility, and peace") on the streets. Alphaethyltryptamine (AET) is another tryptamine hallucinogen that was recently added to the list of Schedule I substances in the Controlled Substance Act (CSA) of the United States.

See also Controlled Substances; Drugs; Lysergic Acid Diethylamide (LSD); Phencyclidine, Phenylcyclhexyl, or Piperidine (PCP)

References

De Forest, P., R. E. Gaensslen, and H. C. Lee. *Forensic Science: An Introduction to Criminalistics.* New York: McGraw-Hill, 1983.

James, S. H., and J. J. Nordby. *Forensic Science: An Introduction to Scientific and Investigative Techniques.* Boca Raton, FL: Taylor and Francis, 2005.

Saferstein, R. *Criminalistics.* 8th ed. Upper Saddle River, NJ: Pearson Education, 2004.

White, P. *Crime Scene to Court: The Essentials of Forensic Science.* Cambridge: Royal Society of Chemistry, 1998.

Winger, G., F. G. Hofmann, and J. H. Woods. *A Handbook on Drug and Alcohol Abuse: The Biomedical Aspects.* 3rd ed. New York: Oxford University Press, 1992.

Handwriting

The most requested type of forensic document examination is handwriting comparison. Handwriting comparison, and hopefully, the identification of the writer, is based on the assumption that no two people have the same handwriting characteristics. Each person's writings reflect the anatomy of the hand; the manner in which he or she learned to write; and the effects of writing habits developed over the years, which eventually become consistently and unconsciously reproduced in his or her writings. Even if a person tries to disguise his or her writing, characteristic peculiarities and traits will still be seen. The careful comparison, analysis, and interpretation of the class and individual characteristics of questioned and known or exemplar writings can often determine whether a questioned writing and known standard were in fact written by the same person.

Class characteristics associated with handwritings are gross features that include the main features and shapes of letters, the relative dimensions of the letters, connections within and between letters, the style of letter capitalization, and styles of punctuation.

Individual characteristics are less obvious features and include the characteristic forms of individual letters, the nature of pen strokes, and flourishes. The individual characteristics of a writer tend to develop over a long period of time and are very difficult to duplicate. In most cases, very similar writings can be differentiated by the comparison of these individual characteristics.

The features of individual and class characteristics can be affected by many factors, such as the speed at which a document is written and the physical condition of the writer. Fast writing speeds tend to result in an increase in the slant of the writing, with an accompanying simplification of the letters. In quickly written documents there may also be visible differences in emphasis and an increased illegibility toward the end of individual words. During a forensic examination, the investigator must also be alert to possible signs of deception.

Signs of deception include broken or deliberate strokes, wavering lines, retouching of individual letters and parts of letters, changes in signs of speed, accentuation at the end of words, and an overemphasis and deliberation associated with initial strokes at the beginning of script or individual words. When the investigator has sufficient and appropriate writing samples available for comparison purposes, attention and consideration should be given to other important individual characteristics such as the spacing and alignment of the script, the use of margins, common spelling, phraseology, and grammatical details.

To provide a correct and valid interpretation of handwriting analyses, the investigator must have sufficient and appropriate control samples for comparison purposes. These control samples are usually obtained from two sources—collected writings and requested writings.

Collected writing refers to written materials collected from a period prior to the onset of the investigation. Collected writings are prepared in the normal course of daily activities and are expected to illustrate the normal writing habits and individualities of the author. Possible sources of collected writings may include business and personal correspondence, canceled checks, school records, and handwritten applications.

Requested writing refers to written materials specifically requested as part of an investigation for use as comparison samples. These kinds of samples ideally will duplicate, as closely as possible, the paper type, ink, and writing implement used in preparation of the questioned document. In obtaining this type of sample, the investigator generally dictates what needs to be written.

To make a positive handwriting identification, the investigator must find sufficient individual characteristics in both the questioned and control scripts. In many cases, inconclusive results may arise when the questioned writing sample is too limited, distorted, or disguised; if there are insufficient or improper standards for the investigator to conduct a thorough comparison; or if a questioned or control sample fails to show sufficient individual characteristics.

See also Document Examination

References
De Forest, P., R. E. Gaensslen, and H. C. Lee. *Forensic Science: An Introduction to Criminalistics.* New York: McGraw-Hill, 1983.
Hilton, O. *Scientific Examination of Questioned Documents.* Boca Raton, FL: CRC, 1993.
James, S. H., and J. J. Nordby. *Forensic Science: An Introduction to Scientific and Investigative Techniques.* Boca Raton, FL: Taylor and Francis, 2005.
Saferstein, R. *Criminalistics.* 8th ed. Upper Saddle River, NJ: Pearson Education, 2004.
Vastrick, T. W. *Forensic Document Examination Techniques.* Institute of Internal Auditors Research Foundation, 2004.
White, P. *Crime Scene to Court: The Essentials of Forensic Science.* Cambridge: Royal Society of Chemistry, 1998.

Haptoglobin

Haptoglobin is a protein found in the bloodstream to which traces of free hemoglobin in the blood are strongly bound. There are two main forms of the haptoglobin molecule, type 1 and type 2. The form found in the blood of an individual is inherited from the person's parents; thus haptoglobin was of interest to forensic serologists as a typing system. Its avidity for hemoglobin provided the basis of a very sensitive detection system, although its discriminatory power was very limited. Like most other traditional systems, haptoglobin is no longer used in forensic science testing.

Reference
Butler, J. *Forensic DNA Typing.* 2nd ed. Burlington, MA: Elsevier, 2005.

Hashish

Hashish is the name given to the sticky resinous material secreted from the plant *Cannabis sativa.* This resin can be extracted from the plant by soaking in alcohol and then dried and compressed into balls, cakes, or cookies for sale on the illicit drug market. Pieces of the cakes are broken off and placed in pipes and smoked. Hashish can also be obtained on the streets in the form of compressed vegetation containing a high percentage of resin. Hashish preparations average about 3.5 percent in tetrahydrocannabinol (THC) content. THC is the pharmacologically active substance that is mainly responsible for the hallucinogenic properties of cannabis. Another commonly encountered form of hashish is liquid hashish; a dark green, viscous tarlike substance. Liquid hashish is also produced by extracting the THC-rich resin from the marijuana plant with an appropriate solvent. Liquid hashish has a variable and higher THC content ranging from 20 percent to 65 percent. This substance is very potent and one drop can cause a "high." Liquid hashish can be used by adding a drop to a regular or marijuana cigarette before smoking.

The main suppliers of hashish include several countries in North Africa, Pakistan, and Afghanistan.

See also Cannabis; Controlled Substances; Drugs; Marijuana

References
De Forest, P., R. E. Gaensslen, and H. C. Lee. *Forensic Science: An Introduction to Criminalistics.* New York: McGraw-Hill, 1983.
James, S. H., and J. J. Nordby. *Forensic Science: An Introduction to Scientific and Investigative Techniques.* Boca Raton, FL: Taylor and Francis, 2005.
Office of National Drug Control Policy, Drug Facts, Marijuana; http://www.whitehousedrugpolicy.gov/drugfact/marijuana/index.html (Referenced July 2005).
Saferstein, R. *Criminalistics.* 8th ed. Upper Saddle River, NJ: Pearson Education, 2004.
U.S. Drug Enforcement Administration, Drug Descriptions, Marijuana; http://www.dea.gov/concern/marijuana.html (Referenced July 2005).
White, P. *Crime Scene to Court: The Essentials of Forensic Science.* Cambridge: Royal Society of Chemistry, 1998.

Hay, Gordon

In 1967 this landmark case was the first in which bite-mark evidence was deemed admissible in court for identification of an offender. The body of a young woman, Linda

Peacock, was found in the local cemetery of Biggar, Scotland. Although it was determined that she had not been raped, the clothing on the top half of her body was in disarray, and a bite mark was noted on her right breast. The bite mark was examined by a forensic odontologist, who was able to determine that the offender had a tooth that was particularly jagged. The investigation focused very much on a detention center in the locality where the girl's body was found. A number of the inmates were asked to give dental impressions so they could be compared to the bite mark. It was found that the impression of one of the inmates, Gordon Hay, matched that of the bite mark on the body. Individual characteristics found on a jagged tooth impression from Hay were found to match the bite on Linda Peacock's breast. When the evidence was presented in court, there was a significant legal battle over its admissibility. It was eventually deemed admissible, and Hay was found guilty of murder.

See also Admissibility of Scientific Evidence; Bite Marks; Odontology

References
Court TV's Crime Library, Criminal Minds and Methods; http://www.crimelibrary.com/criminal_mind/forensics/bitemarks/3.html?sect=14 (Referenced July 2005).
Evans, C. *The Casebook of Forensic Detection: How Science Solved 100 of the World's Most Baffling Crimes.* New York: Wiley, 1998.

Headlight Filaments

Collection of evidence from hit-and-run accidents or vehicle crash scenes includes the collection of all broken headlamp and reflector glass. Examination of a vehicle's headlight filament can provide information as to whether the headlight was on or off at the time of a crash or impact. The presence of oxides on the headlamp filament indicates the filament was hot when it was exposed to air; therefore the filament must have been on prior to the vehicular impact.

See also Bulbs (Automobile, Examination of in Accidents)

References
Curran, J. M., T. N. Hicks, and J. S. Buckleton. *Forensic Interpretation of Glass Evidence.* Boca Raton, FL: CRC, 2000.

Heroin

Heroin is classified as an opiate. Opiates are drugs commonly used by the medical community to relieve moderate to severe pain. Opiates can be naturally occurring, semisynthetic, or synthetic. Naturally occurring opiates are extracted from opium, which is harvested from the opium poppy, also called the *Papaver somniferum*. The opium poppy grows in the Middle East, Southeast Asia, and parts of Central and South America. Semisynthetic opiates, such as heroin, are produced using a naturally occurring opiate as a starting point. Synthetic opiates are manufactured completely in the laboratory. The following are examples of each of these classes.

Naturally Occurring	Semisynthetic	Synthetic
Morphine	Heroin	Methadone
Codeine	Hydrocodone	Propoxyphene
Thebaine	Hydromorphone	Meperidine (Demerol)
Oxycodone	Fentanyl	

Heroin was first synthesized in 1874. It is produced from morphine by a relatively simple chemical process. Heroin remained somewhat obscure until the Bayer Chemical Company of Germany introduced it as an ingredient in some medications in 1898. By 1900 a significant portion of the American population was addicted to opium, morphine, or heroin. In 1956 heroin became a Schedule 1 drug and all stocks of heroin were required to be surrendered to the U.S. government. It should be noted that heroin is still manufactured and administered therapeutically in several countries, including England, France, and Belgium.

Heroin is typically seen in powder form. It ranges in color from pure white to dark brown. The range of colors is attributed to impurities left from the manufacturing process or the

Opium collected from poppies in Afghanistan. Opium is used in both illegal and pharmaceutical drugs such as heroin and morphine. (Ash Sweeting / Courtesy of The Senlis Council)

The overall effect of heroin is that it depresses the central nervous system. Short-term effects include sedation, reduced anxiety, hypothermia, reduced respiration and pulse rate, reduced coughing, disorientation, sweating, dry mouth, and possible death due to overdose, because—as with any illicit drug—the exact purity and content is unknown to the user. An overdose can lead to suppressed respiration, coma, and death.

See also Controlled Substances; Drugs; Methadone; Morphine; Opium

References

De Forest, P., R. E. Gaensslen, and H. C. Lee. *Forensic Science: An Introduction to Criminalistics.* New York: McGraw-Hill, 1983.

James, S. H., and J. J. Nordby. *Forensic Science: An Introduction to Scientific and Investigative Techniques.* Boca Raton, FL: Taylor and Francis, 2005.

Saferstein, R. *Criminalistics.* 8th ed. Upper Saddle River, NJ: Pearson Education, 2004.

U.S. Drug Enforcement Administration, Drug Descriptions, Heroin; http://www.dea.gov/concern/heroin.html (Referenced July 2005).

White, P. *Crime Scene to Court: The Essentials of Forensic Science.* Cambridge: Royal Society of Chemistry, 1998.

addition of adulterants to the final product. "Black tar" is another form of heroin originating in Mexico. It is a tarlike substance varying in color from brown to black.

Heroin can be injected, smoked, or snorted. The most rapid and intense effects are produced by intravenous injection. Effects are felt in seven to eight seconds. Effects from smoking or snorting heroin develop slower, in about ten to fifteen minutes. Snorting or smoking heroin has become more popular in recent years due to the spread of AIDS through needle sharing and the availability of high purity heroin.

After ingestion, heroin users experience a "rush," a sense of relaxation and well-being. The user is driven to take repeated doses of the drug to continue experiencing this euphoria. Effects of heroin last about six hours. The body rapidly develops a tolerance to heroin, requiring larger doses to achieve the same effect.

High Performance Liquid Chromatography (HPLC)

High performance liquid chromatography (HPLC) was first introduced in the late 1960s. It is now a key tool in the examination of many materials and can be used to separate, identify, and quantify unknown substances of forensic significance, including drugs and poisons, inks, and explosive residues.

HPLC is a form of chromatography that employs a column packed with solid particles as the stationary phase and a solvent/buffer mixture as the mobile phase. An HPLC unit consists of an injector, a pump, an analytical column, a detector, and a computer to display the results.

A small amount (around 5 to 10 microliters) of liquid is injected into the system using an injection port or a syringe. Mobile phase is continually being pumped through the instrument at a rate specified by the operator, and so when the sample is injected, it is inserted into the stream of mobile phase and is pumped along to the column. When the sample enters the column, the individual components making up the sample are separated due to each component having a different affinity for the column particles or the mobile phase. Some components will have stronger interactions with the column particles and so be retained in the column for a longer time, while other components will interact more strongly with the mobile phase and so will move at that flow rate and elute from the system more quickly. As components emerge from the system they are detected by the detector. The detector signal is recorded and displayed on the computer so that the results can be interpreted by the analyst. The chromatogram is displayed as a graphical representation with time (in minutes) on the x-axis and peak area or height on the y-axis. The elution of the individual components from the system is viewed as a peak on the chromatogram, and the time at which each component elutes is referred to as the retention time.

There are basically two forms of HPLC to be considered here: normal phase and reverse phase. Both normal and reverse phase chromatography separate components on the basis of polarity.

In normal phase HPLC, the column is packed with an adsorbent material that is polar in nature, usually silica, and the mobile phase is relatively nonpolar (e.g., n-hexane or tetrahydrofuran). The affinity of individual sample components for the column is dependent on their relative polarities. Polar compounds have strong affinities for the column and so are retained for longer periods of time, hence resulting in their having longer retention times. To elute a polar compound, the polarity of the mobile phase must be increased. Normal phase HPLC is mainly used for the separation of nonpolar or moderately polar materials.

Reverse phase HPLC consists of a silica support with a nonpolar phase bonded to it. This is commonly achieved by the covalent binding of n-alkyl chains to the silica. For example, a C-8 column has an octyl chain bonded to the silica, whereas a C-18 column designation signifies an octadecyl ligand. A polar mobile phase is used in this type of system and normally is composed of mixtures of water and polar solvents such as methanol or acetonitrile. Elution is opposite to silica columns, such that polar compounds elute first and nonpolar compounds are more strongly bound to the stationary phase. Reducing polarity increases the eluting strength of the mobile phase.

Significant progress has been made in recent years with regard to HPLC detector types. It is desirable for an LC detector to have a low drift and noise level. This is especially important in trace analysis. A high degree of sensitivity is also desirable. Commonly used detector types are ultraviolet detectors, photodiode array detectors, fluorescence detectors, and more recently, mass spectrometers. Recent advances in liquid chromatography–mass spectrometry have led to this technique becoming increasingly more used in the forensic field.

A major advantage of HPLC in forensic analysis is that the sample does not have to be vaporized as it does during gas chromatography analysis. This makes this an amenable technique for analyzing organic explosives. Some drug samples such as LSD are analyzed using HPLC analysis.

See also Drugs; Gas Chromatography

References

De Forest, P., R. E. Gaensslen, and H. C. Lee. *Forensic Science: An Introduction to Criminalistics.* New York: McGraw-Hill, 1983.

Levine, B. *Principles of Forensic Toxicology.* Washington, DC: American Association for Clinical Chemistry, 1999.

Saferstein, R. *Criminalistics.* 8th ed. Upper Saddle River, NJ: Pearson Education, 2004.

Hitler Diaries

This case is an important illustration of document analysis. Points to note are the importance of document examination for potentially fraudulent materials, as well as the selection of an appropriate examiner and, of course, the importance of ensuring adequate control or "known" samples.

In 1981 the publishers of *Stern* magazine in Germany were approached by one of their senior journalists, who claimed to have a source that had the diaries of Adolf Hitler and also a third previously undiscovered volume of Hitler's autobiography *Mein Kampf.* They purchased all of the materials for a total price of $2.3 million and only later had the diaries examined by scientists for authentication. The diaries were then analyzed by two examiners, who pronounced them genuine. However, the documents were ultimately revealed to be forgeries. There were several reasons the initial analyses failed to reveal the materials as fraudulent. First, many of the "known" samples of Hitler's handwriting actually came from the same source as the diaries. Second, one of the examiners did not speak German and so could only analyze the diaries by simple observation.

The magazine prepared to publish the diaries and recruited international partners for the publication. In the meantime, however, the West German police had conducted an investigation that determined the diaries to be forgeries. They did this by examining the paper on which the diaries were written. It was found to contain blankophor, a paper-whitening agent known not to have been used until 1954! Much of the money for the transaction was found in the bank account of the journalist involved in the purchase of the diaries, who, when faced with criminal charges, gave up the name of his source. The source was a petty criminal named Konrad Kujau and both he and the journalist were given jail sentences for their part in the operation. The cost of the case to *Stern* was estimated to be more than $16 million.

See also Document Examination
References
Court TV's Crime Library, Criminal Minds and Methods; http://www.crimelibrary.com/criminal%5Fmind/scams/hitler%5Fdiaries/ (Referenced July 2005).
Evans, C. *The Casebook of Forensic Detection: How Science Solved 100 of the World's Most Baffling Crimes.* New York: Wiley, 1998.
Saferstein, R. *Criminalistics.* 8th ed. Upper Saddle River, NJ: Pearson Education, 2004.

Human Leukocyte Antigen (HLA)

The human leukocyte antigen (HLA) is an individualizing characteristic, or genetic marker, obtained from blood. HLA markers are histocompatibility antigens of the white blood cells and are composed of a large number of factors. These factors are found on white blood cells and tissues but not on red cells, and play a role in determining the fate of transplanted tissues or organs. The HLA test was routinely performed as part of paternity testing in the days before DNA analysis. If a suspect could not be excluded as the father after this test was performed, then there was a greater than 90 percent chance that he was the father.

See also DNA in Forensic Science
Reference
Butler, J. *Forensic DNA Typing.* 2nd ed. Burlington, MA: Elsevier, 2005.

I

Identification

Pretty well everything that a forensic scientist does can be categorized as comparison or identification. Increasingly, even comparisons are coming to be based on testing that results in identification, with conclusions being based on data that shows that the items have (or do not have) the same identity. Reconstruction of a crime scene and testing of a tire to see if failure was the result of an accident or was its cause are examples of important areas of forensic science that do not depend on establishing identity.

The most common instances of identity testing are those to establish personal identity and those to identify the nature of a material. The entity that is the subject of testing for personal identity can be a person or part of a person—a fingerprint, ear morphology, or a bloodstain for example. Materials tested include chemical or botanical drugs, drugs and poisons in blood, weapons and ammunition, and accelerant traces in fire-scene debris. In some cases the identification is not based on a physical object per se—for example, audio or video tapes and the recovery of data from a computer.

With most laboratory testing, however, the principle of identification is to determine the physical or chemical identity of a substance to the best degree of certainty that existing analytical techniques will allow. This could be the chemical identification of an illicit drug or the identification of blood or biological fluids and hairs for species identification, and it requires the application of stringent testing procedures that provide characteristic results for known standard materials.

For example, to identify a questioned substance as heroin, the analytical data must be identical to the data obtained from a known heroin sample. The number and types of tests administered must be sufficient to give the analyst assurance that they will indeed differentiate the test sample from other substances.

There is an accepted practice of laboratories using at least two tests that depend on different principles for their operation. However, there is no intrinsic justification for that. Some tests, such as fingerprint identification, DNA analysis using multiple STR loci, and drug detection using mass spectrometry, are sufficiently discriminating in their own right. Others, such as blood typing by ABO groups and drug testing by color screening and thin layer chromatography (TLC), will never achieve the degree of differentiation needed for reliable identification unless supported by a battery of much more than two tests.

See also Associative Evidence; DNA in Forensic Science; Drugs; Fingerprints
Reference
Saferstein, R. *Criminalistics.* 8th ed. Upper Saddle River, NJ: Pearson Education, 2004.

Identity

Identity (uniqueness) is defined as the sameness of essential or generic characters in different instances and the distinguishing character or personality of an individual.

DNA or fingerprints are considered identity evidence.

See also DNA in Forensic Science; Fingerprints
Reference
Saferstein, R. *Criminalistics.* 8th ed. Upper Saddle River, NJ: Pearson Education, 2004.

Inceptive Evidence

Inceptive evidence is used to address the question of whether or not an offense has occurred. Controlled substance analysis, identification of accelerants in residues from a fire scene, and alcohol testing are examples of inceptive evidence.

See also Accelerant Residues; Alcohol; Drugs

Indented Writing

Indented writing is the partially indented impression left on a sheet of paper that was present under an original written document when it was made. The indented impressions were made by the pressure of the writing implement on the document above. Indented writings can often be valuable evidence when the original document is unavailable. Indented writing impressions can be observed more clearly by using oblique lighting or using a technique developed by the London College of Printing in collaboration with the Metropolitan Police Forensic Science laboratory in London, England. The electrostatic document analyzer (ESDA), now commercially available, applies an electrostatic charge to the surface of a polymer film placed in contact with the questioned document. Any indented impressions are highlighted when a toner powder is applied to the charged film. This technique has been used to produce readable images from impressions that could not be seen or were barely legible under normal lighting.

ESDA impressions can be obtained from sheets that were several pages below the original when it was made. For example, additions made to an apparently authentic record on a page from a notebook were detected by ESDA development of the notebook. The page was nine sheets below that on which the questioned entries were made. The ESDA impression was a perfect match to the original, other than for the absence of the questioned, and therefore additional, entries.

See also Document Examination
References
Hilton, O. *Scientific Examination of Questioned Documents.* Boca Raton, FL: CRC, 1993.
Vastrick, T. W. *Forensic Document Examination Techniques.* Altamonte Springs, FL: Institute of Internal Auditors Research Foundation, 2004.

Individual Characteristics

Evidence that can be associated with a common source with a very high degree of probability is considered to have individual characteristics. Examples of individual characteristics are the matching ridges of two fingerprints; the comparable random striation markings on bullets; and the comparable random, or mechanical and accidental wear marks on footwear and vehicle tires and their impressions. The common thread in all these cases is that the mark or marks in question were produced by random events and so can be taken as being one of an infinite number of possible patterns with infinitesimally small odds that the marks match by chance.

See also Class Characteristics
References
James, S. H., and J. J. Nordby. *Forensic Science: An Introduction to Scientific and Investigative Techniques.* Boca Raton, FL: Taylor and Francis, 2005.
Saferstein, R. *Criminalistics.* 8th ed. Upper Saddle River, NJ: Pearson Education, 2004.

Infrared Spectroscopy

The visible part of the electromagnetic spectrum can be illustrated by a rainbow (red, orange, yellow, green, blue, indigo, and violet). Just beyond the red side of this lies infrared light, invisible to humans. When this infrared light is directed at molecules, it induces vibrations in the chemical bonds holding them together. The type of vibrations induced by the infrared light depends on the atoms making up the molecule. Thus, this technique allows the forensic scientist to obtain structural information about the substance being analyzed. This makes infrared spectroscopy a confirmation technique in forensic science.

The region of light used in infrared spectroscopy is 2.5 to approximately 15 micrometers in wavelength, and this corresponds to wave numbers of 4,000 to 600 cm^{-1} (proportional to frequency). Wavelength is the length of the wave from peak to peak, and wave number is the number of cycles of the wave in each centimeter. Hence, wavelength is measured in micrometers and wave number is measured in reciprocal centimeters. When light in this range is absorbed by an organic molecule, its chemical bonds undergo vibrations, bending, stretching, or rotation. Molecules undergoing stretching exhibit changes in the length of their bonds—alternately between being stretched to longer than normal and compressed to shorter than normal. Molecules may undergo a vibrational change known as *scissoring* or *rocking,* which results in changes in bond angle. Different vibrations occur with each different wave number. For example, bending and stretching occur at higher wave numbers, while vibrations like scissoring and rocking occur at lower wave numbers.

The infrared spectrum obtained from the analysis may be likened to a fingerprint or "photograph" of a molecule. There are two regions of the spectrum—a fingerprint region and a functional group region. The fingerprint region is specific to each individual molecule and provides the forensic scientist with a unique identification of the substance at hand, while the functional group region is

the same for molecules with similar functional groups. (Functional groups are parts of the chemical structure that determine reactivity. Thus, molecules with similar functional groups have similar chemical reactivity.)

The spectrum is essentially a graph showing the wave numbers on the x-axis and the percentage of light that is transmitted (passes through the sample without being absorbed) on the y-axis. Bands are areas where the transmitted light drops (i.e., the light is being absorbed and inducing vibrational changes), and these can be described as strong, medium, or weak, depending on how far down the chart the band drops. An experienced analyst can read the bands on the spectrum to gain structural information about the identity of a specific sample.

Instrumentation

In its basic form, an infrared spectrometer consists of the following:

- A light source
- A prism to disperse the incident light
- A monochromator to select and vary quickly the wave number of light to be shone on the sample
- A sample to be analyzed
- A detector to detect the amount of light passing through the sample
- A data recorder to provide a graphical representation of the amount of light transmitted of each wave number.

Sample Preparation

Liquid samples can be analyzed using infrared spectrometry by sandwiching them between salt discs. Salts such as sodium chloride used for this purpose do not absorb any infrared light and so do not result in any interference on the spectrum.

Solid samples can be analyzed in two ways. First, they can be finely ground in a mortar and pestle and mixed with liquid paraffin such as nujol. This is often referred to as a nujol mull. The second method by which solids may be analyzed is by grinding them very finely and mixing them with potassium bromide salt. This mixture is then pressed under high pressure to create a disc of the mixture suitable for analysis.

Fourier Transform Infrared Spectroscopy (FTIR)

Fourier transform infrared spectroscopy (FTIR) is the method of infrared spectroscopy that is now more commonly used. It provides the same end result as infrared analysis as described above, but arrives at the spectrum by a slightly different method. It has several advantages over conventional infrared spectroscopy, including increased sensitivity and decreased analysis time. FTIR uses an interferometer in place of a monochromator. This means that the sample is exposed to all wavelengths of light in the selected range and a composite response is recorded. A mathematical transformation known as the *Fourier transform* is performed on the results to produce a spectrum like that described above.

Attenuated Total Reflectance–Fourier Transform Infrared Spectroscopy (ATR-FTIR)

The recent advent of attenuated total reflectance–Fourier transform infrared spectroscopy (ATR–FTIR) has provided forensic scientists the ability to perform infrared spectroscopic analyses on samples without destroying the evidence. This technique eliminates the need for destructive sample preparation because the sample is simply placed on the crystal surface of the ATR (often made of diamond or germanium). An infrared beam is focused on and totally internally reflected by the crystal, creating a wave that extends into the sample contained on the crystal. Because some regions of this wave will be absorbed by the sample, the resultant wave will be altered and detected as an interferogram suitable for mathematical transformation into an infrared spectrum.

Uses in Forensic Science

FTIR as a conformational technique has many uses in forensic science. It may be used in the analysis of drugs. However, it should be noted that it is very difficult to analyze mixtures in this way unless the FTIR is used as a detector for a chromatographic method such as gas chromatography. FTIR analysis is commonly performed if a substance is suspected to be gamma hydroxybutyrate (GHB) because analysis of this drug by the "gold standard" gas chromatography mass spectrometry (GCMS) method requires an additional derivatization method prior to analysis. FTIR is also commonly used to allow drug chemists to differentiate between cocaine hydrochloride and cocaine base. FTIR may also be used in the analysis of fiber, paint, glass, and other trace evidence.

See also Fibers; Gamma Hydroxybutyrate (GHB); Gas Chromatography; Glass; Paint

References

BioForum Applicable Knowledge Center Fourier Transform Infrared; http://www.forumsci.co.il/HPLC/FTIR_page.html (Referenced July 2005).

Saferstein, R. *Criminalistics.* 8th ed. Upper Saddle River, NJ: Pearson Education, 2004.

Ink

The most common writing material is ink. An indefinite number of inks are available with up to twelve different classifications, but only four general classes of ink are regularly encountered. These include ballpoint, liquid, printing, and—less commonly—typewriter inks. The forensic examination of ink can be divided into tests for three characteristics: color, chemical composition, and the age of the writing.

Inks are made by incorporating a number of dye substances into a carrier vehicle. Different ink manufacturers use different ink compositions, but the colored components of ink contain only a few basic substances:

- India ink. The oldest form of ink, it is a suspension of carbon black in a binder solution and is the most permanent of all ink colors.
- Logwood ink. Its main color ingredients are hematoxylin and potassium chromate.

- Iron gallotannate ink. This ink is composed of tannic acid, gallic acid, ferrous sulfate, and an aniline-based dye. This ink penetrates and reacts with paper fibers and is the basis for many commercial writing inks. The aniline dye produces the immediate pigment, but the other components react together with time to produce the permanent dye.
- Nigrosine ink. This was an early black dye ink made from reacting aniline and nitrobenzene. It is now used infrequently due to its water solubility and weathering effects.
- Dye inks. These are composed of a basic, acidic, or neutral organic dye, derived mainly from coal tar or petroleum, which are the main sources of color in most inks. Dyes can be readily removed from an ink mark and the colors separated by thin layer chromatography. The separation components of different dye inks vary sufficiently to allow comparison and determination of whether known or questioned writings were prepared with the same type of ink.

Analysis and Comparison

The first methods applied in the analysis and comparison of questioned inks with known ink standards are nondestructive in nature. These tests include color comparisons; the determination of the type of ink; and creating a profile of its infrared absorption, luminescence, and fluorescence properties. If the unknown ink differs in any of these properties from the standard ink—then it could not have originated from the same source. At this stage questioned and standard inks that exhibit the same properties are further tested.

Chemical tests involve thin layer chromatography to separate the colored and noncolored components of the ink mixtures. The colored substances can be compared under normal lighting and specialized lighting techniques such as ultraviolet illumination

and infrared luminescence. The pattern of separation can easily be compared with those of known standards. The samples are prepared for thin layer chromatography (TLC) or other chromatographic analysis by taking scrapings or microdot punches of the sample and extracting the ink into a suitable solvent system for analysis.

Dating of inks can be of forensic use when trying to determine whether a document is as old as it claimed to be. When questioned and standard inks are found to match, the manufacture's files can be checked to see when the ink was first produced. Current methods of ink aging utilize high performance liquid chromatography and gas chromatography–mass spectroscopy to separate and measure the amounts of volatile components present in ink. As ink on a document ages, the amount of volatile substances in the ink decreases.

See also Document Examination

References

Hilton, O. *Scientific Examination of Questioned Documents.* Boca Raton, FL: CRC, 1993.

Saferstein, R. *Criminalistics.* 8th ed. Upper Saddle River, NJ: Pearson Education, 2004.

Vastrick, T. W. *Forensic Document Examination Techniques.* Altamonte Springs, FL: Institute of Internal Auditors Research Foundation, 2004.

White, P. *Crime Scene to Court: The Essentials of Forensic Science.* Cambridge: Royal Society of Chemistry, 1998.

Integrated Automated Fingerprint Identification System (IAFIS)

Personal identification is an important component of most forensic investigations, from linking a person to a specific crime scene to identification of corpses. As explained in the entry on **fingerprints**, the traditional fingerprint classification is not a very good tool for comparing latent images recovered from crime scenes. However, there are many computer-based systems that can make fast and accurate comparisons between latent prints and database information using ridge characteristics.

Unlike the new IAFIS in place today for easy analysis, fingerprint analysis used to be a laborious and tedious process. ca. 1930–1935. (Library of Congress)

The fact that there were several systems and more than 19,000 law enforcement agencies in the United States compromised the effectiveness of national searches until the Federal Bureau of Investigation developed its integrated automated fingerprint identification system (IAFIS). The IAFIS system now provides law enforcement agencies a rapid, reliable method to identify suspected offenders and a means to obtain all possible information about them. The establishment of this system brought an end to the traditional labor-intensive methods of fingerprint identification: Fingerprint records and evidence were checked and compared manually and involved the review of individual fingerprint cards and latent prints that had been archived in databases for decades. This means of personal identification could take as long as 10 days for completion and involve the review of approximately 34 million criminal fingerprint files in an attempt to establish whether a given individual had an existing criminal record. The IAFIS system is constructed such that the FBI can obtain and accept fingerprints by electronic means from a network of booking stations within any state, county, or local law enforcement agency. By utilizing complex computer systems, the system enables the rapid search of fingerprint files for the identification of suspects and can provide immediate access to their fingerprint records.

See also Fingerprints
Reference
Federal Bureau of Investigation, Integrated Automated Fingerprint Identification System; http://www.fbi.gov/hq/cjisd/iafis.htm (Referenced July 2005).

Integrated Ballistic Identification System (IBIS)

The Integrated Ballistic Identification System (IBIS) is a networkable workstation that combines two databases called Brasscatcher and Bulletproof through links to the Bureau of Alcohol, Tobacco, and Firearms National Tracing Center. The system combines a microscope with illumination-control devices and computer hardware and software to enable firearms examiners and technicians to make digital images of, and automatically sort and categorize, bullet and shell-case markings.

The IBIS program provides numerical rankings or ratings for potential matches from the thousands of evidentiary images in its database. Through specialized software, the system correlates and matches projectile and shell-case evidence, to produce fast, high-probability ballistic matches and to increase the frequency of investigative leads in shooting incidents. Firearms examiners can study and compare more items of evidence in less time and create leads for law enforcement agents by providing information on crimes committed with the same firearm. The IBIS is commercially available to all forensic or investigative laboratories and allows image analysis to be conducted on a local, national, or international level.

See also Bullets; Firearms

Reference
Forensic Technology, IBIS/Ballistic Identification;
    http://www.fti-ibis.com/en/s_4_1.asp
    (Referenced July 2005).

Irving, Clifford

The case of Clifford Irving represents a high-profile document examination case in which the relatively new technique of voice-print analysis was used to uncover fraudulent claims.

Clifford Irving in 1971 made a significant book deal and was paid a $765,000 advance by claiming that the billionaire recluse Howard R. Hughes had provided him with records and documents and authorized him to write his biography. The deal received much publicity and attempts were made to verify the arrangement. Howard Hughes, after spending years as a recluse, held a press teleconference to state that he had never met Irving and that no such authorization had been given for him to write Hughes's biography. Of course, Irving disputed that this was the real Howard Hughes, only to be revealed as an imposter when results of a voiceprint analysis were made public. The analysis involved comparing the caller to that of a taped speech given by Howard Hughes. Comparisons were made with respect to pitch, volume, and tone. The documents Irving claimed had been given to him by Hughes were later shown to be fake also. Irving and his coconspirators were charged with fraud and as a result Irving was sentenced to thirty months in prison.

See also Identification

References
Evans, C. *The Casebook of Forensic Detection: How Science Solved 100 of the World's Most Baffling Crimes.* New York: Wiley, 1998.
Owen, D. *Hidden Evidence: Forty True Crimes and How Forensic Science Helped Solve Them.* Willowdale, Ontario: Firefly, 2000.

J

Jewelry

Jewelry is encountered in forensic science in two entirely different settings. First, the presence of identifiable jewelry on a body can be a valuable tool for victim identification. Obvious examples are rings and bracelets.

The second situation is concerned with the gemstones rather than their cosmetic presentation. Smuggling and stone fraud are crimes that account for many millions of dollars each year. Identification and valuation of smuggled gem stones is readily performed by an expert jeweler.

Fraud can consist of misrepresentation of synthetic stones as being natural or presentation of a stone of lesser value as more valuable than it is, or presentation of an imitation material as a gemstone.

Most gemstones can be created artificially. Stones such as amethyst, rubies, emeralds, and sapphires can be grown as crystals, and synthetic diamonds can be manufactured. Synthetics can be very difficult to distinguish from authentic stones by testing of physical properties such as hardness and refractive index. An authentic history is important, and some of the other disciplines of the forensic science laboratory such as document examination may be called upon.

Imitations are often colored glass and are easy to identify in the laboratory. Hardness and refractive index measurement are effective tests.

Stones can also be treated to alter color by heating or irradiation. These methods are not considered fraud.

Simulated diamonds can be attractive stones in their own right. Common sources are forms of garnet, cubic zirconia, and synthetic moissonite. Synthetic moissonite is a form of silicon carbide and is very close to true diamond in its properties.

Jennings, Thomas

This landmark case was the first murder trial in the United States to rule fingerprint testimony admissible evidence in court.

In September 1910 Clarence Hiller was shot and killed by an intruder in his home. The intruder fled, and there were no living eyewitnesses to the crime. Shortly afterwards Thomas Jennings was apprehended by police. He was found to be carrying a loaded revolver, as well as having blood on his clothing.

Cartridges from Jennings's revolver were examined and found to match unused cartridges that had been found at the Hiller residence. However, of most significance in this case was the finding of four left-hand fingerprints on a newly painted fence close to the window through which the perpetrator had gained access to the house. Fingerprint identification was in its infancy at this time, and so it was a team of fingerprint examiners who agreed that the fingerprints on the railing could only have been left by Thomas Jennings. It was this evidence that led to Jennings receiving a verdict of guilty and being sentenced to death. The defense appealed the ruling based on the fact that fingerprint evidence was new, not "scientific," and therefore not admissible as evidence. However, the ruling was upheld on appeal, and Thomas Jennings was later hanged.

See also Fingerprints
References
Court TV's Crime Library, Criminal Minds and Methods; http://www.crimelibrary.com/criminal_mind/forensics/crimescene/6.html?sect=21 (Referenced July 2005).
Evans, C. *The Casebook of Forensic Detection: How Science Solved 100 of the World's Most Baffling Crimes.* New York: Wiley, 1998.
Owen, D. *Hidden Evidence: Forty True Crimes and How Forensic Science Helped Solve Them.* Willowdale, Ontario: Firefly, 2000.

K

Kerosene

Crude oil is a mixture of aliphatic hydrocarbons, which can be characterized by the length of the carbon chain. Petroleum refining does this by separating the constituents according to boiling point, with the shorter-chain, more volatile hydrocarbons distilling off at lower temperatures. Kerosene is the third fraction recovered (after naphthas and gasoline) and consists of a mixture of C12 to C15 chains. Kerosene is used for lanterns and heating stoves. It is encountered in forensic science as an accelerant in arson fires.

See also Accelerant Residues; Arson
References
Almirall, J., and K. Furton. *Analysis and Interpretation of Fire Scene Evidence.* Boca Raton, FL: CRC, 2004.
Nic Daeid, N. *Fire Investigation.* Boca Raton, FL: CRC, 2004.

Khat

Khat is a naturally occurring stimulant drug produced by the plant *Catha edulis.* The shrub is found in the Arabian peninsula and East Africa. Khat (also referred to as "Abyssinian tea," "African salad," "oat," "kat," "chat," and "catha") use is legal and part of the culture of people in these lands. It is illegal in the United States.

Khat has two active ingredients, cathinone and cathine. Cathinone is a Schedule I drug, and the less potent cathine is a Schedule IV substance. Cathinone activity is lost in the vegetable material unless it is preserved by refrigeration or freezing.

The drugs are similar to amphetamine in their actions. According to the DEA, khat is grown in export quantities in countries such as Kenya and Ethiopia. It is Ethiopia's fourth-largest export according to U.S. embassy reporting, and the recreational use of khat is widely accepted there. Over 33 percent of Yemen's gross national product is associated with the cultivation, consumption, and exportation of khat. The World Health Organization reports that the cultivation and use of khat has profound socioeconomic consequences on countries and individuals.

See also Controlled Substances; Drugs
References
U.S. Drug Enforcement Administration, Drug Intelligence Brief, Khat; http://www.usdoj.gov/dea/pubs/intel/02032/02032.html (Referenced July 2005).
Saferstein, R. *Criminalistics.* 8th ed. Upper Saddle River, NJ: Pearson Education, 2004.
Winger, G., F. G. Hofmann, and J. H. Woods. *A Handbook on Drug and Alcohol Abuse: The Biomedical Aspects.* 3rd ed. New York: Oxford University Press, 1992.

Kicks

Footwear is sometimes submitted to the forensic science laboratory with a request for testing to confirm whether or not the wearer had kicked someone. Identification of blood and typing it to identify the origin is straightforward. However, the question of how the blood got there is another matter.

The physical nature of the stains is a significant guide. Any violent impact—boot to head or bullet to body—results in a splash back of fine blood drops. The drops will leave directional patterns or streaks that show where the blood came from. Thus, a boot could have a mixture of fine spray drops at the toes and streaks running from front to back if the blood had come from kicking. However, the blood may have originated from someone running through a pool of blood, and the deposition of stains could be similar to the novice eye. Experience in interpretation of drop size and directionality is key.

Interpretation is also made more difficult by the likelihood of smearing of stains through contact with the wearer's clothing or other reasons. Taken all in all, it is best to take a conservative view to the interpretation of bloodstain patterns in regard to answering the question of whether or not they originated from kicking.

References
Byard, R., T. Corey, C. Henderson, and
    J. Payne-James. *Encyclopedia of Forensic and Legal Medicine.* London: Elsevier Academic, 2005.
Knight, B. *Simpson's Forensic Medicine.* London and New York: Oxford University Press, 1997.
Payne-James, J., A. Busuttil, and W. Smock. *Forensic Medicine: Clinical and Pathological Aspects.* London: Greenwich Medical Media, 2003.

Knife Wounds

Knife wounds are inflicted in assaults such as stabbing and slashing, in self-inflicted suicide attempts, and occasionally in self-inflicted efforts to simulate assaults.

The injuries made by knife wounds can leave clues as to what happened. First, the shape of a stab wound will reflect the type of knife blade used to make it. The size of the blade, whether it is single or double edged, the force of the stabbing, and whether the knife is sharp or blunt can all influence the physical nature of the wound. For example, a stabbing with a sharp, double-edged knife, but not a blunt or single-edged or thrown knife, will leave a clean, deep wound.

The location of the wound can also be informative. Clearly there are parts of the body where knife wounds cannot be readily self-inflicted. The multiple, shallow wounds on the wrists are well-known as being associated with less-than-determined suicide attempts. Cuts along the palm and inner surfaces of the fingers are associated with defensive actions by the victim of an assault.

References
Byard, R., T. Corey, C. Henderson, and
    J. Payne-James. *Encyclopedia of Forensic and Legal Medicine.* London: Elsevier Academic, 2005.
Knight, B. *Simpson's Forensic Medicine.* London and New York: Oxford University Press, 1997.
Payne-James, J., A. Busuttil, and W. Smock. *Forensic Medicine: Clinical and Pathological Aspects.* London: Greenwich Medical Media, 2003.

Km Blood Typing

See *Gm and Km Typing*

Kumho Tire Ruling

The admissibility of scientific evidence is today largely determined by application of the Supreme Court's decision in *Daubert v. Merrill Dow Pharmaceuticals, Inc.* (see **Daubert Ruling**). *Daubert* was about admissibility of novel scientific evidence and established that the judge is the gatekeeper, and it provided a set of guidelines for evaluation of whether evidence is scientific or not. *Kumho Tire Co. v. Carmichael* extended the findings of *Daubert* to established expert evidence in any area, and confirmed the gatekeeper role of the judge.

See also Daubert Ruling
References
Daubert v. Merrill Dow Pharmaceuticals, Inc.,
    509 U.S. 579 (1993).
Kumho Tire Co. v. Carmichael., 526 U.S. 137 (1999).

L

Latent Prints

The tiny sweat pores on top of the friction skin ridges of the fingers constantly exude perspiration that sticks to the ridge outlines. Substances, like oil from touching one's face or hair, may also be present on these ridges. When an object is touched, an impression of the ridge characteristics may be left on the surface. Impressions that are easily seen are called *patent prints*, while others that can only be seen with specialized techniques are called *latent prints*.

See also Fingerprints

References
James, S. H., and J. J. Nordby. *Forensic Science: An Introduction to Scientific and Investigative Techniques.* Boca Raton, FL: Taylor and Francis, 2005.
Lee, H. C., T. Palmbach, and M. T. Miller. *Henry Lee's Crime Scene Handbook.* London and San Diego, CA: Academic, 2001.
Saferstein, R. *Criminalistics.* 8th ed. Upper Saddle River, NJ: Pearson Education, 2004.

Lead Bullets

Lead is an excellent material for the projectile in ammunition because of its high density. However, projectiles for use in high-velocity ammunition encase the lead in a jacket of copper or steel. This gives better mechanical properties, as well as offering a reduction in the amounts of lead in discharge residues.

See also Ammunition; Bullets; Firearms

Reference
Warlow, T. *Firearms, the Law, and Forensic Ballistics.* 2nd ed. London and Bristol, PA: Taylor and Francis, 1996.

Lead Poisoning

Lead is a cumulative environmental poison. That is, it is found in materials in the environment such as paint and water, and its levels build up in the body with continued exposure. However, awareness has resulted in far fewer instances of lead poisoning today than 50 to 100 years ago. Chronic low-level exposure in children can cause developmental effects. Lead intake can be monitored by measuring its concentration in blood. Acute overdose is readily treated by intravenous infusion of the chelating agent ethylenediaminetetraacetic acid (EDTA). Probably the main concern with lead in forensic science is exposure of firearms examiners to potentially high levels at shooting ranges.

See also Poisoning; Toxicology

Reference
Goldfrank, L., N. Flomenbaum, N. Lewin, M. A. Howland, R. Hoffman, and L. Nelson. *Goldfrank's Toxicologic Emergencies.* 7th ed. New York: McGraw-Hill Professional, 2002.

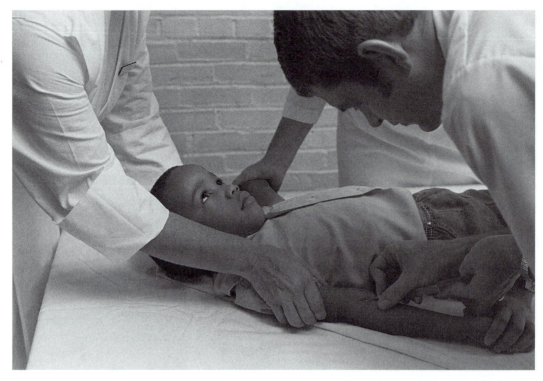

Dr. Joseph Caldwell takes a blood sample to test for lead poisoning; 3-year-old Gregory McClendon doesn't seem to mind the slight pain. Gregory, son of Mrs. Carmen McClendon, is one of hundreds of inner-city children given free tests to detect lead poisoning caused primarily from eating or swallowing chips or flakes of lead based paints still on many walls and interiors of homes in the inner city. September 22, 1970, St. Louis, Missouri. (Bettmann/Corbis)

Lewis Blood Typing

Lewis groups are genetically determined factors present in blood. They are polysaccharides (large molecules made from a variety of simpler sugar molecules linked together) and circulate in the blood bound to proteins and to lipids. There are two main Lewis types: a and b. The polysaccharides that determine the Lewis groups are related to those that establish secretor status. People classified as "secretors" have a gene that results in large amounts of the chemical that determines their blood group (A, B, or O) being found in body secretions such as semen. This chemical is also a polysaccharide. Lewis typing was therefore used in some laboratories to confirm secretor status and support ABO typing of vaginal swabs and of semen stains in rape cases. Nonsecretors are Lewis (a + b −). Lewis typing in stains was not one of the most reliable techniques and is no longer used,

mainly because DNA typing has replaced the need to assign secretor status.

See also ABO Blood Groups; Blood; Blood Grouping; DNA in Forensic Science; Rape

Lie Detector Test (Polygraph)

Polygraph literally means "many writings." Polygraph testing is used for three main purposes: event-specific investigations, employee screening, and preemployment screening. A polygraph instrument is used to measure changes in blood pressure, pulse rate, respiration, muscle movements, and perspiration. By collecting this physiological data from the human body, this data (or "many writings") is used to perform a lie detector test. Respiratory activity is measured by placing rubber tubes over the chest and abdominal area, and two small metal plates attached to the fingers record sweat-gland activity. Blood

Patsy Ramsey and her husband John Ramsey listen during a press conference in Atlanta where they released the results of an independent lie detector test. The Ramseys appeared with their attorneys and the officials who administered the test stating that they had no knowledge of who killed their daughter Jon-Benet Ramsey in Boulder, Colorado. (Reuters / Corbis)

pressure measurements are made to record cardiovascular activity. All of these physiological activities are recorded simultaneously. Originally the equipment used was analog in nature but presently computerized polygraph instruments are used.

The basis of the procedure is the assumption that when a person lies, emotional and physiological reactions occur. Trained polygraph examiners are trained to detect changes and fluctuations in the data recorded caused by the emotional and physiological reactions. The examiner asks the test subject a number of questions and changes in recordings occur as the subject responds. By determining baseline recordings while asking a series of innocuous questions, changes caused by lying are detected by the polygraph. Courtroom objections to the use of the polygraph are to

the same sensitivities measurable by the equipment. It has been argued that emotional changes caused by nervousness or the stress of being tested can also influence the recordings.

Typical polygraph examinations last two to three hours and include a pretest interview in which the examiner talks to the examinee about the test, the questions to be asked, and the procedures that will be used. The chart collection phase is when the examiner asks the questions and collects a number of polygraph charts, and in the test data analysis phase the examiner analyzes and interprets the charts to give an opinion on the truthfulness of the test subject. When appropriate, the test subject is given an opportunity to explain the physiological responses obtained in relation to one or more questions asked during the test. During the test, questions will be asked three to four times, and several charts will be compiled before data is interpreted. After charts are collected, the polygrams are scored by hand and by computer. If there are still further questions after the scoring is concluded, the examiner will seek a second opinion of another certified examiner.

The polygraph technique is accurate, but not infallible. The matter of reliability is addressed in the 2003 report from the National Research Council Committee Division of Behavioral and Social Sciences and Education Committee to Review the Scientific Evidence on the Polygraph. The committee found that there is little scientific basis for the expectation that a polygraph test could have extremely high accuracy. However, the committee did agree that specific-incident polygraph tests can discriminate lying from truth telling at rates well above chance, though well below perfection. However, the committee cautioned that because the studies of acceptable quality all focus on specific incidents, generalization from them to uses for screening is not justified. Because actual screening applications involve considerably more ambiguity for the examinee and in determining truth than arises in specific-incident studies, polygraph accuracy for screening purposes is almost certainly lower than what can be achieved by specific-incident polygraph tests in the field.

The committee was particularly concerned about the paucity of good quality research. Indeed, the report implies that some basic questions may not be answerable. Specifically, laboratory testing in which subjects are tested for their response to staged scenarios cannot mimic the circumstances of the real event because the subject is not exposed to the same serious consequences of being found to be lying. Similar limitations apply to determining the effectiveness of the polygraph to countermeasures.

Although the findings of the committee support the usefulness of the polygraph in criminal investigations, they were very negative about its use for employee screening.

The committee specifically concluded that its accuracy in distinguishing actual or potential security violators from innocent test takers is insufficient to justify reliance on its use in employee screening.

The NRC committee considered alternatives. They concluded that there is no better method for evaluating truthfulness, at this time. Computerized data evaluation may enhance the reliability of the polygraph.

Polygraph results are admissible in some federal circuits and some states. Usually this evidence is admissible where the parties have agreed to its admissibility before the examination is given. However, some jurisdictions have absolute bans on polygraph results as evidence; the suggestion that a polygraph examination is involved is sufficient to cause a mistrial. The requirements of the 1923 *Frye* standard were decided in a trial in which the scientific validity of the polygraph was rejected. The *Daubert* ruling, while not involving polygraph data per se, has had a profound effect on admissibility of polygraph results as evidence.

See also *Daubert* Ruling; *Frye* Rule
References
American Polygraph Association;
   http://www.polygraph.org/journals.htm
   (Referenced July 2005).
Daubert v. Merrill Dow Pharmaceuticals, Inc.,
   509 U.S. 579 (1993).
Frye v. United States, 293 F. 1013 (D.C. Cir. 1923).
Kleiner, M. *Handbook of Polygraph Testing.* San Diego,
   CA: Academic, 2001.

Likelihood Ratio

In theory, DNA profiles are unique. In practice, alleles from a finite number of loci will be identified and so there is a chance that the profile will not be unique. Laboratories have had to develop techniques to account for this. The likelihood ratio is one such technique. The approach owes its origins to the work of the Reverend Thomas Bayes, an eighteenth-century English clergyman. Bayes's theorem expresses the odds on an event in the form of the ratio of two probabilities. Therefore, the probability of an event A given the circumstances B is expressed as the ratio of the odds (probabilities) that B will occur given that A is true, divided by the probability that B will occur randomly. If A is a DNA profile, we can calculate the ratios for the event B, that the profile did indeed come from the suspect during the criminal act, and the event B′, that the semen did not come from the suspect.

Bayes's theorem can be presented in a form that allows us to modify the odds on an event as information relevant to that event becomes available. The presentation is in the form of the ratio of the prior odds (no information) to the posterior odds (we have the information). This is the likelihood ratio, or the probability that we will find A in circumstances B divided by the probability that we will find A by chance (i.e., circumstances B did not occur).

See also DNA Population Frequencies
References
Butler, J. *Forensic DNA Typing.* 2nd ed. Burlington, MA: Elsevier, 2005.
National Research Council. *The Evaluation of Forensic DNA Evidence.* Washington, DC: Academy, 1996.

Locard's Exchange Principle

Edmond Locard, a student of Bertillon in France, was one of the early pioneers of forensic science. He conducted research on the traces of dust and other debris that could be found on clothing, such as in the cuffs of trousers. He showed how the history of where the clothing had been could be reconstructed from the traces that it gathered. The general principle has been adopted as one of the tenets of forensic science, and named the Locard exchange principle. It posits that each contact between two objects results in an exchange of traces between them. This is, of course, not true. There are many dangers in assuming that the Locard principle is in some way a law of nature. For example, traces may indeed be transferred but be retained for such a short time as to be meaningless. It is also a general truth that positive conclusions

(no contact) should never by drawn from negative observations (no traces). Finally, even if there are transferred traces, extreme caution must be exercised in interpretation. There may be no background data on the frequency of the materials, or they may be very common in the environment in which the crime took place.

For example, the English forensic scientist Alan Clift was suspended from his post as a senior scientist in the Home Office forensic science service because of concerns about some of his work. In one case, he found a match between polypropylene twine found at the scene and on the accused. He had never encountered this material before and reported it in terms that would lead one to believe it was a significant indicator of association. However, the environment was a rural farming community, and the twine had widespread use in baling machines. In another example, police in South Australia built a case against Charles Edward Splatt based mainly on trace evidence. The materials included very unusual particles found on Splatt that matched controls removed from a window ledge at the murder victim's home. Entry had been gained through the window. Splatt was convicted, but a royal commission reversed the conviction noting, among other things, that Splatt worked in a nearby factory as a welder and the particles were weld spatters common in his work environment, and that it in turn was close enough to the house of the victim that the minute particles could well have been wind blown onto the exterior of her house.

See also Associative Evidence
References
De Forest, P., R. E. Gaensslen, and H. C. Lee. *Forensic Science: An Introduction to Criminalistics.* New York: McGraw-Hill, 1983.
James, S. H., and J. J. Nordby. *Forensic Science: An Introduction to Scientific and Investigative Techniques.* Boca Raton, FL: Taylor and Francis, 2005.
Lee, H. C., T. Palmbach, and M. T. Miller. *Henry Lee's Crime Scene Handbook.* London and San Diego, CA: Academic, 2001.

Saferstein, R. *Criminalistics*. 8th ed. Upper Saddle River, NJ: Pearson Education, 2004.

White, P. *Crime Scene to Court: The Essentials of Forensic Science*. Cambridge: Royal Society of Chemistry, 1998.

Lysergic Acid Diethylamide (LSD)

Lysergic acid diethylamide, commonly known as LSD, was discovered in 1938 and is one of the most potent hallucinogenic compounds known. It is synthesized from lysergic acid, found in ergot fungus that grows on rye and other grains. LSD, often referred to as "acid," on the streets, is sold as tablets, tabs, capsules, and occasionally in liquid form. Usually taken orally, it is an odorless, colorless, bitter-tasting substance. The tabs that hold a small amount of drug are made of absorbent paper, divided into small perforated squares. Each square, often highly decorative, represents a single dose. Reports from the Drug Enforcement Administration indicate the strength of a single dose of LSD from illicit sources ranges from 20 to 80 micrograms per dose. This is less than the levels reported during the 1960s and early 1970s, when doses ranged from, and often exceeded, 100 to 200 micrograms.

The hallucinogenic effects of LSD are unpredictable and depend on the amount taken; the personality, mood, and expectations of the user; and the environment or surroundings in which the drug is used. First effects of the drug are usually felt thirty to ninety minutes after ingestion. Physical effects include dilated pupils, elevated body temperature, increased heart rate and blood pressure, sweating, loss of appetite, sleeplessness, dry mouth, and tremors.

Changes in sensations and feelings are much more dramatic than physiological changes. The user often experiences several different emotions at once or rapid mood swings. Taken in a large enough dose, the drug produces delusions and visual hallucinations. The user's sense of time and self becomes distorted and stimulatory sensations become crossed. The user feels as though he or she is seeing sounds and hearing colors, causing fear or panic. Such "trips" or "bad trips," if fear and panic ensue, begin to diminish after about twelve hours. LSD users may experience severe terrifying thoughts; fear of losing control; or fear of insanity, death, and despair while under the influence of the drug. Fatal accidents have occurred during states of LSD intoxication. Many users experience flashbacks. Flashbacks are a recurrence of parts of an LSD experience, without actually being under the influence of the drug. Flashbacks can be sudden and without warning and can occur within days or more than a year after LSD use. Flashbacks are associated more with chronic hallucinogen abuse or in persons with underlying personality disturbances. LSD users may also suffer from long-lasting psychoses, such as schizophrenia or severe depression.

LSD is not considered an addictive drug because it does not produce compulsive drug-seeking behavior like cocaine, heroin cocaine, and other substances do. LSD does produce tolerance, and users who take the drug repeatedly may have to take higher doses to achieve the comparable states of intoxication previously achieved.

See also Controlled Substances; Drugs; Hallucinogens

References

De Forest, P., R. E. Gaensslen, and H. C. Lee. *Forensic Science: An Introduction to Criminalistics*. New York: McGraw-Hill, 1983.

James, S. H., and J. J. Nordby. *Forensic Science: An Introduction to Scientific and Investigative Techniques*. Boca Raton, FL: Taylor and Francis, 2005.

Saferstein, R. *Criminalistics*. 8th ed. Upper Saddle River, NJ: Pearson Education, 2004.

White, P. *Crime Scene to Court: The Essentials of Forensic Science*. Cambridge: Royal Society of Chemistry, 1998.

Winger, G., F. G. Hofmann, and J. H. Woods. *A Handbook on Drug and Alcohol Abuse: The Biomedical Aspects*. 3rd ed. New York: Oxford University Press, 1992.

M

Maggots

In some circumstances maggots can be used to estimate the period that has elapsed since death. Flies lay their eggs in the tissue of the corpse, and the eggs hatch into maggots. Experienced entomologists can estimate the time since the eggs were laid from examination of the maggots. Hence, an estimation of time of death can be made.

See also Entomology; Time of Death
Reference
Byrd, J. H., and J. L. Castner. *Forensic Entomology: The Utility of Arthropods in Legal Investigations.* Boca Raton, FL: CRC, 2000.

Magic Mushrooms

Magic mushrooms are hallucinogenic mushrooms. Although several different species of mushroom have hallucinogenic properties, liberty cap mushrooms are the most commonly used. Liberty cap (psilocybin) mushrooms contain traces of the hallucinogenic compound psilocin. The use of these mushrooms dates back for thousands of years to when people ate them in search of a heightened state of consciousness and spiritual insight.

The effects experienced with magic mushrooms depend on their strength, expectations of the user, and the physical and social environment in which they are used. The usual dose is in the range of ten to thirty liberty caps. With a large dose, the effects are often comparable to those experienced with LSD. Effects experienced thirty to sixty minutes after ingestion include numbness, muscle weakness, tension, drowsiness, distortion of senses and thought processes, and unpleasant reactions including confusion, fearfulness, paranoia, and depression. Elevated temperatures and seizures have been reported in children. Common effects that last for six to twelve hours include powerful changes in perception and temporal and visual disturbances whereby familiar objects and surroundings appear strangely and fantastically distorted. Everyday situations or objects seem funny or frightening. Flashbacks have been reported up to three months after use.

Magic mushrooms are not physically addictive, but they are potentially psychologically addictive. Repeated use results in tolerance and reduced hallucinogenic effects. More serious problems such as kidney failure and death occur with mushroom poisoning by picking and ingesting the wrong kind of mushroom. Possession of mushrooms (in any form) that contain psilocybin is illegal in the United Kingdom, and varies from state to state in the United States.

See also Controlled Substances; Drugs; Hallucinogens

References

De Forest, P., R. E. Gaensslen, and H. C. Lee. *Forensic Science: An Introduction to Criminalistics*. New York: McGraw-Hill, 1983.

James, S. H., and J. J. Nordby. *Forensic Science: An Introduction to Scientific and Investigative Techniques*. Boca Raton, FL: Taylor and Francis, 2005.

Saferstein, R. *Criminalistics*. 8th ed. Upper Saddle River, NJ: Pearson Education, 2004.

White, P. *Crime Scene to Court: The Essentials of Forensic Science*. Cambridge: Royal Society of Chemistry, 1998.

Winger, G., F. G. Hofmann, and J. H. Woods. *A Handbook on Drug and Alcohol Abuse: The Biomedical Aspects*. 3rd ed. New York: Oxford University Press, 1992.

Marijuana

Marijuana is a green, brown, or gray mixture of dried, shredded leaves, stems, seeds, and flowers of the hemp plant *Cannabis sativa*. Street names include "pot," "herb," "weed," "grass," "boom," "Mary Jane," or "chronic," but there are more than 200 slang terms for marijuana. Marijuana is a mind-altering substance. All forms of marijuana, sensemilla, hashish, and hash oil, contain the main active chemical delta–9-tetrahydrocannabinol (THC). The effects experienced by a marijuana user depend on the strength or potency of the THC the material contains. The overall THC potency of marijuana has increased since the 1970s but has been about the same since the mid-1980s. Marijuana is usually smoked as a cigarette (*joint*) or in a pipe or a bong, and more recently, in cigars called *blunts*. Once ingested, THC is strongly absorbed by fatty tissues in various organs. Metabolites of THC can be detected by standard urine-testing methods several days after a smoking session, and in heavy chronic users, traces can often be detected weeks after they have stopped using the drug.

The pharmacological effects of the drug on each person, like many other hallucinogenic compounds, depends on the user's experience, the strength of the marijuana, the expectations of the user, the environment in which it's used, the route of administration, and whether the drug is being combined with alcohol or other drugs or pharmaceutical agents.

Although some people feel no effects when they smoke marijuana, others experience feelings of euphoria or relaxation. Other effects include increased thirst and hunger commonly referred to as "the munchies." Some users may have bad experiences, with sudden feelings of anxiety and paranoia. The likelihood of this happening increases with the potency of the marijuana is used.

The short-term effects of marijuana include:

- Problems with memory and learning
- Distorted perception
- Trouble with thinking and problem solving
- Loss of coordination
- Increased heart rate or anxiety.

Signs that someone is high on marijuana include:

- Dizziness and trouble walking
- Silliness and giggling for no apparent reason
- Very red, bloodshot eyes
- Problems remembering things that just happened.

When the early effects fade, the user can become very sleepy.

Marijuana has serious effects on driving skills including decreased alertness, concentration, coordination, and response time. The effects can last up to twenty-four hours after smoking and make it difficult to judge distances and react to signals and sounds on the road.

Forensic scientists are frequently called upon to determine whether a substance contains marijuana. Although presumptive color tests can be conducted on extracts of herbal material, a quick way to determine if an unknown herbal material is marijuana is to microscopically determine the presence of specific cellular features (glandular trichomes). Although several cellular features are present, such as covering and cystolithic hairs, these features are common to many other plant materials. It is the presence of glandular trichomes containing orangish cannabis oil that distinguish the material as being derived from *Cannabis sativa*. This test is particularly useful for testing herbal samples that are mixed with tobacco and other nonrestricted herbal material. The most frequently encountered structures are three types of hairs or trichomes.

See also Cannabis; Controlled Substances; Drugs; Hashish

References
De Forest, P., R. E. Gaensslen, and H. C. Lee. *Forensic Science: An Introduction to Criminalistics.* New York: McGraw-Hill, 1983.
James, S. H., and J. J. Nordby. *Forensic Science: An Introduction to Scientific and Investigative Techniques.* Boca Raton, FL: Taylor and Francis, 2005.
Office of National Drug Control Policy, Drug Facts, Marijuana; http:// www.whitehousedrugpolicy.gov/drugfact/ marijuana/index.html (Referenced July 2005).
Saferstein, R. *Criminalistics.* 8th ed. Upper Saddle River, NJ: Pearson Education, 2004.
U.S. Drug Enforcement Administration, Drug Descriptions, Marijuana; http://www.dea.gov/ concern/marijuana.html (Referenced July 2005).
White, P. *Crime Scene to Court: The Essentials of Forensic Science.* Cambridge: Royal Society of Chemistry, 1998.

Markov, Georgi

This case of poisoning illustrates the difficulties associated with identifying poisons, as well as presents a suspected assassination.

Georgi Markov was a Bulgarian dissident who, through the BBC World Service, broadcast antigovernment programs to his country during the cold war years. In 1979, several years after defecting, he felt a sharp pain in his leg while waiting at a bus stop. When he turned around, a man with an umbrella apologized to him (Markov later recalled that the man had a very thick foreign accent) and walked away. Later that evening, the wound became red and swollen, and Markov became very ill with a high fever. Following three days of illness, Markov died. The postmortem

revealed that in his calf at the site of the wound was a small platinum pellet (around 1.5 mm diameter) with two holes in it. The pellet was suspected to contain ricin, an extremely toxic substance for which there is no known antidote.

It was later admitted by several high-profile KGB defectors that Markov's death was an assassination carried out by the Bulgarian secret police with the assistance of the Russian KGB.

See also Poisoning; Toxicology

References
Evans, C. *The Casebook of Forensic Detection: How Science Solved 100 of the World's Most Baffling Crimes.* New York: Wiley, 1998.
Owen, D. *Hidden Evidence: Forty True Crimes and How Forensic Science Helped Solve Them.* Willowdale, Ontario: Firefly, 2000.

Mass Disaster Victim Identification

Identification of the dead is a requirement of society, and most jurisdictions have clear legislative requirements allocating the responsibilities to certain groups that include police, coroners, and forensic specialists. These

The two towers of the World Trade Center burn after being struck by airliners on September 11, 2001. (http://www.bigfoto.com)

requirements ensure that any situation in which a person dies suddenly, violently, or unnaturally is investigated thoroughly. The process of identification of bodies resulting from mass disasters should not be any different than that of identifying a single individual. However, the process should provide double checks, as any error in the identification of a body may create doubt as to the identification of all the others. Applying at least two methods of identification is one way of preventing this kind of errors. The techniques applied for disaster victim identification (DVI) include:

Visual Recognition

Visual recognition is the most common, but probably the most unreliable, of identifying techniques. Difficulties in accurate identification can occur because of mutilation, time lapse since the witness last saw the deceased, and emotional trauma associated with the identifying procedure.

Fingerprints

Where the skin is undamaged and when exemplars are available, fingerprints are the fastest and most reliable means of identification. Sources of exemplars include records for military and law enforcement personnel and criminal identification records.

Odontology

Teeth are extremely hard wearing and durable; they are usually protected by soft tissue and may therefore be available for comparison purposes when other body parts are missing or partly destroyed. Forensic odontologists consider several parameters for identification purposes. These include the dentition (number, shape, and position of the teeth), any restoration work that has been carried out, and any diseases that can be identified. In some situations it may be necessary to conduct cranio-facial superimposition.

Medical Records or Medical Condition

Comparison of the condition of the body with the medical history of the missing person can produce both confirmations and eliminations. Congenital conditions and conditions associated with chronic disease can be very useful in the identification process. Repaired trauma, such as scarring and broken bones, can be used two ways. Any scars can be compared, and the x-rays used at the time can be compared with x-rays of the unidentified body. X-rays can also be useful in comparing other features such as the shape of sinuses.

Tattooing and Other Identifying Marks

The presence of tattoos and other identifying marks such as birthmarks can be used when they are present and the skin is not damaged. Some marks and tattoos are very characteristic and detailed.

DNA Testing

DNA testing is now an accepted first-line tool for identification of disaster victims. The same properties of robustness and discriminating power that make it a valuable tool in investigation of crimes such as rape make it valuable for identification of remains. It is possible to obtain reference material from personal possessions or by inference from blood samples taken from parents or other immediate relatives.

Mitochondrial DNA (mtDNA) typing has been especially valuable in some cases, because it can be carried out on bone and hair remains and because mtDNA is inherited solely from the mother. The maternal inheritance makes it easier to obtain exemplar samples.

Identification of remains at the scene of the September 11, 2001, attack on the World Trade Center presented the greatest challenge to mass disaster victim identification in recent years. According to a press release from the New York Academy of Science, researchers worked to match DNA samples from 14,249 body parts recovered from the wreckage of the twin towers to the 2,795 people who were thought to have died in the attacks. Of that number, just over half—or 1,508 people—were identified by DNA analysis as of late June 2003. Forensic scientists took up the monumental task of scrutinizing at least 20,000 remains, testing some of them up to six times for usable DNA. Unfortunately, the extreme heat and moisture inside the mountain of debris presented a highly hostile environment to this process.

Among the new approaches to DNA evidence to emerge are "MiniSTR" tests, a variation of the traditional DNA tool, short tandem repeat tests (STR), which can provide results from shorter, more degraded pieces of DNA. Genetic differences between individuals also can be detected through the analysis of single nucleotide polymorphisms (SNPs), as well as mitochondrial DNA, which is somewhat hardier and more plentiful than nuclear DNA. The latter helps the forensic team group remains into the families that they could belong to and may prove even more useful as new techniques to analyze this genome become available.

In addition, the data from these remains could be matched using software such as the Mass Fatality Identification System (M-FISys), which compares the results gleaned from forensic analysis against the DNA information collected by thousands of toothbrushes, razors, hairbrushes, and other personal items provided by the relatives of the victims. CODIS (Combined DNA Index System), a national DNA database, also can be used.

With millions of dollars already spent on the identification process, the rapid advances in forensic analysis reflect "the really compelling need for victims and families to have some closure," says Sarah Hart, director of the National Institute of Justice, the research arm of the Department of Justice. "We were seeing the agony of families on a mass scale here, but throughout the country, families go through this all the time when someone disappears."

Personal Effects and Possessions

If possessions are found in reasonable condition, they may be unique and easily identifiable. The group includes clothing, jewelry, wallets, identity papers, and so on.

Circumstances

The facts surrounding the incident can also be useful in identifying an unknown body. Airline listings, passenger listings, and travel documents can be used to link a person to a place.

Administrative Tracking

The International Criminal Police Organization, Interpol, provides a manual of disaster victim identification to all its member countries, and structured teams for DVI are in place. To facilitate the identification process Interpol provides a disaster victim identification form (Interpol DVI) that consists of two parts. A pink section is used to document information about the deceased; the yellow section documents information about the missing person. They are designed to be easily comparable. In a mass disaster situation, a pink form would be completed as much as possible for each body at the scene. The body and the form would be transported to the morgue where the form would be completed further during the postmortem examination. Police would obtain missing person information on the yellow form, and the two sets of forms for all bodies and missing persons would then be compared and matched in a process of reconciliation.

Scene

When intentional mayhem is suspected, a disaster scene for all intents and purposes is a large crime scene and should be treated as such in order to maximize the information obtained. Usually, scene investigators will not begin until any living persons have been evacuated and hazards removed. The scene would then be secured and the investigation begun. The usual approach is to search for bodies, mark their location with pegs or flags, photograph and record the body and the situation in which it was found, and then bag and remove the body to refrigerated transport.

Mortuary

The systematic recording that began at the scene continues at the mortuary. Forensic pathologists undertake postmortem examinations, and all the property that was collected with the bodies (clothing, jewelry, etc.) is cleaned, examined, and documented. Once the information obtained from the body is maximized, the reconciliation process is started. This involves checking matches, or conducting further tests as needed. The Interpol DVI form has been invaluable in these situations for facilitating the requests for specific medical or dental records from foreign countries for comparison purposes, and as a result, minimizing time delays, errors, or failures in communication.

Reconciliation

Once it is determined how many persons are missing and likely dead, the process of matching bodies to those reports can begin. The method by which missing person information is collected varies between organizations and countries. The information must always be checked and verified against the body being examined.

To maximize the information obtainable at the scene, bodies should be left in situ whenever possible. This can be very stressful to emergency services personnel. Parts of bodies are usually treated as separate bodies. In aircraft accidents, the impact forces can propel whole or dismembered bodies very large distances in all directions.

The time required for the DVI process can be measured in weeks or months. This depends on the size and type of event. Some or all the bodies may not be released until all have been identified.

See also Autopsy; Combined DNA Index System (CODIS); DNA in Forensic Science; Fingerprints; Identification; Odontology; Pathology

References

Butler, J. *Forensic DNA Typing.* 2nd ed. Burlington, MA: Elsevier, 2005.

Byard, R., T. Corey, C. Henderson, and J. Payne-James. *Encyclopedia of Forensic and Legal Medicine.* London: Elsevier Academic, 2005.

James, S. H., and J. J. Nordby. *Forensic Science: An Introduction to Scientific and Investigative Techniques.* Boca Raton, FL: Taylor and Francis, 2005.

Mass Spectrometry (MS)

Mass spectrometry (MS) is a detection method used in the analysis of chemical compounds. It is an adjunct to the techniques of high performance liquid chromatography (HPLC) and gas chromatography. In drug analysis the positive identification of the compound by chemical structure is required. Gas chromatography–mass spectrometry is the gold standard in structural identification.

Samples are vaporized and then separated by the gas chromatograph component of the instrument. Upon entrance to the mass spectrometer, the sample molecules are bombarded with a high-energy beam of electrons. The result of this is that ions are formed and some fragmentation of the molecules occurs. After this, the ions and fragments are sorted according to their "mass to charge" ratio, referred to as the *m/z ratio*. This is done by a part of the mass spectrometer known as the analyzer. Different types of mass spectrometer have different types of analyzers. Common types include quadrupole mass filter, quadrupole ion trap, and time-of-flight mass analyzers. This separation according to m/z value is achieved using magnetic fields because larger fragments will take a different path than smaller fragments in an electric field. By changing the strength of the magnetic field, it is possible in some analyzers (such as the quadrupole) to select which analytes will be allowed to pass through the analyzer to the detector. Following separation by the analyzer, the ions are then detected and a mass spectrum generated. Computer systems are essential for capturing, manipulating, interpreting, and storing mass spectrometry data.

A mass spectrum shows the m/z values of the separated fragments along the x-axis and their relative abundances on the y-axis. Mass spectrometry data can be used to qualitatively identify substances because the data provides structural information, allowing a definitive and specific identification to be made. MS software packages also provide means by which substances can be quantified, that is, a determination of the amount of substance present in a sample can be made.

Mass spectrometry is used widely in forensic science and particularly in the fields of drug chemistry, toxicology, and fire debris analysis. The technique is so commonly used because of its ability to provide both qualitative and quantitative data. It is a very sensitive technique and can detect limits of substances at the nanogram level (and as technology improves, it can detect lower and lower levels). It is also a very specific technique; because the molecules fragment in a predictable way, structural information is provided to allow the specific identification of the substance being analyzed.

It should be noted, however, that mass spectrometry is an expensive technique that requires skilled analysts able to perform complex analysis and interpretation, as well as the high maintenance often required by the instrument.

See also Gas Chromatography
References
De Forest, P., R. E. Gaensslen, and H. C. Lee. *Forensic Science: An Introduction to Criminalistics.* New York: McGraw-Hill, 1983.
James, S. H., and J. J. Nordby. *Forensic Science: An Introduction to Scientific and Investigative Techniques.* Boca Raton, FL: Taylor and Francis, 2005.
Saferstein, R. *Criminalistics.* 8th ed. Upper Saddle River, NJ: Pearson Education, 2004.
White, P. *Crime Scene to Court: The Essentials of Forensic Science.* Cambridge: Royal Society of Chemistry, 1998.
Yinon, J. *Forensic Applications of Mass Spectrometry.* Boca Raton, FL: CRC, 1994.
———. *Advances in Forensic Applications of Mass Spectrometry.* Boca Raton, FL: CRC, 2004.

MDMA

See *Ecstacy*

Mengele, Josef

Dr. Josef Mengele was the chief medical officer at the Nazi concentration camp Auschwitz and was held responsible for the cruel deaths of at least 400,000 innocent people. On his orders, prisoners entering the camp were separated into those fit to work and those not. Those who were not were killed either in the gas chamber or by lethal injection into the heart. Mengele frequently carried out these acts himself. His cruelty has been unrivaled. He was known to dissect live infants, castrate men with no anesthetic, and torture women.

He frequently carried out "medical experiments" in order to perfect the Aryan race. He frequently performed work on twins and dwarfs. One of the experiments he was known to perform was injecting dye into the eyes of children to determine if eye color could be changed. It was reported that if any of these children died, he took their eyes and pinned them to the wall of his office.

After the war, Mengele escaped capture by fleeing to South America and apparently working as a farmhand under an assumed name. In 1985 the United States vowed that

Josef Mengele in Paraguay in 1960. The doctor from the death camp at Auschwitz escaped prosecution and died in 1979. (Bettmann/Corbis)

Josef Mengele would be found and brought to justice. However, authorities were informed by a German couple who lived in Brazil of a grave purported to be that of Josef Mengele. The body was exhumed and examined by forensic teams from the United States, Germany, and Austria. Their examination revealed that the bones were those of a Caucasian male, probably right-handed, and between sixty and seventy years old. The decedent would have been 5 feet 8.5 inches tall (as was Mengele), and dental examination correlated with hand-drawn dental records of Mengele from 1938. One German examiner used the technique of superimposition to compare identifying features of the skull with a photograph of Mengele. At least thirty points were shown to match. The most definitive evidence was provided in 1992 when DNA analysis confirmed that the exhumed body was that of Josef Mengele.

See also DNA in Forensic Science; Identification
References
  Court TV's Crime Library, Criminal Minds and Methods; http://www.crimelibrary.com/serial_killers/history/mengele/index_1.html (Referenced July 2005).
Evans, C. *The Casebook of Forensic Detection: How Science Solved 100 of the World's Most Baffling Crimes.* New York: Wiley, 1998.
Holocaust, The, Crimes, Heroes and Villains, Victims of Mengele; http://auschwitz.dk/Mengele/id17.htm (Referenced July 2005).
Owen, D. *Hidden Evidence: Forty True Crimes and How Forensic Science Helped Solve Them.* Willowdale, Ontario: Firefly, 2000.

Methadone

Methadone was initially developed by German scientists during World War II due to a shortage of morphine and is chemically dissimilar to heroin or morphine.

The 1960s heralded methadone as a cure for heroin addiction, and presently it is primarily used to control narcotic addiction and to ease the discomfort of the accompanying withdrawal symptoms. The effects of a single methadone dose can last up to twenty-four hours. In maintenance, it is administered orally under controlled conditions and is usually accompanied by some form of rehabilitation program. Although in some cases methadone can be helpful, it can also convey tolerance and addiction. In certain situations this has resulted in the heroin addict trading addictions from one drug to the other. Methadone is normally ingested orally in liquid form, but it can also be injected.

Methadone is a DEA Schedule II controlled substance. Drugs in this class have current medical use and high potential for abuse. Other examples of Schedule II drugs are hydromorphone, meperidine, cocaine, phencyclidine (PCP), morphine, and certain cannabis, amphetamine, and barbiturate types.

Methadone users show the common signs of narcotics abuse seen with other narcotic drugs like heroin, codeine, and morphine. These drugs cause the following pharmacological and behavioral effects:

- Lethargy and drowsiness
- Constricted pupils that fail to respond to light
- Scars (tracks) on inner arms or other parts of body from needle injections
- Use or possession of paraphernalia, including syringes, bent spoons, bottle caps, eyedroppers, rubber tubing, cotton, and needles
- Slurred speech

See also Controlled Substances; Drugs; Heroin; Morphine
References
Levine, B. *Principles of Forensic Toxicology.* Washington, DC: American Association for Clinical Chemistry, 1999.
Saferstein, R. *Criminalistics.* 8th ed. Upper Saddle River, NJ: Pearson Education, 2004.
U.S. Drug Enforcement Administration, Drug Descriptions, Methadone; http://www.whitehousedrugpolicy.gov/publications/factsht/methadone/ (Referenced July 2005).

Methamphetamine

Methamphetamine is a synthetically produced central nervous system stimulant. Chemically, it is very similar to amphetamine, but meth-amphetamine has a greater effect on the central nervous system. Both drugs are Schedule II substances, meaning they have limited medical applications coupled with a very high potential for abuse.

Effects

Methamphetamine is taken intravenously or orally or is smoked or snorted. After intravenous injection or smoking, the user feels an intense sensation known as a "rush" that lasts only a few minutes. Oral ingestion or snorting produces a high or euphoria, but not the "rush" experience. Users can rapidly become addi-cted to methamphetamine. The body will develop a tolerance to the drug, requiring increasingly larger doses to achieve the same effect.

A person who takes methamphetamine may experience some or all of the following symptoms:

- Increased heart rate and respiration
- Damage to blood vessels
- Decreased appetite
- Insomnia
- Irritability
- Anxiety
- Tremors
- Paranoia
- Aggressiveness

Preparations

Methamphetamine is accessible in a variety of forms. Illicit methamphetamine is most commonly encountered as a powder. It is bitter in taste, water soluble, and ranges in color from white to reddish brown, depending on the manufacturing process. Powder methamphetamine is typically injected, snorted, or orally ingested.

In the 1980s the smokable form of methamphetamine came into vogue. Sometimes referred to as "ice" or "rock candy," this version of methamphetamine is usually a clear crystal-like material with greater than 90 percent purity.

Methamphetamine tablets, referred to as "Yaba," are seen most frequently in the western states. These tablets are manufactured in Southeast and East Asia, and arrive in the United States by mail, courier, air, and maritime cargo.

Cuts

As is true for any illicit drug, methamphetamine is "cut" or diluted with a variety of substances that may include the following:

- Sugars—mannitol, inositol
- Caffeine
- Nicotinamide
- Niacin
- Dimethylsulfone—a nonregulated veterinary food supplement
- Heroin
- Cocaine

Paraphernalia

Paraphernalia associated with methamphetamine use includes:

- Syringes
- Spoons
- Glass pipes
- Glass tubes with a glass bulb at one end fashioned from lightbulbs
- Hollow tubes (empty pen casings, straws, or aluminum foil)

Street Names

Methamphetamine is known by a number of street names, including "meth," "speed," "crank," "crystal," "ice," "glass," and "zip."

Manufacture

Methamphetamine was developed during the first part of the twentieth century from its parent drug, amphetamine. Initially, methamphetamine was manufactured for use in nasal decongestants, bronchial inhalers, in the treatment of narcolepsy, and for weight control. In the 1970s methamphetamine became a Schedule II drug.

Although some legitimately produced methamphetamine is diverted for illicit use, most street methamphetamine is manufactured clandestinely. The illicit production and distribution of methamphetamine was originally controlled by outlaw motorcycle gangs. In the early 1990s, Mexican drug-trafficking organizations gained control of the methamphetamine market. They manufacture methamphetamine in large clandestine labs in Mexico and California and distribute it across the United States. Small, independent laboratories also play a role in methamphetamine production. Simple, makeshift clandestine laboratories, mostly in the Midwest and southeastern part of the United States, produce and distribute methamphetamine on a smaller scale.

Clandestine methamphetamine labs are found in a variety of locations, including apartments, hotel rooms, rented storage units, and vehicles. Many meth labs are portable and easily dismantled, stored, or moved. This mobility facilitates evasion of law enforcement.

At one time, the P2P amalgam method was the most common synthesis used to manufacture illicit methamphetamine. Phenyl–2-propanone (P2P) and methylamine are the primary precursors. In 1980 P2P became a Schedule II substance, compelling the meth "cooks" to manufacture their own P2P or obtain it on the black market. Due to this difficulty, clandestine lab operators switched to other methods requiring more readily available precursors.

The ephedrine/pseudoephedrine reduction synthesis is easier than the P2P method and produces a higher yield of methamphetamine. It is the method most commonly associated with Mexican drug traffickers. The required starting compound is ephedrine or pseudoephedrine. Because of its popularity as a precursor to methamphetamine, ephedrine has become controlled and difficult to obtain. Pseudoephedrine is more easily acquired by extracting it from over-the-counter cold medications. Other necessary chemicals for the synthesis are red phosphorous and iodine.

Red phosphorus is purchased at chemical supply stores or scraped from the striker plates of road flares or matchbooks. Iodine crystals are available at veterinary supply stores.

The Nazi or Birch method for producing methamphetamine was developed during World War II in Germany. This method is very popular among independent manufacturers and dealers. The process is quick, there is little setup, and it produces a very high yield. Ephedrine or pseudoephedrine is again the primary precursor, along with lithium metal and anhydrous ammonium. Often used as fertilizer, anhydrous ammonium is purchased at industrial supply companies or stolen from farms. Lithium metal is found in batteries.

Regardless of the synthesis process used, all clandestine labs are dangerous and should be approached only by specially trained personnel. A variety of booby traps are commonly employed by clandestine lab operators. Many of the chemicals used are extremely hazardous to people and the environment.

See also Amphetamines; Clandestine Drug Laboratories; Controlled Substances; Drugs
References
Christian, D. R. *Forensic Investigation of Clandestine Laboratories.* Boca Raton, FL: CRC, 2004.
Drummer, O.H. *The Forensic Pharmacology of Drugs of Abuse.* New York: Oxford University Press, 2001.
James, S. H., and J. J. Nordby. *Forensic Science: An Introduction to Scientific and Investigative Techniques.* Boca Raton, FL: Taylor and Francis, 2005.
Office of National Drug Control Policy, Drug Facts, Methamphetamine; http://www.whitehousedrugpolicy.gov/drugfact/methamphetamine/index.html (Referenced July 2005).
Saferstein, R. *Criminalistics.* 8th ed. Upper Saddle River, NJ: Pearson Education, 2004.
U.S. Drug Enforcement Administration, Drug Descriptions, Methamphetamine/Amphetamines; http://www.usdoj.gov/dea/concern/amphetamines.html (Referenced July 2005).

Methanol and Methanol Poisoning

Methanol is an industrial solvent that is sometime consumed by people believing it is "alcohol." Unfortunately, methanol is extremely toxic. It is metabolized by alcohol dehydrogenase to produce formic acid, which is responsible for most of the toxic effects. These include headache, confusion, and symptoms arising from the action of the formic acid on the optic nerve. Treatment consists of intravenous infusion of ethanol to saturate the alcohol dehydrogenase and so prevent formation of formic acid by oxidation of the methanol.

See also Poisoning; Toxicology
Reference
Goldfrank, L., N. Flomenbaum, N. Lewin,
    M. A. Howland, R. Hoffman, and L. Nelson.
    Goldfrank's Toxicologic Emergencies. 7th ed.
    New York: McGraw-Hill Professional, 2002.

Microcrystal Tests

Crystal tests are used to determine the presence of a chemical substance. The basis of the test is that the chemical of interest combines with a reagent to produce a unique crystal form.

The main application of microcrystal tests today is in the examination of controlled substances. These tests are an effective means to deal with high numbers of drug cases encountered by laboratories. The test produces crystal derivatives from simple molecular addition of compounds of the substance tested and a test reagent. The crystal derivatives are then observed using a polarizing microscope. Many microcrystal tests have been considered to be so characteristic that they could be employed as confirmatory tests. However, few analysts have such microscopy or crystallographic expertise that they can unequivocally identify a drug microcrystal using microscopy alone. Hence, there is some controversy as to their reliability other than in the hands of well-trained and experienced personnel. Microcrystal tests are useful for the identification of compounds that contain basic nitrogen and readily precipitate from aqueous solution.

Tests applied for chemical identification purposes can be classified into direct and indirect tests. Direct tests include precipitation tests and color tests; the substance being tested is responsible for the crystals produced. Indirect tests include negative tests, reagent tests, and derivative tests. The types of test are described below:

- Negative tests. The test is the failure of a specific action or reaction to occur.
- Reagent tests. A color change or precipitate formation is due to some change in the reagent caused by the test substance.
- Color tests. Largely reagent tests, these involve a change in color of the reagent by an oxidation or reduction reaction.
- Derivative tests. The drug-reagent derivative formed becomes the compound to be tested.

Microcrystal test procedure involves adding a drop of reagent to a drop of a solution of the test substance on a flat glass microscope slide and observing the product at frequent intervals until crystals form or the solution evaporates and microscopic information can be acquired. These tests are useful for quick reactions in which the crystals are expected to form immediately or within a matter of minutes. When the test substance has dispersed or dissolved in the test reagent and the mixture becomes saturated, crystal formation occurs. Supersaturation may occur as a result of evaporation of the solvent, and any impurities present in the test sample will also be concentrated. This will hinder or distort crystal development.

Forensically, the microcrystal test is an efficient means of preliminary identification due to the directness, simplicity, sensitivity, and rapidity of the technique, as well as the ability to conserve evidence. Quantities needed for identification of alkaloids are in the range of 1 microgram (µg) to 100 nanograms (ng).

Alkaloid precipitating agents are aqueous solutions of acids, salts, or bases with sufficient acid to prevent basic hydrolysis. The agents precipitate ptomaines, proteins, basic nitrogenous compounds, or amines. The most common reagents are gold and platinic chloride.

Microcrystal reaction products are usually examined by polarized light microscopy. Without a cover slip, at 100x magnification, crystals can be seen forming at the edge of the reaction drop. Microscopic examination can be used to determine the size, form, color, and aggregation characteristics of these crystals. Distortions of crystal form resulting in crystal di- or polymorphism can be caused by changes in ambient temperature and humidity and the concentration of the reagent or substance tested. It should be noted that a colored test reagent may result in the formation of colored crystals.

See also Drugs; Microscope

References

Fulton, C. C. *Modern Microcrystal Tests for Drugs.* New York: Wiley, 1969.

Saferstein, R. *Criminalistics* 8th ed. Upper Saddle River, NJ: Pearson Education, 2004.

Microscope

A microscope is an instrument that is used to magnify an object, making fine details visible. The simplest microscope is a magnifying glass. Light from the object is bent, or refracted, as it passes through the curved lens and into the eye, where it forms a magnified virtual image. Magnifying glasses have a limit to their useful magnification—about five to ten times the true size of the object. Obtaining higher magnifications requires the use of a compound microscope, constructed with two or more lenses organized in the correct manner. The image is viewed through the upper lens, or eyepiece; the lower lens is the objective. In light microscopy the radiation is collected by the lenses and converted to a visible image of the magnified object.

The optical and mechanical components of the microscope, including the specimen mounted on a glass slide, are aligned along the central axis of the microscope. These components are mounted on a stable base that enables them to be centered, aligned, or changed easily to provide the best magnification image.

The objective lenses are the most important components of an optical microscope because they determine the quality of images that the microscope is capable of producing. The eyepieces are designed to work in combination with the objectives to reduce chromatic aberrations and to magnify an intermediate image, allowing the details of the specimen to be clearly observed. There are a wide range of objectives designed to provide excellent optical performance and to eliminate most optical aberrations. Standard bright field objectives, corrected for varying degrees of optical aberration, are most common and are useful for examining specimens with traditional illumination techniques. The numerical aperture of a microscope objective is a measure of its ability to gather light and resolve fine detail at a fixed object distance. The resolution of a microscope objective can be defined as the smallest distance between two points on a specimen that can still be distinguished as two separate entities. Although the numerical aperture determines the resolving power of an objective, the total resolution of a microscope system also depends on the numerical aperture of the substage condenser. As the numerical aperture of a total system increases, so does the resolution. Most low-power objectives can be used with air as the imaging medium, and higher magnification objectives generally use liquid immersion media to minimize aberrations and increase the numerical aperture.

The condenser is important for obtaining high-quality images. The substage condenser gathers light from the microscope light source and concentrates it into a cone of light that illuminates the specimen uniformly over the entire field of view. The condenser light cone can be adjusted to optimize the intensity and angle of light entering the objective front lens.

Microscopes usually have a moveable stage to hold the specimen slide in place for observation that can translate the slide back and forth and from side to side. Other stages may be rotated through 360 degrees or may have anchors for auxiliary light sources and accessories.

Illumination of the specimen is the most important variable in achieving high-quality images in microscopy and critical photomicrography. The optimum illumination method for both practices is Kohler illumination (named after a German scientist, August Kohler) that produces even, glare-free, specimen illumination throughout the field of view.

Modern microscopes are designed in a way that the collector lens and any other optical components built into the base of the microscope will project an enlarged and focused image of the lamp filament onto the plane of the aperture diaphragm of a properly positioned substage condenser. Closing or opening the condenser diaphragm controls the angle of the light rays emerging from the condenser and reaching the specimen. Because the light source is not focused at the level of the specimen, the light at specimen level is essentially grainless and extended and does not suffer deterioration from dust and imperfections on the glass surfaces of the condenser. The setting of the condenser's aperture diaphragm, along with the aperture of the objective, determines the realized numerical aperture of the microscope "system." As the condenser diaphragm is opened, the working numerical aperture of the microscope increases, resulting in greater resolution and light transmittance. Parallel light rays that pass through and illuminate the specimen are brought to focus at the back focal plane of the objective, where the image of the variable condenser aperture diaphragm and the image of the light source will be seen in focus.

Some specialized versions of compound microscopes are used in forensic science. They include the comparison microscope, which consists of two compound microscopes joined by a bridge and arranged so that the images on each can be seen side by side. The low-power stereomicroscope is useful for screening debris. The polarized light microscope is useful in fiber identification and microcrystal analyses.

See also Document Examination; Fibers; Glass; Hair; Microspectrophotometry; Paint; Trace Evidence

References

De Forest, P., R. E. Gaensslen, and H. C. Lee. *Forensic Science: An Introduction to Criminalistics.* New York: McGraw-Hill, 1983.

Florida State University, Molecular Expressions, Exploring the World of Optics and Microscopy; http://micro.magnet.fsu.edu/ (Referenced July 2005).

Saferstein, R. *Criminalistics.* 8th ed. Upper Saddle River, NJ: Pearson Education, 2004.

White, P. *Crime Scene to Court: The Essentials of Forensic Science.* Cambridge: Royal Society of Chemistry, 1998.

Microspectrophotometry

Color analysis has been applied for decades in the pigment, paint, dyestuff, and fabric industries. There are numerous methods of color measurement and description. Colors can be described by systems such as those of Munsell and the Commission International de l'Eclairage (CIE). These systems can be used to classify colors for database systems, but usually absorption spectra of known and questioned samples are directly compared in forensic color comparisons. Microspectrophotometry can provide objective color data for paint comparison from small samples. The technique can be applied to the outer surfaces of paint films by diffuse reflectance (DR) measurements with visible spectrum illumination. Diffuse reflectance measurements of paint surfaces are affected significantly by surface conditions such as weathering, abrasion, contamination, and texture, and can provide useful discriminating information.

Comparison of paint layers by transmission microspectroscopy of thin cross sections offers a more definite form of color analysis for these samples compared to reflectance techniques. Transmission microspectroscopy requires consistent sample thickness and choice of measurement size and location for meaningful comparisons.

See also Fibers
References
James, S. H., and J. J. Nordby. *Forensic Science: An Introduction to Scientific and Investigative Techniques.* Boca Raton, FL: Taylor and Francis, 2005.

Saferstein, R. *Criminalistics.* 8th ed. Upper Saddle River, NJ: Pearson Education, 2004.

White, P. *Crime Scene to Court: The Essentials of Forensic Science.* Cambridge: Royal Society of Chemistry, 1998.

Morphine

Morphine is a natural opium alkaloid with strong analgesic properties. Like all opiate drugs, morphine produces euphoria and respiratory and physical depression. The combination of its analgesic, sedative, and cardiovascular effects makes this Schedule II drug extremely useful in emergency care. Morphine is marketed in a variety of forms, including oral solutions (Roxanol), sustained-release tablets (MSIR and MS-Contin), suppositories, and injectable preparations.

Morphine exerts its principal pharmacological effect on the central nervous system (CNS) and gastrointestinal tract. Morphine increases the patient's tolerance to pain and decreases discomfort. Alterations in mood, euphoria, dysphoria, and drowsiness commonly occur. CNS respiratory centers and the cough reflex are depressed, pupils become constricted, and nausea and vomiting may occur through direct stimulation of the chemoreceptor trigger zone. Morphine increases the tone and decreases the propulsive contractions of the smooth muscle of the gastrointestinal tract, causing constipation. Other commonly experienced effects include dizziness, sedation, dysphoria, euphoria, and sweating.

Once absorbed, morphine is distributed to the skeletal muscles, kidneys, liver, intestinal tract, lungs, spleen, and brain. Morphine also crosses the placental membrane and has been found in breast milk. Once in the circulation system, morphine is converted to glucuronide metabolites, which also have some analgesic activity; less than 5 percent of absorbed morphine is demethylated. Following oral administration, most morphine is metabolized to morphine–3-glucuronide; a smaller fraction is converted to morphine–6-glucuronide, which has the greater analgesic activity of these two metabolites. Morphine overdose is characterized by respiratory depression. The narcotic antagonist, naloxone, is a specific antidote. Other adverse reactions caused by morphine are typical of those observed with other opiate drugs and include respiratory depression, circulatory depression, apnea, shock, and cardiac arrest secondary to respiratory and/or circulatory depression.

Opioid analgesics may cause psychological and physical dependence. Physical dependence results in withdrawal symptoms in patients who abruptly discontinue the drug, or via the administration of drugs with antagonistic activity, such as naloxone. Physical dependence usually results from several weeks of continued opiate drug use. Tolerance to the drug is also a common occurrence, in which increasingly larger doses are required to produce the same degree of analgesia or euphoria. Withdrawal symptoms include yawning, sweating, lacrimation, rhinorrhea, restless sleep, dilated pupils, gooseflesh, irritability, tremor, nausea, vomiting, and diarrhea. Depressant effects of morphine are accentuated by the presence of other CNS depressants such as alcohol, sedatives, antihistamines, or psychotropic drugs.

See also Controlled Substances; Drugs; Heroin; Methadone; Morphine
References
Drummer, O.H. *The Forensic Pharmacology of Drugs of Abuse.* New York: Oxford University Press, 2001.

Levine, B. *Principles of Forensic Toxicology.* Washington, DC: American Association for Clinical Chemistry, 1999.

Saferstein, Richard. *Criminalistics.* 8th ed. Upper Saddle River, NJ: Pearson Education, 2004.

White, P. *Crime Scene to Court: The Essentials of Forensic Science.* Cambridge: Royal Society of Chemistry, 1998.

N

Napalm

Napalm is the generic name for incendiary devices that burn with an intense heat and are difficult to extinguish. Originally a mixture of naphthalene and palmitate (hence the name), napalm was then applied to any mixture of gasoline and oil in a gel matrix. The gel or thickener caused the flammable material to stick to its target. Apart from military uses, napalmlike materials are used by criminals in bombs intended to destroy and maim. The laboratory investigation is directed to identification of the accelerant and thickener (such as polystyrene).

See also Accelerant Residues; Arson

References
Almirall, J., and K. Furton. *Analysis and Interpretation of Fire Scene Evidence.* Boca Raton, FL: CRC, 2004.
Nic Daeid, N. *Fire Investigation.* Boca Raton, FL: CRC, 2004.

Narcotics

Narcotic, derived from the Greek word for "stupor," was a term used to classify a variety of substances that induced sleep. In a legal context, *narcotic* refers to opium, opium derivatives, and their semisynthetic or synthetic substitutes. Cocaine and coca leaves, classified as narcotics in the Controlled Substances Act (CSA), are not actually narcotic or sleep-inducing in nature.

Narcotic drugs can be administered orally, transdermally (as skin patches), intravenously, or by suppository. As drugs of abuse, they are smoked, snorted, and self-administered subcutaneously ("skin popping") and intravenously ("mainlining").

The ensuing drug effects depend on the dose, route of administration, previous exposure to the drug, and the expectation of the user. These substances have clinical applications in the treatment of pain, cough suppression, and acute diarrhea, and produce a general sense of well-being by reducing tension, anxiety, and aggression.

Narcotic use comes with a variety of unwanted effects, including drowsiness, inability to concentrate, apathy, lessened physical activity, constriction of the pupils, dilation of the subcutaneous blood vessels causing flushing of the face and neck, constipation, nausea, vomiting, and respiratory depression.

As with most forms of illicit drug use there is a high risk of infection, disease, and overdose. Medical complications arise mainly from the adulterants found in street drugs and nonsterile injecting habits. Skin, lung, and brain abscesses, hepatitis, and AIDS are common among narcotic abusers. As there is no simple way to determine the purity of a street drug, the subsequent pharmacological effects are unpredictable and often fatal.

Repeated use of narcotics leads to tolerance and physical dependence. Generally, narcotics that exhibit shorter durations of action tend to produce shorter, more intense withdrawal symptoms, while drugs that produce longer narcotic effects have prolonged symptoms that tend to be less severe.

Early withdrawal symptoms include watery eyes, runny nose, yawning, and sweating. Restlessness, irritability, loss of appetite, tremors, and severe sneezing appear as the syndrome progresses, followed by severe depression and vomiting. Chills alternating with flushing and excessive sweating are also characteristic. Without intervention, the syndrome will run its course, and most of the physical symptoms will disappear within seven to ten days.

Long after the physical need for the drug has passed, an addict may continue to think and talk about the use of drugs. There is a high probability that relapse will occur after narcotic withdrawal when neither the physical environment nor the behavioral motivators that contributed to the abuse have been altered.

Narcotic abuse may involve individuals initially prescribed narcotics for therapeutic purposes, and who subsequently increased their dosage through further medical intervention, but the most common pattern of abuse begins with experimental or recreational use of narcotics.

See also Controlled Substances; Drugs
References

Drummer, O. H. *The Forensic Pharmacology of Drugs of Abuse.* New York: Oxford University Press, 2001.

Levine, B. *Principles of Forensic Toxicology.* Washington, DC: American Association for Clinical Chemistry, 1999.

Saferstein, R. *Criminalistics.* 8th ed. Upper Saddle River, NJ: Pearson Education, 2004.

White, P. *Crime Scene to Court: The Essentials of Forensic Science.* Cambridge: Royal Society of Chemistry, 1998.

National Crime Information Center (NCIC)

The National Crime Information Center, which was established by the Federal Bureau of Investigation, houses a computerized database containing documented criminal justice information. Easily accessed by any authorized criminal justice or law enforcement agency, the system rapidly provides information about crimes and criminals. The purpose of this information is to assist law enforcement agencies in apprehending fugitives, finding missing persons, locating stolen property, and helping protect the law enforcement agents who encounter and apprehend the individuals described in the system. The data contained in the NCIC is provided by the FBI; other federal, state, and local law enforcement and criminal agencies; and authorized foreign agencies.

Reference
Federal Bureau of Investigation, National Crime Information Center; http://www.fas.org/irp/agency/doj/fbi/is/ncic.htm (Referenced July 2005).

National Integrated Ballistic Information Network (NIBIN)

The increasing incidence of firearms-related crimes has perpetuated the development of more efficient means of sharing crime data and important information between investigating agencies. In efforts to maintain the accurate flow and exchange of information, the Federal Bureau of Investigation and the Bureau of Alcohol, Tobacco, and Firearms (BATF) have combined their individual firearms databases, Drugfire and Ceasefire, into the National Integrated Ballistic Information Network (NIBIN). The NIBIN system collates digital images of bullets and cartridges that are then analyzed using computer software and stored in an electronic database for matching against bullets or cartridges fired from the same gun. By combining both systems, evidence analysis time is reduced from weeks to minutes. It is now possible to compare bullets recovered from a crime scene

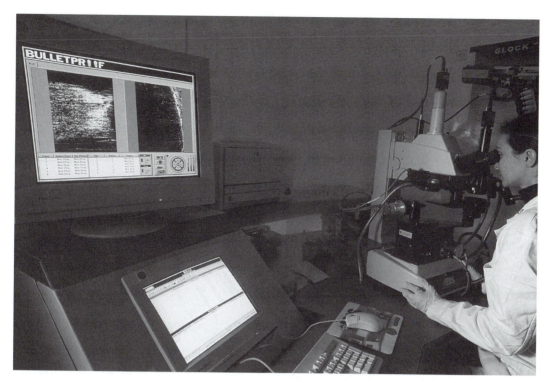

Forensic scientist using a microscope to compare markings on bullets. The barrel of a gun has grooves on the inside that makes a bullet spin to improve accuracy when firing. These grooves leave marks on bullets fired from the gun, and are unique to each gun. If a bullet can be recovered from a crime scene, the markings can be analyzed and compared to markings on bullets test-fired from suspect guns. If these markings match, then the gun was used to fire that bullet. The markings (seen on the screen) can also be kept on file in a national database network and used if the gun is recovered at a later date. (Mauro Fermariello / Photo Researchers Inc.)

with evidence from crime scenes across the country and to identify suspects and leads by quickly generating the most likely candidates for matching crime evidence. The database contains images of criminal evidence directly connected with gun crimes and images from test firings with crime- and noncrime-related guns. The BATF is now working closely with gun manufacturers in making digital images of their stock before it is distributed to commercial retailers, allowing easier tracking of hardware and serial numbers.

See also Ammunition; Bullets; Firearms
Reference
Bureau of Alcohol, Tobacco, Firearms and Explosives, National Integrated Ballistic Information Network; http://www.nibin.gov/ (Referenced July 2005).

National Tracing Center (NTC)

The National Tracing Center (NTC) of the Bureau of Alcohol, Tobacco, and Firearms (BATF) provides a system to trace firearms recovered from crimes worldwide. The system currently holds more than 100 million firearms records, including the records of firearms dealers who are no longer in business. The system assists law enforcement agencies by allowing firearms examiners access to the NTC's reference library for the identification of firearms, firearms manufacturers, and importers. It also provides information to confirm the accuracy of firearms traces and identifies trends and trafficking patterns for investigators. The center is also responsible for collecting, collating, and generating firearms statistics for each state in the United States and can provide investigative leads for the law enforcement community by identifying persons such as unlicensed firearms dealers.

Reference
U.S. Department of State, International Information Programs; http://usinfo.org/usia/ usinfo.state.gov/topical/pol/arms/stories/ 01061323.htm (Referenced July 2005).

Nitrocellulose

Nitrocellulose is a constituent of modern gun powder, usually termed *smokeless powder*. It may be the sole active ingredient ("single base") or combined with nitroglycerin ("double base"). Traces of unburned nitrocellulose can be detected by testing for nitrates. Tests used include the Griess test and the dermal nitrate test. The dermal nitrate test is not used today. It was the active step in the paraffin wax test whereby a mold was taken of the hand of someone suspected of firing a handgun. The mold was peeled off and sprayed with diphenylamine, which turns blue in the presence of nitrates. However, nitrate traces can be deposited on the hands from many sources, including tobacco and fertilizer. In an infamous British case—the Birmingham Six—a review of evidence that led to the conviction of suspected IRA terrorist bombers concluded that nitrate test results should not have been considered, as it was possible that nitrate traces had been deposited on the hands of some of the suspects from their handling of playing cards.

Unlike the dermal nitrate test, the Griess test is still used today in measurements of discharge distance. The usual form of the test is to press photographic paper onto the garment over the bullet hole. The paper is then sprayed with Griess reagent and nitrites produced from nitrocellulose residues to produce a pink color. The diameter of the residue-containing area around the bullet hole is used to estimate discharge distance.

Nitrocellulose is also encountered in another quite different setting in forensic science: Some of the manipulations in DNA analysis involve transfer of an immobilization of DNA fragments from an electrophoresis gel to a stable support medium. The support medium is usually a nitrocellulose membrane.

See also Black Powder; Color Tests; Explosions and Explosives

References
Saferstein, R. *Criminalistics.* 8th ed. Upper Saddle River, NJ: Pearson Education, 2004.
White, P. *Crime Scene to Court: The Essentials of Forensic Science.* Cambridge: Royal Society of Chemistry, 1998.

Nitroglycerin

Nitroglycerin (also spelled nitroglycerine) is the component found with nitrocellulose in double-base smokeless powder. It was the result of Alfred Nobel's research into explosives. Nitroglycerin is still used in explosives today, although it is not found in most modern material bearing the name *dynamite*.

Nitroglycerin has an important use as a medicine for treating heart diseases such as angina. It can thus be encountered in forensic science in the **firearms** section, in the **trace evidence** section, and in the **toxicology** section.

See also Dynamite; Explosions and Explosives; Nitrocellulose

References
Saferstein, R. *Criminalistics.* 8th ed. Upper Saddle River, NJ: Pearson Education, 2004.
White, P. *Crime Scene to Court: The Essentials of Forensic Science.* Cambridge: Royal Society of Chemistry, 1998.

O

Objective Test

There are several definitions of *expert witness* but they all include recognition that the witness gives opinion evidence, unlike lay witnesses. This is somewhat ironic given that forensic science is also under attack for being too dependent on the subjective views of the scientists. For example, trace evidence involving comparison of hairs is based entirely on microscopic comparison of test and control hairs and the conclusion reflects the judgment of the examiner. The attacks have been countered by an increasing emphasis on quality systems and the recognition of practitioners, for example by registration or board certification, and by the accreditation of facilities.

The major international accreditation programs comply with the standards set out by the International Organization for Standardization in ISO 17025. These standards are generic, but there are several guidance resources for acceptable practice in different industries. Chief among them is the International Laboratory Accreditation Cooperation or ILAC. ILAC publishes guides for different testing areas, including one for forensic science. The forensic science guide is built around the concept of objective tests, and the definition and discussion below reflect the thinking of the ILAC guide.

An objective test is one that, having been documented and validated, is under control so that it can be demonstrated that all appropriately trained staff will obtain the same results within defined limits. Each element of the definition is important, but the part that addresses the concern of critics as to what they regard as subjective testing is the requirement that all appropriately trained staff will obtain the same results within defined limits.

Objective tests are controlled tests. Procedures used to ensure control include:

- Documentation of the test
- Validation of the test
- Training and authorization of staff
- Maintenance of equipment
- Calibration of equipment
- Use of appropriate reference materials
- Provision of guidance for interpretation
- Checking of results
- Testing of staff proficiency
- Recording of equipment/test performance

If these rules are correctly applied, then tests such as microscopy, tool marks, questioned documents, and even polygraph examinations can all be controlled and objective.

See also Accreditation

Odontology

Forensic odontology is the use of dental information and records in personal identification. This can be applied in the identification of living individuals, sources of bite marks, and in the identification of human remains.

Odontology can be a powerful tool for identification of human remains, whether at the site of a mass disaster, a highly decomposed body recovered from a river, incinerated remains at a fire scene, or a skeleton. The basis of the identification is comparison of the dentition in the remains with dental records. There is a reasonable degree of individualization in dentition due to the variations in size and structure of jaws and teeth. If we add environmental changes, such as fillings, extractions, and even wear patterns in pipe smokers and teeth grinders, we find that there is a very high degree of individualization in adults. Dental information can be used for identification of children, with the age-dependent changes to adult dentition and, in some cases, records of orthodontic interventions to deal with malformed or cosmetically undesired patterns of tooth eruption and alignment. However, all instances depend on the existence of dental records, whether they are charts or x-rays.

If records do exist, then forensic odontology offers a rapid and reliable means of identification of remains. The jaws and teeth are hard materials not subject to easy decomposition and degradation. Dentition patterns do not change after death. In contrast, DNA testing can take several months to complete, especially in a mass disaster with many body parts to type. In some cases, postmortem degradation can make DNA testing more

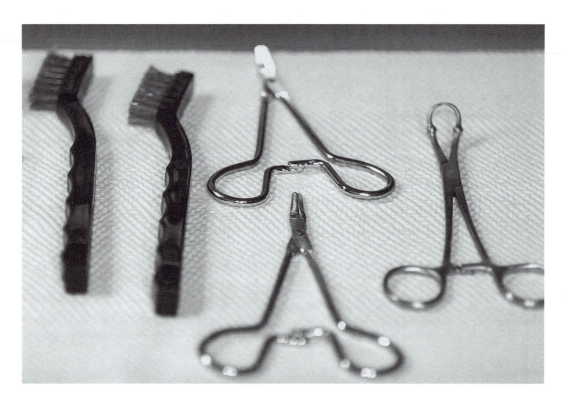

The tools used for the identification of bodies found in the New Orleans region are displayed on a tabletop at the Disaster Mortuary Operational Response Team (DMORT) base camp and morgue facilities in St Gabriel, Louisiana. Depending on the condition of the body, teeth records and examination can play a large role in the identification of a body. The St. Gabriel morgue site was created to perform the identification and reunification of bodies left in the wake of Hurricanes Katrina and Rita. October 14, 2005. (Lucas Jackson/Reuters/Corbis)

difficult. However, it is usually possible to obtain reference samples after the event for DNA identification. Personal effects and family-tree studies can be worked to provide the expected DNA profile of the victim.

Radiographs and Photographs

Dental x-ray records are an extremely valuable resource in identification. They contain detail that is not found in charts, such as the physical presence of, and patterns associated with, conditions such as root canals and impacted teeth, as well as exact replications of the visible tooth structures. Another source of x-ray records is hospital records—for example, X-rays taken in investigation of sinus problems may include tooth and jaw images. Occasionally, photographs may provide reference data. For example, someone may have some characteristic attribute on their front teeth—such as a gold filling—that is reproduced in a photograph.

Decomposed Bodies

Teeth resist decomposition but may become dislodged from the jaw in a highly decomposed body. This would make identification more difficult but not impossible, because the teeth themselves will be intact and may have sufficient individualizing features to permit identification.

Severely Burned Bodies

Even teeth are affected by burning if the temperature is high enough and the exposure long enough. A particular problem is that teeth may appear intact but be sufficiently affected by the heat that they will disintegrate on handling. It is possible to reinforce the dental structures to prevent fracture by spraying an adhesive over the teeth.

Bite Mark Evidence

A bite mark is a physical alteration in a medium caused by the contact of teeth, or a representative pattern left in an object or tissue by the dental structures of an animal or human.

A bite mark will reflect the physical shape of the jaws of the biter and the characteristics of the medium in which the mark was made. Jaws are U-shaped, but human bite marks are usually a shallow oval. If the medium in which the bite mark is found is appropriate, then the size, shape, arrangement, and distribution of the contacting surfaces of the teeth will be recorded in the mark. Some bite marks—for example, in foods such as cheese—can be preserved directly. Others will need to be recorded by photography or casting. A number of different techniques can be applied in the analysis and comparison of bite marks including the use of acetate overlays on life-size photos of wounds, models of teeth and life-size photos of wounds, special lighting techniques to enhance marks on the tissue, computer enhancement and/or digitization of mark and/or teeth, and stereomicroscopy and scanning electron microscopy.

The impressions and records of a bite mark can be compared with dental information and photographs of the suspect's dentition, and when the bite site is accessible to the victim's dentition, impressions of the victim's teeth should also be obtained and compared.

Important points for comparison would include the presence of any damage or trauma such as:

- Missing and misaligned teeth
- Broken and restored teeth
- Periodontal condition and tooth mobility

Forensic odontology is one of the most highly criticized areas of forensic science in regard to concerns about whether it is an objective test or too subjective to be reliable. Bite marks are certainly a contrast with the reliable and useful application of forensic odontology to identification of remains. Today, the best evidence that can be obtained from a bite mark is through DNA typing of the saliva traces, recovered by swabbing.

Forensic odontologists are trying to rectify their image of nonobjectivity. There are societies and even registration boards for odontologists, for example, the American Board of Forensic Odontology (ABFO) and the British Association for Forensic Odontology. As an example of how the societies are dealing with reliability issues, the ABFO recognizes that a range of conclusions can be reached when reporting a dental identification. The board recommends that these be limited to the following four conclusions:

- Positive identification: The antemortem and postmortem data match in sufficient detail, with no unexplainable discrepancies, to establish that they are from the same individual.
- Possible identification: The antemortem and postmortem data have consistent features but, because of the quality of either the postmortem remains or the antemortem evidence, it is not possible to establish identity positively.
- Insufficient evidence: The available information is insufficient to form the basis for a conclusion.
- Exclusion: The antemortem and postmortem data are clearly inconsistent.

Some of the evidence in the Chamberlain ("dingo baby") case illustrates the concerns about forensic odontology, especially bite marks. The general circumstances of the case are, briefly, that on the night of August 17, 1980, at a campsite near Ayers Rock in central Australia, Lindy Chamberlain cried out that her daughter Azaria was missing and had been taken by a dingo. The child was never seen again, but various articles of clothing were recovered in the country surrounding the campsite. Blood was found in the tent. A police inquiry followed, including extensive forensic science testing. A lengthy set of legal proceedings ensued, with the findings of the initial inquest that the baby had been taken by a dingo reversed and a verdict of guilty returned at the subsequent murder trial of Lindy and her husband Michael.

Part of the initial evidence dealt with damage to the jumpsuit worn by the baby and found outside the tent. The issue was whether or not it was caused by dingo teeth or by tearing or cutting. This was a critical question: If the damage could reasonably have been attributed to a dingo, Lindy's story was strongly corroborated. If not, then the inference would be that the parents had produced the damage to make it seem that a dog had taken the baby when, in fact, they were the perpetrators. The clothing was examined by Dr. Kenneth Brown of the Department of Odontology at the University of Adelaide and Sergeant Barry Cocks from the Technical Services Unit of the South Australian Police Department. They concluded that dingo teeth could not have caused the damage but that scissors could. Their experiments and conclusions were strongly criticized at the first inquest. For example, Brown had weighted a baby's jumpsuit to about ten pounds (Azaria's weight) and suspended it overnight from the incisor in a dingo skull. It did not tear and he concluded that a dog could not have caused the damage.

The matter did not end with the trial but proceeded through a retrial and a royal commission investigation. Commissioner Justice Morling made it clear that he believed prejudice had affected the objectivity of the scientists involved in the examinations of the clothing.

Morling found that in regard to the damage on the jumpsuit, the only substantial evidence that could be accepted was that it was a dingo's dentition that had caused it.

See also Bite Marks; Mass Disaster Victim Identification

References

Bowers, C. M. *Forensic Dental Evidence: An Investigator's Handbook.* Amsterdam and Boston, MA: Academic, 2004.

Bowers, C. M., and G. Bell. *Manual of Forensic Odontology.* 3rd ed. Saratoga Springs, NY: American Society of Forensic Odontology, 1995.

Dorion, R. B. J. *Bitemark Evidence.* New York: Marcel Dekker, 2004.

Opium

Nonsynthetic narcotic substances are obtained from the *Papaver somniferum* poppy. This plant has been grown in Mediterranean regions since 5000 BC and has now been cultivated in a number of countries throughout the world. Incisions made in the unripe seedpod of the plant ooze milky fluid that can be collected by hand and air dried to produce opium.

Industrial harvesting of opium involves the extraction of alkaloids from the mature dried poppy straw. This extract may be in liquid, solid, or powder form, but most commercially available poppy straw concentrate comes in the form of a fine brown powder. More than 500 tons of opium is legally imported into the United States each year for legitimate medical use.

Opium is still used in the form of paregoric to treat diarrhea, but most of the substance imported into the United States is already broken down into its alkaloid constituents. These alkaloids can be divided into two distinct chemical classes, phenanthrenes and isoquinolines. The main pharmacologically active phenanthrenes include morphine, codeine, and thebaine, but the isoquinolines have no significant central nervous system effects and are not regulated under the Controlled Substance Act (CSA).

See also Controlled Substances; Drugs; Heroin; Methadone; Morphine

References

Drummer, O. H. *The Forensic Pharmacology of Drugs of Abuse.* New York: Oxford University Press, 2001.

Levine, B. *Principles of Forensic Toxicology.* Washington, DC: American Association for Clinical Chemistry, 1999.

Saferstein, R. *Criminalistics.* 8th ed. Upper Saddle River, NJ: Pearson Education, 2004.

White, P. *Crime Scene to Court: The Essentials of Forensic Science.* Cambridge: Royal Society of Chemistry, 1998.

P

Paint

Forensic paint examination can often be difficult because of the physical size of the evidence that needs to be examined. This usually prevents the use of the more standard analytical techniques used in the paint industry. During examination, a forensic paint examiner must take into consideration any pertinent issues of the case; the sample's size, complexity, and condition; any environmental effects the sample has encountered; and the sample collection methods. This requires the examiner to select test methods, sample preparation schemes, test sequences, and the degree of sample alteration and consumption allowable to each specific case.

Paint is typically encountered as evidence during the investigation of automobile accidents (vehicle—vehicle or vehicle—person contact) or house breaking.

Paint is a mixture of pigment, which gives color and opacity to the paint, and a binder dissolved or dispersed in a solvent. When fresh paint is applied to a surface, the solvent evaporates, leaving behind a hard polymeric film with an appearance determined by the pigment composition.

Automobile Paints

Vehicle paint is usually made up of at least four coats, or layers. The first layer is known as the electrocoat primer, an epoxy resin applied by electroplating directly to the steel that provides protection against rust. It is usually gray or black in color.

The second coat is the primer. This is an epoxy-modified polyester that functions to smooth fine imperfections in the surface and is colored the same as the topcoat.

The topcoat (also called the basecoat or colorcoat) provides the visual finish, and must be able to withstand weathering effects from ultraviolet light and environmental insults. This is usually an acrylic-based polymer. Organic-based color pigments have now replaced lead- and chrome-based color pigments because they are less toxic. Luster or mica pigments added to the topcoat provide a pearl-like finish. The fourth layer is a nonpigmented acrylic- or polyurethane-based clear coat, which improves gloss and durability.

Analysis of Paint Evidence

During the forensic examination of paint evidence, the possibility of an exact physical match between a recovered chip and a possible source should always be considered. This type of match would provide proof of origin. Such a match may be accomplished by a physical fit of the fragment, or a match of the layers within a chip.

It is not common for there to be sufficiently large fragments recovered to offer the opportunity for a physical match. However, because of the multilayered nature of paint application in new and respray paint jobs, layer structure is more frequently used. During layer examination, it is important to consider whether or not the source vehicle is new, as there may be similarities in composition with other models produced around the same time. Vehicles that have been repainted provide better opportunities for individualization.

Obviously, more layers in the sample and source result in more evidential value to a match. The number and thickness of layers may vary at different sites on a vehicle; therefore reference samples should be obtained from as near the damaged area as possible without causing further damage.

More often than not, a layer structure is not complex enough to uniquely associate the recovered material with a suspect source, and analytical techniques must be used to help identify the source. These techniques aim to give a specific identification of the paint by the chemical characterization of its components. The tests that may be used are:

- Solubility tests
- Color characterization using microspectrophotometry
- Fourier transform infrared spectroscopy with microscope attachment
- Pyrolysis-gas chromatography for trace polymer characteristics
- X-ray fluorescence, neutron activation, and inductively coupled plasma—mass spectrometry for characterization of inorganic pigments

The first steps in forensic paint analysis involve visual and stereomicroscopic evaluation and a detailed documented description of the paint fragments and other trace material including the general condition, weathering characteristics, size, shape, exterior colors, and visible layers present in the sample. The most conclusive type of examination that can be conducted on paint evidence is the physical matching of the sample.

Layer Analysis

Layers in a paint sample can be seen under magnification along the sample edges. Obvious layers are generally visible without sample preparation, but some manual or microtome sectioning, edge mounting, and polishing might be required. Manual separation of individual layers can be carried out to facilitate solubility tests.

Microscopic examination may show subtle differences in color, pigment appearance, surface details, inclusions, metallic and pearlescent flake size and distribution, and layer defects between known and questioned paint samples. These comparisons are usually made by viewing the samples side by side under a stereomicroscope.

Solvent/Microchemical Tests

Solvent/microchemical tests can be used to distinguish between paint films with different pigment and binder compositions that are indistinguishable visually and macroscopically. These destructive tests are based on the dissolution of paint binders and on pigment and binder color reactions with oxidizing, dehydrating, or reducing agents. Reactions such as softening, swelling, curling, or wrinkling, layer dissolution, pigment filler effervescence, flocculation, and color changes should be observed and documented.

Polarized Light Microscopy (PLM)

Polarized light microscopy (PLM) can be used to examine the layer structure for the comparison or identification of pigments, extenders, additives, and contaminants.

Infrared Spectroscopy (IR)

Infrared spectroscopy (IR) and fourier transform infrared spectroscopy (FTIR) may be used to obtain information about binders, pigments, and additives. An infrared microscope attachment can also be used to sequentially analyze individual layers of a multiple-layer coating system. A related technique, Raman spectroscopy, can be used to obtain information about binders, pigments, and additives used in coatings. This technique, based on light scattering rather than absorption, can provide information that is complementary to that produced by infrared spectroscopy.

Pyrolysis Gas Chromatography (PGC)

Pyrolysis gas chromatography (PGC) is a destructive technique that uses pyrolytic breakdown products to compare paints and to identify binder types, enabling discrimination of chemically similar paints.

Microspectrophotometry

Microspectrophotometry can provide objective color data for paint comparison from small samples and so may be used in the analysis of paint. This technique provides a means for comparison of the absorption spectra of known and questioned samples.

Scanning Electron Microscopy

Scanning electron microscopy–energy dispersive x-ray analysis (SEM-EDX) can be used to characterize the morphology and elemental composition of paint samples. Emitted x-rays produce information regarding the presence of specific elements, and the electron signals produce composition and topographical information about a sample.

Architectural Paints

Architectural or household paints sometimes present as evidence in house-breaking or burglary cases. The general approach to their laboratory examination is similar to that for automobile paints, except that there are hardly ever sufficient layers for layer matching to be of value.

See also Fracture Matching; Individual Characteristics; Microscope; Microspectrophotometry

References

Caddy, B. *Forensic Examination of Glass and Paint: Analysis and Interpretation.* London and New York: Taylor and Francis, 2001.

James, S. H., and J. J. Nordby. *Forensic Science: An Introduction to Scientific and Investigative Techniques.* Boca Raton, FL: Taylor and Francis, 2005.

Saferstein, R. *Criminalistics.* 8th ed. Upper Saddle River, NJ: Pearson Education, 2004.

White, P. *Crime Scene to Court: The Essentials of Forensic Science.* Cambridge: Royal Society of Chemistry, 1998.

Pan Am 103

Pan Am Flight 103, which originated in Frankfurt, Germany, took off from London, at 6:25 p.m. on December 21, 1988, bound for New York. At 7:02 p.m. it was traveling at 31,000 feet when it disappeared from the radar—all that could be seen was several vanishing traces. The airliner had exploded, and the remaining pieces hit the small town of Lockerbie, Scotland. All 259 passengers and crew as well as 11 people on the ground died as a result of the explosion. A large crater was left where the main part of the aircraft fell, and 21 houses in Lockerbie were destroyed. An investigation was carried out by the Air Accident Investigation Branch (AAIB) in Scotland. In order to determine the cause of the disintegration of the aircraft, all of the pieces of debris were collected and reconstructed, and from this it was determined that the crash originated between the cockpit and the fuselage. Examination of the flight data recorder revealed that all of the control settings had been normal and correct, and investigators also concluded that there had been no problems with the engine or body of the aircraft that could account for its disintegration. It was revealed that the explosion had happened in the forward luggage hold, and the finding of a circuit board revealed that a Toshiba cassette player had been used for the bomb. Examination of fiber evidence on this device revealed that it had been inside a brown suitcase. Semtex (a plastic explosive) was also found on the device. Investigators determined that the explosion had caused the nose of the aircraft to break away, causing it to go into a steep dive and break apart. By reconstructing the explosion, investigators found that the suitcase would have been at the bottom of the baggage container, indicating that it was loaded onto the aircraft in Frankfurt. Analysis of clothing fibers traced the baggage to a Libyan who did not board the flight from London, but whose baggage did.

See also Crime Scene; Explosions and Explosives
References

Emerson, S., and B. Duffy. *The Fall of Pan Am 103: Inside the Lockerbie Investigation.* New York: Putnam, 1990.

Evans, C. *The Casebook of Forensic Detection: How Science Solved 100 of the World's Most Baffling Crimes.* New York: Wiley, 1998.

Pan Am 103; http://plane-truth.com/pan_am_103.htm (Referenced July 2005).

Parentage Testing

Each human cell nucleus contains forty-six chromosomes: twenty-three chromosomes of genetic material from the biological mother and twenty-three chromosomes from the biological father, which combine at conception to provide a unique DNA profile. Every person's DNA is unique, except for that of identical twins.

Parentage testing compares the genetic material of the child with that from its biological mother and putative father. Almost always, the issue being addressed is the identity of the father—paternity. Paternity tests can be performed on a variety of samples, including blood cells, cheek cells, tissue samples, and semen. However, the more traditional blood groups, such as ABO and Rhesus, are common throughout the population, making the power to differentiate between individuals not nearly as high as with DNA testing. A DNA paternity test is the most accurate form of paternity testing available.

All the DNA techniques used for typing of blood and semen stains can be applied to parentage testing.

When comparisons are made of the DNA patterns of mother, child, and alleged father, the child will share DNA information with the biological mother and with the biological father. There are two different ways to establish

nonparentage. If the child has an allele not found in the mother and not found in the putative father, then either the true father is someone else or there has been a mutation at that locus producing a different allele. Because there is always a very small but real chance of a mutation, laboratories should not call the finding as nonpaternity unless at least one more nonmatching allele is found. Most laboratories require at least three nonmatching loci. The second way to identify nonpaternity is if some alleles present in the putative father are absent in the results from the child. A laboratory would not normally report nonpaternity based on the absence of alleles.

When the DNA patterns between the mother, child, and the alleged father match at every DNA locus, then the probability of paternity is 99.9 percent or more, indicating that the alleged father is the biological father of the child. Most courts in the United States accept 99.0 percent as proof of paternity.

Paternity testing using DNA requires only a few drops of blood or cheek cells collected by swabbing the mouth. This enables the noninvasive testing of newborns and infants. Paternity tests can also be performed on an unborn child through chorionic villi sampling (CVS) or amniocentesis, and on postmortem specimens, or missing or deceased persons by reconstructing their DNA patterns from samples obtained from biological relatives.

See also DNA in Forensic Science; Exclusion of Paternity
Reference

Butler, J. *Forensic DNA Typing.* 2nd ed. Burlington, MA: Elsevier, 2005.

Pathology

The roles and practices of a forensic pathologist, coroner, and medical examiner are often confused or unclear and the terms used interchangeably, albeit incorrectly.

A coroner is an appointed or elected public official within in a particular jurisdiction whose duty involves the inquiry into certain categories of death. The coroner assigns the cause of death and completes the death certificate. The cause of death may be a disease, injury, or poison. A coroner decides whether a death occurred under natural circumstances or was the result of accident, homicide, or suicide. Some jurisdictions require the coroner to be a qualified physician, but in others the coroner need not be a physician or be trained in medicine. The coroner often employs physicians, pathologists, and forensic pathologists to conduct autopsies. A coroner without medical training may not immediately recognize subtle, nonviolent and violent causes of death.

A medical examiner is a physician responsible for the investigation and examination of persons dying a sudden, unexpected, or violent death within a given jurisdiction. The medical examiner is expected to use medical expertise in evaluating the medical history and physical examination of the deceased. A medical examiner is not necessarily required to be a specialist in death investigation or pathology, and may practice any branch of medicine.

A pathologist is a trained physician in the area of medicine that deals with the diagnosis of disease and causes of death by means of laboratory examination of body fluids, cytology, and tissues. The method of examination is autopsy, a systematic external and internal examination for the purposes of diagnosing disease and determining the presence or absence of injury. This includes the chemical analysis of body fluids for medical information, as well as analysis for drugs and poisons.

Forensic pathology is the examination of persons who die sudden, unexpected, or violent deaths. A forensic pathologist is trained to perform autopsies to determine the presence or absence of disease, injury, or poisoning and to evaluate investigative information relating to the manner of death. During the autopsy the forensic pathologist collects medical evidence, which may include trace evidence and bodily secretions to document sexual assault, and often reconstructs how a person received injuries. He or she usually has in-depth knowledge of toxicology, firearms examination, trace evidence, forensic serology, and DNA analysis. The forensic pathologist coordinates the medical and forensic scientific assessment of a given death, ensuring that appropriate procedures and evidence collection techniques are conducted. In this role he or she uses his or her expertise in interpretation of the scene of death, in assessing the consistency of witness statements regarding injuries, and the interpretation of injury patterns.

A forensic pathologist may need to identify the deceased, establish a time of death, determine the manner of death, the cause of death, and, if the death was by injury, the type of instrument used to cause the death. When the results of the autopsy and the laboratory tests are completed, the forensic pathologist correlates all the information and draws conclusions as to the cause and manner of death. Subsequently, the pathologist may then be subpoenaed to testify before courts and other tribunals about the pathologic findings and conclusions.

See also Autopsy; Toxicology

References

DiMaio, V. J. M., and D. DiMaio. *Forensic Pathology.* 2nd ed. Boca Raton, FL: CRC, 2001.

Dolinak, D., E. Matshes, and E. O. Lew. *Forensic Pathology: Principles and Practice.* Amsterdam and Boston: Elsevier Academic Press, 2005.

James, S. H., and J. J. Nordby. *Forensic Science: An Introduction to Scientific and Investigative Techniques.* Boca Raton, FL: Taylor and Francis, 2005.

Saukko, P., and B. Knight. *Knight's Forensic Pathology.* 3rd ed. London: Arnold, 2004.

Phencyclidine, Phenylcyclhexyl, or Piperidine (PCP)

Phencyclidine (phenylcyclhexyl, piperidine, PCP), was originally used as an animal tranquilizer in the United States. Since its legitimate production has ceased, it is now produced illegally in clandestine labs as a white powder that is taken orally, smoked (often in combination with marijuana), or snorted (in combination with cocaine). Pharmacological effects have a rapid onset and resemble the symptoms of schizophrenia. Euphoria, depression, agitation, violence, hallucinations, paranoia, panic, and suicidal tendencies are commonly experienced. PCP is a dissociative anesthetic, meaning the user is aware of what is happening, but does not feel involved. This results in insensitivity to pain, which can allow a person under the influence to run on two broken legs, or continue to fight after being shot. Hypertension, tachycardia, coma, cardiac failure, and psychosis are reported adverse effects of this drug.

See also Controlled Substances; Drugs; Hallucinogens
References
Saferstein, R. *Criminalistics.* 8th ed. Upper Saddle River, NJ: Pearson Education, 2004.
White, P. *Crime Scene to Court: The Essentials of Forensic Science.* Cambridge: Royal Society of Chemistry, 1998.
Winger, G., F. G. Hofmann, and J. H. Woods. *A Handbook on Drug and Alcohol Abuse: The Biomedical Aspects.* 3rd ed. New York: Oxford University Press, 1992.

Phenotypes

A phenotype can be defined as the physical appearance of a genetic trait such as eye color, hair color, and blood type.

See also DNA in Forensic Science
Reference
Butler, J. *Forensic DNA Typing.* 2nd edition. Burlington, MA: Elsevier, 2005.

Physical Evidence

Physical evidence is a term sometimes used to describe all physical materials examined in the forensic laboratory (including biological materials and drugs). Sometimes the term is used as a synonym for trace evidence.

Examination of physical evidence is used to identify or compare samples. Identification of unknown substances may be required in situations involving drugs or accelerants in fire debris. How close the analyst approaches absolute identification depends on the techniques used and the nature of the material being examined.

There is no hard-and-fast rule about what constitutes identification. Some laboratories require at least one molecular confirmation technique, such as mass spectrometry. Where less discriminating techniques have to be used, the laboratory may require a minimum of three independent tests. Assuming identification on the basis of the number of tests without reference to the individual discriminating power is not warranted.

Class characteristics enable a piece of physical evidence to be associated with a class of objects or items but cannot provide information on the exact source of the evidence—for example a piece of evidence can have the class characteristics of glass or paint, but the class characteristics cannot identify the glass or paint as originating from a particular source. Individualizing characteristics, such as the matching of glass fracture mark striations or the composition of a paint fragment (layering structure and pigment/additive composition), can individualize the sample to an extent that the probable source of the evidence can be identified. Often, examination of physical evidence is conducted on the basis of comparisons, to see if materials originated from the same source. With biological evidence, DNA testing may be used to compare a recovered semen stain with blood from a suspect. Relating the test results to frequencies in a population database can give groupings so rare that they are effectively unique. Latent print examinations are also used to compare crime-scene material with knowns from a possible source. Again, the comparison can produce associations that are accepted as being unique. Depending on the nature of the evidence, comparisons may be conducted using specialized techniques such as comparison microscopy by which two samples can be compared side by side, or by the application of sophisticated analytical techniques whereby compositional or structural

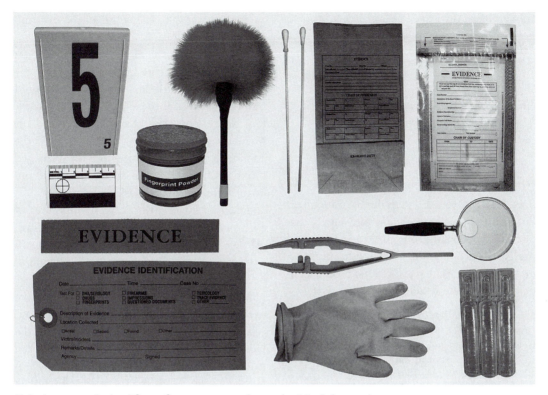

Multiple items are displayed from a forensic investigator's crime kit. (iStockphoto.com)

data obtained from the sample can be compared with similar data from a suspect source.

See also Accelerant Residues; DNA in Forensic Science; Drugs; Fibers; Glass; Hair; Paint

References

James, S. H., and J. J. Nordby. *Forensic Science: An Introduction to Scientific and Investigative Techniques.* Boca Raton, FL: Taylor and Francis, 2005.

Saferstein, R. *Criminalistics.* 8th ed. Upper Saddle River, NJ: Pearson Education, 2004.

White, P. *Crime Scene to Court: The Essentials of Forensic Science.* Cambridge: Royal Society of Chemistry, 1998.

Pitchfork, Colin

This landmark case was the first in which DNA profiling was used to eliminate a suspect and also the first in which DNA was used to identify the suspect.

In 1983 the body of raped and murdered fifteen-year-old Lynda Mann was found in rural England. Forensic analysis of traces of semen revealed that her attacker was a Group A secretor with a strong PGM 1+ enzyme, a combination that is known to occur in approximately 10 percent of the population. This was then linked to a second murder when a victim was found close to the same area and the semen type was found to be the same. Initially, the suspect in the case was a "slow" man employed at a local mental institution and, in fact, he admitted to the murder of the second victim, Dawn Ashworth, but denied attacking Lynda Mann.

At the time, the technique of restriction fragment length polymorphisms (RFLP) had recently been discovered, and analysis of a blood sample from the "slow" man revealed his innocence of both crimes. Because it was suspected that the attacker was a local man, all men in the area were asked to submit a blood sample for the purposes of elimination from the inquiry. It was soon discovered that Colin Pitchfork had asked a coworker to submit a sample on his behalf. Analysis of Pitchfork's DNA revealed him to be the attacker in both cases, for which he pleaded guilty and was sentenced to life in prison.

See also Blood Grouping; DNA in Forensic Science; Restriction Fragment Length Polymorphisms (RFLP)
References
Evans, C. *The Casebook of Forensic Detection: How Science Solved 100 of the World's Most Baffling Crimes.* New York: Wiley, 1998.
Owen, D. *Hidden Evidence: Forty True Crimes and How Forensic Science Helped Solve Them.* Willowdale, Ontario: Firefly, 2000.

Poisoning

Poisoning can be defined as the harmful effect on a living organism of ingestion, inhalation, or contact with a poison. There are many naturally occurring and man-made poisons, including bacterial toxins, plants, pharmaceutical agents, venoms, chemicals, and metals.

See also Poisons; Toxicology
References
Goldfrank, L., N. Flomenbaum, N. Lewin, M. A. Howland, R. Hoffman, and L. Nelson. *Goldfrank's Toxicologic Emergencies.* 7th ed. New York: McGraw-Hill Professional, 2002.
Levine, B. *Principles of Forensic Toxicology.* Washington, DC: American Association for Clinical Chemistry, 1999.

Poisons

There are thousands upon thousands of poisonous agents. The following provides an overview of poisons commonly encountered in forensic toxicology.

Organic Liquids
Ethanol: Ethanol is the most commonly encountered toxic substance in forensic toxicology. This is an organic liquid with a low boiling point and is present in beverages, some household cleaning agents, and cosmetics.

Methanol is another type of commonly encountered household alcohol, also known as wood alcohol. It is commonly used as a solvent and reagent in the chemical industry and is extremely toxic. Methanol is metabolized first to formaldehyde then to formic acid, which can lead to lowered blood pH, causing potentially fatal metabolic acidosis. Formic acid is damaging to the retina of the eye, and may cause blindness.

Isopropyl alcohol is also known as rubbing alcohol or isopropanol and is a commonly used antiseptic. It is not as toxic as methanol but causes more intense central nervous system depression than alcohol. Isopropanol is metabolized to acetone, which requires the application of analytical methods for suspected intoxication to be able to differentiate and quantify both isopropanol and acetone.

Ethylene glycol is the active component of antifreeze. Unlike isopropanol and methanol, ethylene glycol is not as volatile, but it is still extremely toxic. Ethylene glycol is metabolized to oxalic acid, which reacts with calcium in the body to form insoluble calcium oxalate. As well as general calcium depletion, calcium oxalate deposits may form in the kidney and brain. Ethylene glycol deaths can usually be determined without analytical testing as the deposits can be seen by polarized light microscopy in tissue sections.

Volatile Substances and Gases
Death due to volatile substances is generally associated with solvent abuse, although fatal accidents do occur in workers with halogenated compounds in poorly ventilated rooms. Solvents such as toluene and alcohols are inhaled by teenagers for central nervous system depression effects, causing intoxication and euphoria similar to that produced by alcohol intoxication. Death is usually due to cardiac failure. The majority of the abused solvents are used commercially or in industry as cleaning and degreasing agents. These agents commonly produce well-recognized toxic effects of cardiac arrhythmia, coma, and death, although the mechanisms of action are not fully understood. Their presence in blood can be readily determined using gas chromatography and appropriate handling techniques.

Gas inhalation. A number of pressurized gases with narcotic properties have become agents of abuse over recent years. Gases such as nitrous oxide and butane, legally used as propellants in aerosol spray products and anesthetics, are among these. Abuse of these compounds often results in death.

Death may also be caused by inhalation of gases released by burning synthetic materials. Most fire-related deaths are caused by inhalation of carbon monoxide and hydrogen cyanide. These compounds are readily released from blood and body tissues, enabling analyses by gas chromatography.

Metals

Exposure to elevated levels of essential and nonessential metals can result in toxicity and even death. People can be exposed to metals as part of their daily lives through their occupations, living environments, contaminated food and water, or intentional and unintentional poisoning. As with any poisoning, the chemical form and dosage of the poison, along with the age and physical condition of the person exposed, play a large role in the resultant toxic response.

Aluminum is present in drinking water, foods, and dust. It may be ingested from eating utensils and cookware and is found in antiperspirants, cosmetics, internal analgesics, antiulcer medications, antidiarrhea medications, and food-packing materials. Blood and tissue levels rise as exposure increases. The central nervous system and bone tissue are the target organs. Toxicity causes impaired cognitive function or central nervous system effects.

Arsenic is found naturally in soil, water, and air. It is used in wood preservatives, insecticides, herbicides, sheep dip, flypaper, arsenical soaps, germicides, rat poison, and growth promoters in poultry and livestock. Arsenical compounds such as arsenate and arsenite are extremely toxic. Although the frequency of arsenic poisonings has decreased, it is still known to occur due to the availability of arsenic-containing herbicides and pesticides. Intentional poisonings occur using arsenic in rodenticide or pesticide products. Arsenic is deposited in hair and nails and accumulates in muscle tissue, the liver, lungs, intestinal wall, spleen, and bone.

Nails and hair can be used to determine the occurrence of arsenic exposure using chemical/elemental analysis techniques. Arsenic exposure causes pain, vomiting, diarrhea, bloody stools, renal damage, proteinuria, hematuria, cardiovascular effects, muscle pain, and severe thirst. Encephalopathy and peripheral neuropathy are common in both acute and chronic exposure. With high enough doses, spasms, stupor, convulsions, and death ensue. Chronic arsenic poisoning produces symptoms that are similar to or mimic other medical conditions. These include muscle weakness, hyperkeratosis on the feet and hands, garlic breath, anemia, hematological disturbances, and neuropathy.

Arsine toxicity produces different symptoms, as the mechanism of toxicity is different, resulting in hemoglobinemia, anemia, and kidney damage.

Iron. Iron deficiency is actually more common than toxicity. Toxicity usually involves children ingesting prenatal iron supplements or vitamins. There are five phases of iron toxicity:

Phase I occurs one-half hour to two hours after ingestion, with symptoms manifesting as lethargy, bloody vomiting, abdominal pain, and restlessness.
Phase II occurs at variable times after ingestion. The symptoms disappear and the patient appears to recover.
Phase III occurs two to twelve hours after ingestion. Symptoms manifest as shock, metabolic acidosis, cyanosis, and fever.
Phase IV occurs two to four days after ingestion. Symptoms manifest as hepatic necrosis.
Phase V occurs two to four weeks after ingestion. Symptoms manifest as gastrointestinal obstruction due to gastric or pyloric scarring.

The mortality rate from acute poisoning is about 1 percent.

Mercury exposure is usually from mercury disposal sites. It is used in thermometers, barometers, manometers, gauges, valves, and in industries making lamps, batteries, wiring devices, switches, and in the production of chorine and caustic soda. Mercuric compounds are used in pigments, refining, lubricating oils, paint preservative, water-based paints, and dental amalgams. Alkyl mercury compounds are used as fungicides and preservatives for wood, paper, pulp, textiles, and leather.

Mercury itself poses little danger. Organic mercury is much more toxic and percutaneous exposure can be lethal.

Acute toxicity results in flulike symptoms, interstitial pneumonitis, bronchitis, chills, aching muscles, and dry mouth and throat. Orally ingested mercuric chloride causes gastrointestinal disturbances, burning sensations of the mouth and throat, gingivitis, and esophageal destruction. There has been a recent focus on mercury toxicity as a result of mercury poisoning in children that has been attributed to mercury in the fish they consume. Almost all fish and shellfish contains traces of mercury and so it is currently recommended that children, pregnant women, and nursing mothers limit their intake of fish to two meals of low mercury-containing fish (such as shrimp, salmon or catfish) and avoid fish known to have high mercury content such as shark, swordfish or tilefish.

Chronic poisoning manifests as neurological symptoms, with low concentrations causing tremors. Pink disease develops with reddening of the extremities, chest, and face, with photophobia, diaphoresis, anorexia, and tachycardia that lead to renal injury. Classic symptoms include excitability, gingivitis, and hallucinations.

Lithium is an anticonvulsant, sedative, and gout treatment. Small amounts are found in soil, tobacco, grain, coffee, seaweed, and milk. Lithium is also used for the treatment of manic depression. Lithium hydride is used industrially as a desiccant, a source of hydrogen, a nuclear-shielding material, and a condensing agent in organic synthesis. In the 1940s it was used as a table salt substitute, resulting in many deaths. Acute exposure results in polydipsia, polyuria, sedation, tremor, ataxia, diarrhea, coma, and seizures. Other side effects include cardiac arrhythmias, hypotension, and albuminuria.

Lead. A common source of exposure is from lead traces in the foods we consume. Lead carbonate and lead oxide are commonly used in paint pigments. Lead-based paints, which are now banned from use in homes in the United States, resulted in many toddler poisonings, caused by chewing the sweet-tasting chips of paint found in older houses. These paints are still used on bridges and street surfaces. Lead is a component of solder, brass and bronze, plastics, glass, and ceramics. Lead arsenates are used in pesticides.

Lead is gradually deposited in soft tissues, mainly the tubular epithelium of the liver and the kidney. Eventually, it is redistributed to bone, teeth, and hair with 95 percent of the body burden being in bone, deposited as a tertiary lead phosphate in the same way as calcium.

Thallium is commonly used in the semiconductor industry and is present in mineralogical solutions, optic systems, photoelectric cells, and cement. Due to its high toxicity, thallium is regulated in many countries. Thallium can be found in red blood cells, brain, lung, gut, skeletal and cardiac muscle, salivary glands, testes, spleen, and pancreas, and is accumulated in the kidneys, liver, and bone. Acute symptoms include gastrointestinal distress, paralysis, and respiratory failure. Chronic symptoms include alopecia, which is the most characteristic symptom of intoxication—especially if the patient lives for twenty days after exposure. Other symptoms include lobster-red skin, paresthesis of hands and feet, psychosis, delirium and convulsion, optic cataracts and neuritis, incoordination, paralysis of the extremities, hepatic and renal disorders, endocrine disorders, and psychosis. Death usually ensues from respiratory and cardiovascular collapse.

Carbon monoxide is a colorless, odorless gas that is produced by incomplete combustion and is often associated with old heaters and stoves. Carbon monoxide is also produced in vehicle exhaust. The toxicity of carbon monoxide is due to the higher affinity of hemoglobin for carbon monoxide than for oxygen. Death is caused by a lack of oxygen.

Initial symptoms of carbon monoxide toxicity include headache and fatigue. This eventually progresses to confusion, coma, and death from lack of oxygen if the victim does not get fresh air. A sign of carbon monoxide poisoning is that the victim takes on a bright pink color caused by the binding of carbon monoxide to the hemoglobin in the blood.

In postmortem forensic cases such as poisonings from a faulty heater, suicide using fumes from car exhaust, or death associated with smoke inhalation in fire, carbon monoxide poisoning should be a consideration.

See also Poisoning; Toxicology
References
Goldfrank, L., N. Flomenbaum, N. Lewin, M. A. Howland, R. Hoffman, and L. Nelson. *Goldfrank's Toxicologic Emergencies.* 7th ed. New York: McGraw-Hill Professional, 2002.
Klaassen, C. *Casarett and Doull's Toxicology: The Basic Science of Poisons.* 6th ed. New York: McGraw-Hill Professional, 2001.
Levine, B. *Principles of Forensic Toxicology.* Washington, DC: American Association for Clinical Chemistry, 1999.

Polygraph
See *Lie Detector Test (Polygraph)*

Polymerase Chain Reaction (PCR)
Polymerase chain reaction, or PCR, is a technique that is used to increase the amount of DNA available for typing. In forensic DNA testing, PCR is used to identify and copy one or more variable regions of DNA. The tests use "primers" that identify constant regions adjacent to the variable region of interest. These primers, short pieces of DNA complementary to their target sequences, serve as the starting points for copying the variable regions of the DNA sequence. The process occurs in a test tube placed in a thermal cycler, which goes through a series of heating and cooling cycles. During each cycle, DNA fragments from the target region are duplicated, increasing the quantity of target DNA exponentially with each cycle. Millions of copies of a single DNA molecule can be produced in a few hours. The copying process is very efficient if the original sample is in good condition. However, DNA introduced into the system erroneously will also be multiplied and cause contamination, producing false and misleading results. Prevention of false results involves the use of carefully applied controls and techniques.

See also DNA in Forensic Science
Reference
Butler, J. *Forensic DNA Typing.* 2nd ed. Burlington, MA: Elsevier, 2005.

Population Genetics
The characteristics of any organism are determined by their genetic makeup or DNA. Any random mutations causing changes within that DNA sequence can result in slight changes in those characteristics. With time, these changes may accumulate, resulting in adaptive or evolutionary changes in the organisms. Population genetics is the mathematical analysis of this process. Approximately 90 percent of an organism's DNA is considered to be nonfunctional, and mutations in the nonfunctional DNA generally have no effect. The remaining functional DNA influences the physical and biochemical properties of the organism, and mutations can be neutral, beneficial, or harmful. Of these mutations, it's likely that less than half have no effect. The majority of the rest of the mutations are harmful or fatal, though occasionally one may be actually beneficial to the organism. Population genetics can help to explain the adaptation of organisms to their environment through changes in frequency of existing genetic material (alleles) even without mutations. Harmful mutations tend to be eliminated from the population because the

organism usually doesn't survive. Beneficial mutations also may be eliminated by chance, but are generally preserved. As these beneficial mutations accumulate, the species gradually adapts to its environment. Neutral mutations may be eliminated or can spread to the whole population. When this occurs, the mutation has become fixed in the population. The rate at which beneficial or neutral mutations fix in the population is the rate of evolution. Rates of evolution are calculated using the likelihood that various mutations will be passed on to offspring. If each individual with a mutation has two offspring also with the mutation, then that mutation will rapidly spread through the population. If less than one offspring has it, the mutation is usually eliminated. This is known as selective advantage, and can be defined as any favorable structural, functional, or behavioral characteristic that increases an organism's opportunity and likelihood of survival and reproduction, compared to individuals in the same population who lack this characteristic. Selective advantage is usually reported as the ratio of the number of offspring with the mutation to the number of offspring without the mutation. If this ratio is 0.01 then the chance this mutation will be passed on to an offspring is about 1.01 times higher than the chance that DNA without this mutation will be passed on to an offspring. After n generations, the frequency of this mutation in the population should increase by a ratio of $(1.01)^n$. If several mutations exist in the same individual, they have an additive effect.

See also DNA in Forensic Science; DNA Population Frequencies

References

Butler, J. *Forensic DNA Typing*. 2nd ed. Burlington, MA: Elsevier, 2005.

National Research Council. *The Evaluation of Forensic DNA Evidence*. Washington, DC: Academy, 1996.

Probability and Interpretation of Evidence

Probability can be defined as the frequency of the occurrence of an event. To appreciate the significance of a piece of evidence, or the evidential value of an item, one must be aware of the probability of the origin of the two specimens under comparison being the same. Evidence that can be associated with a common source with an extremely high degree of probability is said to possess individual characteristics. An example would be the jigsaw match of glass fracture striations between a glass fragment and a suspect source, or the matching ridges of two fingerprints. In this type of situation, it is not always possible to state the exact mathematical probability that the specimens are of common origin, but it is concluded that the probability is so high that it defies mathematical calculations. In all cases the conclusion of common origin also must be substantiated by the experience of the examiner.

When an evidentiary item does not possess characteristics that relate it to a single source but has only class characteristics, the probability of finding a common source with a suspect item is far lower. Such is the case with blood specimens. The ABO blood groups are distributed throughout the population and the frequencies of each group, although known, do not offer a basis for establishing common origin. This is the situation for most class-associated physical evidence and is a major problem for the investigator, who is often unable to assign exact or even approximate probability values to his or her findings. Although it would be beneficial to be able to assign an approximate probability that a particular fiber came from a particular carpet or that a glass fragment came from a particular window, there is currently little statistical information available to the forensic scientist to derive this information. The development and maintenance of numerous statistical databases would be necessary for the evaluation of the significance of class physical evidence.

References

Butler, J. *Forensic DNA Typing*. 2nd ed. Burlington, MA: Elsevier, 2005.

National Research Council. *The Evaluation of Forensic DNA Evidence*. Washington, DC: Academy, 1996.

Propoxyphene

Propoxyphene is a centrally acting, mild narcotic analgesic that is metabolized in the liver to yield the active metabolite norpropoxyphene. Propoxyphene has a half-life of six to twelve hours, whereas that of norpropoxyphene is thirty to thirty-six hours. Norpropoxyphene has much fewer central nervous system depressant effects than propoxyphene but a much greater effect as a local anesthetic. Propoxyphene is structurally related to methadone, with less potency than codeine. When extensively taken in high doses, propoxyphene can produce drug dependence, characterized by psychic dependence, physical dependence, and tolerance. Propoxyphene only partially suppresses the withdrawal symptoms in morphine or narcotic addicts, but the liability of abuse is similar to that of codeine.

See also Controlled Substances; Drugs

Psilocybin

Psilocin and psilocybin are naturally occurring hallucinogens present in the psilocybe mushroom. They have an action similar to that of LSD but are less active. Ten to fifteen psilocybe mushrooms will induce a hallucinogenic effect when taken orally. Psilocybe mushrooms are also dried and smoked with tobacco and extracted with boiling water to produce infusions.

See also Controlled Substances; Drugs; Hallucinogens
References
Saferstein, R. *Criminalistics.* 8th ed. Upper Saddle River, NJ: Pearson Education, 2004.
Winger, G., F. G. Hofmann, and J. H. Woods. *A Handbook on Drug and Alcohol Abuse: The Biomedical Aspects.* 3rd ed. New York: Oxford University Press, 1992.

Psychiatry

A forensic psychiatrist is a physician who combines clinical experience, knowledge of medicine, mental health, and the neurosciences to form an independent, objective opinion. This expert opinion is presented when necessary by written report, deposition, or courtroom testimony. The applications of forensic psychiatry are often requested in health care, the workplace, and in matters relating to criminal justice and public safety.

References
Gutheil, T. G., and P. S. Appelbaum. *Clinical Handbook of Psychiatry and the Law.* 3rd ed. Philadelphia, PA: Lippincott, Williams & Wilkins, 2000.
Rosner, R. *Principles and Practice of Forensic Psychiatry.* 2nd ed. New York: Chapman & Hall, 2003.

Q

Qualitative and Quantitative Analysis

Testing and *analysis* are used as synonyms when discussing laboratory examinations in forensic science. In everyday language, therefore, *analysis* means examining samples or objects. The outcomes of the analysis can vary considerably. Where the examination is conducted solely to identify the nature or origin of the material, it is best to describe the examination as "qualitative analysis." Examples include identification of a white powder as methamphetamine or a red-brown stain as blood. When the examination is conducted with the objective not only of identifying the nature of the object but also of measuring the amount present, then the examination is a "quantitative analysis." Examples include blood- and breath-alcohol testing and examination of white powder to discover the amount of methamphetamine that it contains.

The reliability, or error rate, in a qualitative test is usually expressed as the incidence of false negatives and false positives. Some degree of value has to be exerted in considering the reliability of a qualitative test. This can be illustrated using the simple examples presented above. A white powder, believed to contain methamphetamine, can be tested by several techniques. Marquis reagent (2 percent formaldehyde in sulfuric acid) will turn orange-brown when the reagent is spotted onto a small sample of the powder, if the powder contains methamphetamine. However, it will also change color with a wide range of amphetamine-like chemical substances. Using the Marquis reagent as a spray to visualize spots on a thin-layer plate will be a little more specific, but not entirely so. A completely certain identification requires a different technique to be used. Fourier transform infrared spectroscopy (FTIR) and gas chromatography–mass spectrometry (GC-MS) are more appropriate qualitative tests for identification of drugs. Likewise, testing the red-brown stain with benzidine or leukomalachite green (see **Blood**) will give a color change if the stain is indeed blood. However, other substances, such as vegetable tissues, can also give a positive response. For this reason, qualitative tests must always be first considered as screening tests, unless used in combination or are of a very highly discriminating nature.

The reliability of quantitative tests is usually described in terms of accuracy and precision. The accuracy of the test is how close the result is to the true amount of analyte present. The precision is a measure of the spread of results obtained on repeated testing. Precision is usually expressed statistically as the standard deviation of a number of repeated tests. Sometimes the standard deviation is presented as the coefficient of variation (CV), which is the standard deviation as a percentage of the mean.

CV can range from 1 or 2 percent in a highly precise test such as blood-alcohol determination, to 30 percent or more for low levels of drug in putrefied materials tested in forensic toxicology.

Questioned Document Examination

See *Document Examination*

R

Ramirez, Richard

This case presents one of the first convictions resulting from a computerized fingerprint database.

A series of attacks by someone termed "The Nightstalker" took place in Los Angeles in the summer of 1984. The attacker broke into homes, raped the women, and shot any males present dead. He was described by victims who survived as a Hispanic with bad breath and bad teeth. Finally, in 1985, one victim managed to get the license number of the vehicle he left in. This was found to be a stolen vehicle, but police were able to get a partial fingerprint from it. They ran the print through the LAPD's new computerized fingerprint database and got a positive match for twenty-five-year-old Richard Ramirez, who had been fingerprinted after being arrested for a traffic violation. Police circulated pictures of Ramirez, and he was soon recognized by members of the public in a liquor store, leading to his arrest. In 1989, despite the fact that Ramirez improved his appearance for court (including having dental work done), he was convicted and sentenced to death for his violent crimes.

See also Databases; Fingerprints
References
Court TV's Crime Library, Criminal Minds and
    Methods; http://www.crimelibrary.com/
    ramirez/ (Referenced July 2005).

Evans, C. *The Casebook of Forensic Detection: How Science
    Solved 100 of the World's Most Baffling Crimes.*
    New York: Wiley, 1998.
Owen, D. *Hidden Evidence: Forty True Crimes and How
    Forensic Science Helped Solve Them.* Willowdale,
    Ontario: Firefly, 2000.

Rape

The precise legal definition of rape varies from jurisdiction to jurisdiction, but all share the common features of penetration (not necessarily by the penis) without consent. Forensic science can play a role in providing inceptive evidence (that there was indeed penetration), corroborative evidence (that consent was denied), and identity evidence (the DNA of the perpetrator). These three types of evidence will now be considered one by one.

Inceptive evidence relies mainly on the presence of the assailant's seminal fluid or its components in the victim. Identification of tissues from the victim on the penetrating organ—penis or finger—of the assailant also provides strong evidence of penetration. However, it is equally important to note that the absence of seminal components does not mean that rape did not occur. The most obvious situation is when a condom was worn by the assailant and discarded. However, if the condom is recovered, then identification of cellular material from the victim on the outside, and semen from the assailant on the inside, can produce very compelling evidence. The screening test for semen and the genetic material producing DNA profiles are all detectable in sperm-free semen samples.

Corroborative evidence can be furnished by the forensic medical examination of the victim, and by laboratory examination of clothing and, in some circumstances, typing of blood shed from injuries. A violent assault may result in bruising and bleeding and physical trauma to the victim. Forceful removal of clothing can produce damage. However, there is little that the laboratory can do to distinguish between physical passion and physical violence. On the other hand, just as with the limitations of inceptive evidence, the absence of physical damage to the person of the victim or to the victim's clothing does not prove the absence of force. The most obvious instance is the psychological force arising from the threat of violence.

Estimation of time since intercourse is a potentially powerful area of corroborative evidence. The presence of spermatozoa in the vaginal cavity provides evidence not only that intercourse did occur, but the motility of the sperm can also indicate the time of assault, as sperm can survive four to six hours in the vaginal cavity. However, the search for motile sperm is generally uncommon, because the vaginal smear would have to be microscopically examined immediately after obtaining the sample from the victim.

Vaginal swabs and smears are more commonly examined in the forensic laboratory. Nonmotile sperm may be found for three to six days after intercourse. Intact sperm are rarely found after sixteen hours, but have been found as long as seventy-two hours after the event. The presence of seminal acid phosphatase in the vaginal cavity decreases with time following intercourse, and is often undetectable within forty-eight hours. In interpreting these findings correctly it is also imperative that the investigator seek information regarding the victim's voluntary sexual history prior to the assault. However, it has to be strongly emphasized that all these estimates of changes with time are just that—estimates. There is no hard relationship, but there is a wide range of normal findings.

In contrast to the interpretative limitations that may accompany inceptive and corroborative evidence, the power of DNA to identify the source of biological material such as semen and epithelial cells is as good as it gets for identity. DNA typing has had a major impact on the scientific investigation of alleged rapes. It takes all doubt out of identification of the donor of the semen, and the application of DNA databases has had a major impact on serial offenses. For example, more than 10 percent of no-suspect rape kit samples analyzed in the United States are producing evidence linking the samples to a known offender or to another rape.

As with all forensic science investigations, care must be taken to ensure that all trace evidence is collected and that cross-contamination of evidence does not occur. As well as collecting the appropriate clothing from both suspect and victim, it may also be pertinent to collect items of bedding or the object (or representative samples from the object) on which the assault took place.

The main source of evidence items in a rape investigation is the so-called rape kit that is used during the medical examination of the victim. The kit allows for collection of pubic hair combings, pubic hair control samples (usually plucked), vaginal swabs, rectal swabs (depends on case history), oral swabs, pulled head hair, a saliva sample, a blood sample, and fingernail scrapings. Clothing is also collected into clean brown paper bags that are sealed to prevent loss and contamination. The paper permits the clothing to air dry, thereby preventing degradation of the biological materials.

Similar samples are taken from a suspect on apprehension, namely all clothing, pubic hair combings, pulled head and pubic hair controls, a saliva sample, and a blood sample.

Some Illustrative Cases

A nine-month-old child was left quietly sleeping in her baby carriage outside the family home. Around noon the mother went out to check on her daughter and found both the baby and carriage missing. The parents reported the abduction to the police, who began a search, interviewing neighbors and family members. Shortly after dusk, the baby, alive and apparently well, was discovered on the doorstep. Initial inquiries had raised some concerns about a male neighbor who sometimes babysat for the family. Police reinterviewed him and took samples, including fingernail scrapings, and sent them to the lab for analysis. Results showed the presence of female cells in the fingernail scrapings with a DNA profile that matched that of the baby. The man pled guilty at the subsequent trial.

One cold winter Sunday morning in 1972, the partially clothed body of an attractive young woman was found partly immersed by a riverbank near a large industrial town in southwest Scotland. Police interviews traced her actions and placed her in a bar with a group of friends and associates on the previous Saturday night. One of the group was a distant cousin who had only recently been released from prison for attempted rape. He was reported to have left the bar with the deceased at about 11 p.m. He obviously became a strong suspect. The postmortem examination showed that the woman had been raped, partially strangled, and that death was due to drowning. This was the same modus operandi that had been used by the suspect in the earlier case. Semen in the vagina of the victim was of the same type as the suspect. Unfortunately, at that time the power of traditional blood-group testing was very limited and almost 25 percent of the population had the same groups as the suspect. The suspect claimed that as soon as he and the victim left the bar, they had separated and he had gone home while the victim had gone in the opposite direction to take a shortcut home by way of the riverbank. He had an alibi from about midnight when his parents returned from a night out and found him at home. Estimation of time of death and time since intercourse became of major significance. The fact that the body was thrown into cold water on a cold winter's day complicated the time-of-death estimation but made the estimation of time since intercourse more reliable. The time-dependent change in the nature and number of spermatozoa found in the vagina after intercourse is due to mechanical drainage and biochemical alterations. A death very shortly after the ejaculate is deposited and rapid cooling will stop the biochemical clock. In this case, the tests revealed a considerable quantity of intact spermatozoa in the vaginal samples. The accused had no alibi for the time of death. That and the grouping results, such as they were, corroborated other evidence and a guilty verdict resulted.

See also Date Rape; Sexual Offenses; Time since Intercourse

References

Butler, J. *Forensic DNA Typing.* 2nd ed. Burlington, MA: Elsevier, 2005.

Hazelwood, R. R., and A. W. Burgess. *Practical Aspects of Rape Investigation: A Multidisciplinary Approach.* 3rd ed. Boca Raton, FL: CRC, 2001.

Turvey, B. E., and J. Savino. *Rape Investigation Handbook.* Amsterdam and Boston: Elsevier Academic Press, 2004.

RDX

See *Cyclotrimethylenetrinitramine*

Red Cell Enzymes

Numerous red cell enzymes have been shown to exhibit polymorphism. Only the principal ones of forensic value will be mentioned.

Phosphoglucomutase (PGM)

Phosphoglucomutase (PGM) has three loci showing polymorphism. The PGM1 locus is widely used in forensic typing. It is found on red cells and in body fluids such as semen and vaginal secretions. Separation on starch gel shows two alleles giving rise to three phenotypes: PGM1 1; PGM1 2; and PGM1 2–1.

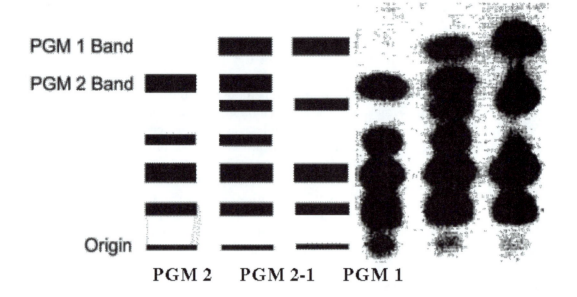

PGM 1 Band

PGM 2 Band

Origin

PGM 2   PGM 2-1   PGM 1

Isoelectric focusing shows four alleles, 1+ and 1−, and 2+ and 2−, giving ten phenotypes in all. This is often referred to as PGM subtyping.

The diagram above shows the results from a conventional electrophoresis of PGM. The bands on the left are schematic representations of the actual results on the right.

Other Enzymes

Other enzyme systems include glyoxylase (GLO), esterase D (EsD), erythrocyte acid phosphatase (EAP), adenylate kinase (AK), and adenosine deaminase (ADA).

See also Blood Grouping
Reference
Saferstein, R. *Criminalistics.* 8th ed. Upper Saddle River, NJ: Pearson Education, 2004.

Refractive Index

Light travels through a homogeneous medium in a straight line. At the boundary between two media (air and glass) some light is reflected and some is transmitted. The amount of reflected light versus transmitted light is a function of the angle of incident (incoming) light and the nature of the media. Light penetrating the second medium is partially absorbed and partially passes through to a second boundary where again reflection and transmission occur. Light crossing a boundary at any angle other than perpendicular to the boundary will change direction if the speed of the light wave in the two media is different. This change of direction is called refraction.

The change in the direction of light as it passes from one medium to the next is a function of the speed of light in that medium. Light crossing into a region of higher refractive index (going from air to glass) is bent toward normal. Light crossing into a region of lower refractive index (glass to air) is bent away from normal.

Because refraction is a function of the velocities of light in the two media, an index of refraction (nr) may be expressed as follows:
nr = speed of light in the first medium / speed of light in the second medium.

Refractive indices (RIs) of a material are dependent on the composition and on particular directions within the crystal. Some materials, glasses and those belonging to the cubic crystal structure, have one refractive index regardless of orientation and are described as *isotropic* in character. Materials with two or three refractive indices are *anisotropic* in character. Refractive index is the main method used in comparision of glasses.

A Case Example

Hooligans were causing regular acts of vandalism in a moderately wealthy suburb of Adelaide, in South Australia. Law enforcement officers had a reasonable idea of the persons involved but could not catch them either in the act or with any usable evidence to associate them with the vandalism. One night the vandals ran along a line of parked cars breaking the windows as they went. The alarm of one of the cars went off and the owner called the police. They arrived at the scene quickly and picked up a youth. Examination of his clothing in the laboratory showed many small glass fragments with refractive index values in the range expected for automotive window glass. Not only that, but the RI values fell into two clusters, one of which encompassed the glass used by Ford and one that encompassed the range of glass used by Volvo. A Ford and a Volvo were the last two cars in the line of damaged vehicles.

See also Glass
References

Curran, J. M., T. N. Hicks, and J. S. Buckleton. *Forensic Interpretation of Glass Evidence.* Boca Raton, FL: CRC, 2000.

Saferstein, R. *Criminalistics.* 8th ed. Upper Saddle River, NJ: Pearson Education, 2004.

White, P. *Crime Scene to Court: The Essentials of Forensic Science.* Cambridge: Royal Society of Chemistry, 1998.

Restriction Fragment Length Polymorphisms (RFLP)

Restriction fragment length polymorphisms, or RFLP, are different fragment lengths of base pairs resulting from cutting a DNA molecule with restriction enzymes.

DNA can be cut into smaller lengths by restriction enzymes, such as Hae III. When this happens, the cut fragments can be separated by length using electrophoresis and detected by probing with chemically or radioisotopically labeled nucleic acids that search for complementary sequences. There have been hundreds of probes identified. People display variations in the length of fragments that carry the same nucleotide sequences. The target DNA fragments that are recognized by the probe are then measured for size. These target fragments come from places in the DNA where there is length polymorphism, which means that the length of the fragments tends to vary from person to person. This is called restriction length fragment polymorphism. The fragments of interest are composed of sequences repeated head to tail a variable number of times, the number of repeats being genetically determined. This is a variable number tandem repeat (VNTR). RFLP fragments are characterized by the number of bases in their length, and analysis involves a sizing procedure. The loci examined by RFLP analysis are highly polymorphic.

If two samples are tested and the target DNA fragments from a particular locus are of different lengths, then the samples must be from different people. If they are the same length, then the samples are either from the same person or from two people who by coincidence have fragments of the same length.

RFLP was the first type of DNA testing conducted in forensic science. It is now used only for legacy systems where comparison may have to be made with data from an old case (such as cold cases). The problems compared to the nearly universal PCR systems used today were sensitivity, the long time required to complete an analysis, a degree of susceptibility to degradation, and data handling of results that is more complex than that for the discrete alleles detected in PCR systems such as STR analysis.

See also DNA in Forensic Science; Polymerase Chain Reaction (PCR)
References

Butler, J. *Forensic DNA Typing.* 2nd ed. Burlington, MA: Elsevier, 2005.

Saferstein, R. *Criminalistics.* 8th ed. Upper Saddle River, NJ: Pearson Education, 2004.

Romanov Family

On July 17, 1918, on the orders of Vladimir Lenin (the new leader of the Bolshevik revolution), Czar Nicholas II of Russia, together

Russian czar Nicholas II and his family, ca. 1900. (The Illustrated London News Picture Library)

with his wife, five children, and four of their staff, was executed, and ever since there has been much speculation about their fates. At the time, a Russian investigator examined the house where the executions took place and found no evidence of the remains of the family. His conclusion (though often disputed) was that the remains of the family had been dissolved in acid and burned.

In the 1970s, a Russian filmmaker investigated the deaths of the family and was able to interview the children of the guard who had overseen the executions. They provided him with a note stating where the bodies were apparently buried. The filmmaker, together with a geologist, went to this site and was able to find the grave as it was detailed in the note. He found bones in very fragile condition, as well as expensive clothing that corresponded to the age and gender of those who had been executed. Some ten years later, on the authority of President Boris Yeltsin, the bones were exhumed and an examination

conducted of them. The examination of the bones compared to photographs of the family led Russian forensic scientists to the conclusion that the remains of all the family were present except the Czar and his daughter Marie. A group of U.S. forensic scientists, headed by William Maples of the University of Florida, were then asked to examine the evidence. This group concluded from their examinations that it was actually the remains of the daughter Anastasia that were missing. DNA analysis was then done in an attempt to confirm the identities, but due to the poor state of the bones, and the difficulties associated with extracting DNA from them, blood relatives of the Romanovs were sought and their DNA compared with the mitochondrial DNA of the remains. This resulted in the confirmation of the remains belonging to the Romanovs. However, no bones were found belonging to Prince Alexei and his sister Anastasia. Since then, there have been many people coming forward (some very convincingly) claiming to be Anastasia or Alexei Romanov.

See also DNA in Forensic Science; Identification

References
Butler, J. *Forensic DNA Typing.* 2nd ed. Burlington, MA: Elsevier, 2005.
Evans, C. *The Casebook of Forensic Detection: How Science Solved 100 of the World's Most Baffling Crimes.* New York: Wiley, 1998.

Ruxton, Buck

The case of Buck Ruxton illustrates several important forensic concepts. This case was the first time in which a photographic overlay was successfully used in the identification of a victim—a photograph of Buck Ruxton's wife was superimposed on her skull and found to be a perfect match. This was one of the most compelling pieces of evidence used in the trial of Buck Ruxton, which resulted in his being found guilty of murdering his wife and the family nursemaid.

The case began when two dismembered bodies were found in a rural area of Scotland in 1935. The bodies had been dismembered with some skill, and care had been taken to ensure that identifying features such as fingertips were obliterated. Some of the body parts had been wrapped in a newspaper that was found to be a local edition from the city of Lancaster, where it was known that a nursemaid and her employer, Mrs. Ruxton, had recently disappeared. A search of the Ruxton house turned up numerous blood stains, and further investigation revealed that Buck Ruxton, a surgeon and the husband of the missing woman, had even tried to give a blood-stained suit to one of his patients in return for helping him to clean up the bloody mess!

Of further forensic significance in this trial was the use of entomology. Maggots were found on the bodies, and the life cycles of these insects were used in the estimation of the time of death of the victims.

Buck Ruxton received a death sentence and was executed on May 12, 1936. A full confession from him was later published in a newspaper. He admitted that while killing his wife, he was disturbed by the nursemaid and so killed her also.

References
Evans, C. *The Casebook of Forensic Detection: How Science Solved 100 of the World's Most Baffling Crimes.* New York: Wiley, 1998.
Owen, D. *Hidden Evidence: Forty True Crimes and How Forensic Science Helped Solve Them.* Willowdale, Ontario: Firefly, 2000.
University of Glasgow, Forensic Medicine Archives Project, Case against Dr. Buck Ruxton, Lancaster, Moffet and Manchester (1935); http://www.fmap.archives.gla.ac.uk/Case%20Files/Ruxton/Case_File9.htm (Referenced July 2005).

S

Safe Insulation

Safes are intended to protect the contents, the main threats to which are burglary and fire. Protection against fire is obtained by a safe construction with one or more layers of insulation inside the case. Many years ago, this insulation often contained sawdust, which, in combination with the widespread style of wearing trousers with cuffs, presented the forensic botanist with entertaining challenges in safe-cracking cases. Debris from the trouser cuffs of suspects could be recovered and the wood identified and compared to the type and range in the insulation of the safe that was burgled. By this means, a perpetrator could be caught.

Today, different materials are used for insulation. Some safes have a lining of poured concrete. Others use asbestos or vermiculite. Examination of safe insulation is now even more specialized, but is still conducted in some laboratories, such as those of the U.S. Postal Inspection Service.

An old safe in a bank in Richmond, Maine. (Library of Congress)

Saliva

Saliva is the fluid that moistens the mouth. It is secreted from three sets of glands—the submaxillary, parotid, and buccal glands. The saliva from the parotid glands contains amylases, which aid in the digestion of carbohydrates.

Saliva is of some significance in forensic biology. It can be the source of evidence, for example, in sexual offenses where oral contact is alleged, or on cigarette butts found at crime scenes.

Before the advent of DNA testing, saliva samples were subjected to ABO typing. Although saliva does not contain red blood cells, about 80 percent of the population has saliva with high levels of a water-soluble form of the chemicals that determine ABO blood types. This is under genetic control, and persons with the attribute are termed *secretors*. Indeed, saliva samples are used in one of the tests for secretor status. Using ABO to characterize the source of a saliva sample suffers from all the drawbacks of ABO testing in forensic science. It is not a very discriminating test, and mixtures can be very difficult to interpret.

Today, the sensitivity and discriminating power of DNA testing mean that the identity of the source of a saliva sample can be determined. Indeed, saliva samples in the form of swabs from inside the mouth are often used as reference samples for DNA analysis.

The S family were accustomed to having the brother of Mr. S baby-sit for them. One evening after returning home, when the mother went to check on her nine-year-old daughter, the child was still awake. The goodnight conversation took a dramatic turn when the girl told her mother about the game she played with uncle, which involved him licking her external genitalia. The parents called the police, and the nurses from the sexual referral center took swabs from the girl's vulva and surrounding areas. The laboratory found traces of material that screened positive for saliva and went on to perform DNA testing. However, interpretation of results was very complex because many of the alleles in the reference sample from the suspect were also found in the girl, inherited from her father. There were alleles at three loci that could not have been from the girl but did match those of the uncle. He was subsequently charged and convicted.

See also ABO Blood Groups; DNA in Forensic Science

References
Butler, J. *Forensic DNA Typing.* 2nd ed. Burlington, MA: Elsevier, 2005.
De Forest, P., R. E. Gaensslen, and H. C. Lee. *Forensic Science: An Introduction to Criminalistics.* New York: McGraw-Hill, 1983.
James, S. H., and J. J. Nordby. *Forensic Science: An Introduction to Scientific and Investigative Techniques.* Boca Raton, FL: Taylor and Francis, 2005.
Saferstein, R. *Criminalistics.* 8th ed. Upper Saddle River, NJ: Pearson Education, 2004.
White, P. *Crime Scene to Court: The Essentials of Forensic Science.* Cambridge: Royal Society of Chemistry, 1998.

Scientific Evidence

By definition, forensic science is about scientific evidence. However, there is no generally accepted answer to the question, "What is scientific evidence?" There may be a tendency to an answer along the lines of "applying the scientific method to analysis of materials to produce evidence for a court." Probing further on the understanding of what makes a method "scientific" will often lead to very unsatisfactory answers, such as one that is done "very carefully." Although care is certainly an essential attribute of the work of a forensic scientist, it is not one of the elements that define "scientific method."

The U.S. Supreme Court wrestled with the concept of "scientific method" in *Daubert v. Merrell Dow Pharmaceuticals, Inc.* (1993). The majority opinion included confirming that the subject matter of the testing did indeed satisfy the requirements for "scientific method," which was defined as being "based on generating hypotheses and testing them to see if they can be falsified; indeed, this methodology is what distinguishes science from other fields of human inquiry."

The concept of falsifiability owes its origins to the philosopher Karl Popper, and may briefly be explained as follows: Science is a search for truth based on formulating hypotheses. The hypothesis must prohibit something and that prohibition must be capable of testing by an experiment. If the experiment produces a result that is in conflict with the hypothesis, then the hypothesis has been falsified and fails. No hypothesis can be proved true by testing, it can only be proved false or "falsified." Hypotheses that have stood rigorous testing over a long time become accepted laws of nature. This is a simple and effective means of identifying a theory as "scientific" and one that is well accepted in the scientific community. (As an aside, the concept was too much for Chief Justice Rehnquist, who wrote in a dissenting opinion, "I am at a loss to know what is meant when it is said that the scientific status of a theory depends on its falsifiability").

Probably the most important aspect of *Daubert* was not the flawed attempts to define "scientific" but the clear enunciation of the principle that the judge is the gatekeeper of what is or is not to be admitted as evidence. A later decision (1999) extended the scope of gatekeeper to all evidence of a technical nature requiring specialized knowledge. Between *Daubert* and *Kumho Tire* came *General Electric Co. v. Joiner* (1997), which made a very significant contribution to the debate on "scientific evidence." The *Joiner* ruling declared that there must be a demonstrable link between the conclusions of the expert and the facts of the case.

References
Daubert v. Merrell Dow. 509 U.S. 579, 596 (1993).
General Electric Co. v. Joiner, 522 U.S. 136 (1997).
Kumho Tire Co. v. Carmichael, 526 U.S. 137 (1999).

Semen Identification

Semen is a fluid of complex composition produced by the male sex organs. There is a cellular component, spermatozoa, and a fluid component, seminal plasma. An average ejaculate is 3 to 4 ml and contains 70 to 150 million sperm. Sperm are the male reproductive cells, each consisting of a head, tail, and middle. In humans, the head is a tiny disc, about 4.5 micrometers (μm, or 1/1 000 000 meter) long and 2.5 μm wide. The tail is about 40 μm long, and is rapidly lost in ejaculates. The head is where the DNA is preserved. It is interesting to note that ape sperm are similar in size and shape, and while dogs have similar-shaped sperm, they are about one-third the size of human sperm. Other animals have different-shaped sperm.

Seminal plasma contains proteins, salts, organics (including flavins, which are the source of its ultraviolet fluorescence, and choline), and some cellular material, and has components originating from several sources, including seminal vesicles and the prostate (the source of its acid phosphatase and p30 protein).

Vasectomy severs or ligates the ducts carrying sperm to the penis. Thus, vasectomized men will have no sperm but will have the plasma components present in their ejaculate.

After ejaculation, semen deposited during intercourse is lost by drainage and by biochemical change. Tails are lost first—the damage begins immediately and about 25 percent will have no tails by six hours following ejaculation. By twelve hours, there will be few sperm with intact tails and by twenty-four hours there will be only a few heads left. It should be remembered that these proportions and times are highly variable. Sperm survival in stains outside the body depends on environmental conditions, but a small stain that has dried quickly may have intact sperm preserved for months or even years.

Human semen contains unusually high concentrations of acid phosphatase. This is exploited in presumptive screening tests for the presence of semen. Screening tests are based on the hydrolysis of phosphate esters and detection of the liberated organic moiety by production of a color complex. Acid phosphatase, present in seminal fluid, reacts with sodium alphanapthylphosphate and Fast Blue B to produce a purple-blue coloration. As with the screening test for blood, a positive result is the rapid formation of the intensely colored product—within twenty seconds or so, or thirty seconds at most. Also note that a number of vegetable and fruit juices can result in the generation of false positive results with this test.

The best identification of semen is from its microscopy. The morphology and dimensions of the human spermatozoon are unique. The small sperm, particularly if they have lost their tails, can be difficult to locate microscopically, especially in samples that have bacterial or yeast infection. Detection is simplified by histopathological staining. The most usual stain is popularly known as "Christmas tree stain" because of the bright colors. It utilizes Nuclear Fast Red, which stains the DNA-containing head bright crimson, and a counterstain of picric acid—indigocarmine (PIC)—which stains the tails blue-gray. The traditional histological staining of hematoxylin and eosin (H and E) is also used, as is Giemsa stain, another dye widely used in histopathology.

Problems may be encountered if the seminal fluid is from a man who has a low sperm count (oligospermia) or who has no spermatozoa present in his seminal fluid (aspermia). In situations where the presumptive alkaline phosphate test indicates the presence of semen, but the microscopical analysis yields no detectable spermatozoa, tests are carried out to determine the presence of a protein, P30 or prostate specific antigen (PSA), which is only found in high concentration in human semen. It can be detected by precipitin reaction with a specific antiserum using the Ouchterlony process. It is also possible to perform a quantitative immunological test utilizing an enzyme-linked reaction (ELISA).

See also Color Tests; DNA in Forensic Science; Ejaculate; Rape; Species Identification

References
Butler, J. *Forensic DNA Typing.* 2nd ed. Burlington, MA: Elsevier, 2005.
De Forest, P., R. E. Gaensslen, and H. C. Lee. *Forensic Science: An Introduction to Criminalistics.* New York: McGraw-Hill, 1983.
James, S. H., and J. J. Nordby. *Forensic Science: An Introduction to Scientific and Investigative Techniques.* Boca Raton, FL: Taylor and Francis, 2005.
Saferstein, R. *Criminalistics.* 8th ed. Upper Saddle River, NJ: Pearson Education, 2004.
White, P. *Crime Scene to Court: The Essentials of Forensic Science.* Cambridge: Royal Society of Chemistry, 1998.

Serial Number Restoration

Metallic objects such as automobile engine blocks and firearms bear unique identifying marks or numbers stamped in by a hard steel die. When the automobile or firearm is stolen, the mark or number is often erased (usually by grinding) to prevent tracing. A false number may be stamped in place of the original.

It is often possible to restore an original number because the die stamp process produces changes in the metal that go beyond the layers of visible indentation. The simplest and most effective restoration process is chemical etching of the metal surface. The metal crystals distorted in the original stamping will dissolve more rapidly than unaltered metal when an etching solution, such as copper chloride in hydrochloric acid, is applied. It is possible to use several other methods, including electrolytic processes, ultrasonic etching, and magnetic particle methods, to reveal the original number.

References
De Forest, P., R. E. Gaensslen, and H. C. Lee. *Forensic Science: An Introduction to Criminalistics.* New York: McGraw-Hill, 1983.
Saferstein, R. *Criminalistics.* 8th ed. Upper Saddle River, NJ: Pearson Education, 2004.
White, P. *Crime Scene to Court: The Essentials of Forensic Science.* Cambridge: Royal Society of Chemistry, 1998.

Serology

Serology is the detection, identification, and typing of body tissues, either in native form or as stains or residues left at a scene. Examples are blood, semen, and hair.

The detection and identification process begins with a physical examination and proceeds to a screening test. Confirmatory identification then follows, and finally the sample is typed.

Typing is the detection of genetically determined characters in the sample. A difference in type between stain and reference sample from the postulated source provides a positive elimination. If there is no elimination, the results can be evaluated against population frequency data to give a measure of the significance of the failure to eliminate.

The features of a good typing system are that:

- It shows variability from person to person but is constant within one individual.
- It is stable in shed form.
- It can be detected reliably at the concentrations found in forensic samples.
- Its frequency of occurrence within the population is known and stable.

The characteristic does not have to be inherited.

See also Blood; Saliva; Semen Identification

References

Butler, J. *Forensic DNA Typing.* 2nd ed. Burlington, MA: Elsevier, 2005.

De Forest, P., R. E. Gaensslen, and H. C. Lee. *Forensic Science: An Introduction to Criminalistics.* New York: McGraw-Hill, 1983.

James, S. H., and J. J. Nordby. *Forensic Science: An Introduction to Scientific and Investigative Techniques.* Boca Raton, FL: Taylor and Francis, 2005.

Saferstein, R. *Criminalistics.* 8th ed. Upper Saddle River, NJ: Pearson Education, 2004.

White, P. *Crime Scene to Court: The Essentials of Forensic Science.* Cambridge: Royal Society of Chemistry, 1998.

Sexual Offenses

The forensic science aspects of unlawful sexual intercourse are dealt with in the section **rape**. However, sexual assault can take many different forms and need not be penis to vagina, nor does there have to be ejaculation. Many women and young boys are victims of sexual abuse. This type of sexual assault is generally defined as unwanted touching for the purpose of sexual arousal.

Lack of consent is a factor in almost all sexual offenses. Lack of consent usually results from the threat or use of physical force, incapacity to consent, or—in the case of sexual abuse—any circumstances in which the victim does not agree with the perpetrator's conduct. Victims may or may not actively resist the attack.

Lists of sexual offenses can be located in the statutes of each state. Although the classification and name of the offense may differ, the list of offenses for many states include the following:

- Rape
- Sodomy
- Statutory rape
- Child molestation
- Enticing a child for indecent purposes
- Sexual assault against persons in custody
- Bestiality
- Necrophilia
- Public indecency
- Prostitution
- Keeping a place of prostitution
- Pimping
- Pandering
- Incest
- Sexual battery
- Aggravated sexual battery
- Distributing obscene materials

Illustrative Case

The forensic science aspects tend to be similar to those described for rape. However, sexual assault cases are full of surprises. One of the very first sexual assault cases involving DNA that was processed in the author's laboratory involved an alleged homosexual rape. A man reported that he had been forced into anal sex by a casual acquaintance. An anal swab from the victim did indeed contain semen. Conventional grouping gave results that did not exclude the suspect but that could also have come from the accuser. DNA testing had only just become available in the laboratory and this looked like an ideal early case. The case worker was stunned when the results came out with the same groups as the accuser and different from those of the suspect. The samples were reanalyzed, and fresh reference materials tested, but with the same result. Finally, the decision was made to inform the police of the results. The police interviewed the parties once more, and, supported by the DNA results, obtained a somewhat different set of statements. It turned out that the two men were established lovers, but that the suspect had left the relationship. The accuser had masturbated into a condom and inserted the semen into his rectum and reported the "rape." But for the DNA tests, an innocent man may have received a substantial jail sentence.

See also Date Rape; Rape
References
Butler, J. *Forensic DNA Typing.* 2nd ed. Burlington, MA: Elsevier, 2005.
Hazelwood, R. R., and A. W. Burgess. *Practical Aspects of Rape Investigation: A Multidisciplinary Approach.* 3rd ed. Boca Raton, FL: CRC, 2001.
Seneski, P. C., N. Whelan, D. K. Faugno, L. Slaughter, and B. W. Girardin. *Color Atlas of Sexual Assault.* St. Louis: Mosby, 1997.
Turvey, B. E., and J. Savino. *Rape Investigation Handbook.* Amsterdam and Boston, MA: Elsevier Academic, 2004.

Shoe Examination

Shoe prints may be left at a crime scene on floors and carpets, in the form of bloody prints or even tread marks on paper. When a shoeprint is considered to be evidence it is photographed for preservation and as a record of all the observable details of the print.

When possible to do so, the surface on which the print is found should be submitted to the laboratory for examination. This is easy if the print is on a tile, piece of paper, or piece of glass. When the item is too large to move, or the print is present in dust or dirt, then the print has to be recovered in a similar manner to that in which fingerprints are recovered. Recovery techniques use sticky tapes and electrostatic lifting devices. If the print is impressed into soft soil, then photography and casting with gypsum dental stone is the best means of documentation and recovery. Wax-based sprays are recommended for the recovery of prints in snow. There are also a variety of chemicals available for the enhancement and recovery of prints left in blood.

Back at the laboratory, a shoe print found at a crime scene is compared and matched with test shoe print impressions made with a suspect shoe. During the examination, investigators

A reconstructed crime scene with a forensic officer taking photos for the permanent record. Photographs will record the layout of the room (including points of entry and exit) and the location of other evidence (labeled), such as the bloody footprints and a spent cartridge shell (D) from a handgun. The protective clothing is intended to prevent contamination of the crime scene. (Mauro Fermariello / Photo Researchers, Inc.)

look for a sufficient number of matched points on the suspect's shoe and the shoe print found at the crime scene. The more points on the suspect shoe that match the shoe print found at the crime scene, the more evidence exists to support the conclusion that the shoe print at the crime scene was made by that particular shoe.

In the absence of a suspect shoe for comparison, class characteristics such as size, track pattern, shape, and design can be used to establish that a particular type and brand of shoe may have been used to make a particular print.

There are a number of shoe print databases under construction that use specialized software to assist in making shoe print comparisons. A British automated shoe print identification system called Shoeprint Image Capture and Retrieval (SICAR) uses multiple databases to search known and unknown footwear files for comparison against footwear specimens.

Forensic podiatry is the subdiscipline in which the podiatrist tries to demonstrate a relationship between a person and specific footwear, a foot impression, or gait evidence recovered from a crime scene. This discipline requires knowledge of podiatric biomechanics, foot pathology, and footwear. The obtained results can range from exclusion to identifying the source of presented evidence with a high percentage of probability.

The comparison process involves the examination of impression evidence made by the weight-bearing contact of a foot with a "recording surface." This evidence can be in the form of a two-dimensional footprint or shoe impression, or a three-dimensional foot impression. The quality of the impression is dependent on the nature of the contacting surfaces and the age, type, and condition of the footwear, as well as the activities of the individual. Footprints can be made from bare or sock-covered feet, and shoeprints can be made from outsole and in-shoe characteristics.

The podiatric examination requires the examination of the individual and the associated footwear, and may extend to the comparison between a questioned impression and gait characteristics and data collected from a suspect individual and his or her footwear and/or foot impressions. This detailed examination requires a working knowledge of lower-limb biomechanics and human movement. For a thorough and complete examination and comparison, consideration should be given to:

- The presence of class and individual characteristics
- The presence of structural differences
- Whether there are features distinguishable within the impressions that can help identify structural and mechanical features of a foot
- Whether characteristics are common to a broad range of individuals, such as shoe size, foot size, and structural or developmental characteristics (such as clawed toes)

Individual characteristics, such as footprint features or a foot impression specific to the individual, are often limited in podiatry. However, soft tissue or bony injuries and scars may be transferred to foot impressions or result in unusual alterations in weight-bearing characteristics.

The examination of a suspect requires a detailed examination of foot structure and biomechanics. Impression casts can be made of the dorsal and plantar surfaces using foam or plaster bandages. Inked impressions are useful for recording weight-bearing characteristics and size and for comparison with questioned footprints. The feet are also photographed from all angles and views. All parts of the footwear are examined, and detailed records and photographs of the wear of uppers, insoles, and outsoles are important for correlation of physical features of wear with structural features.

See also Footwear; Individual Characteristics
References

De Forest, P., R. E. Gaensslen, and H. C. Lee. *Forensic Science: An Introduction to Criminalistics.* New York: McGraw-Hill, 1983.

James, S. H., and J. J. Nordby. *Forensic Science: An Introduction to Scientific and Investigative Techniques.* Boca Raton, FL: Taylor and Francis, 2005.

Lee, H. C., T. Palmbach, and M. T. Miller. *Henry Lee's Crime Scene Handbook.* London and San Diego, CA: Academic, 2001.

Saferstein, R. *Criminalistics.* 8th ed. Upper Saddle River, NJ: Pearson Education, 2004.

White, P. *Crime Scene to Court: The Essentials of Forensic Science.* Cambridge: Royal Society of Chemistry, 1998.

Shotguns

See *Firearms*

Skeletal Remains

Forensic anthropology is the application of physical anthropology principles to the legal process. The identification of skeletal, decomposed, or unidentified human remains is important legally, as well as for humanitarian reasons. In this discipline, forensic anthropologists apply standard scientific techniques used routinely in physical anthropology to identify human remains, to assist in the detection of crime, and to help establish or estimate the postmortem interval. As well as assisting in locating and the subsequent recovery of skeletal remains, forensic anthropologists use their skills to determine the approximate age, sex, ancestry, stature, and unique features from examination of the recovered skeleton.

Once skeletal remains are collected and packaged appropriately, they are transferred from the site of discovery to the laboratory or morgue. The bones are then cleaned and arranged in anatomical position for examination purposes.

One of the most important aspects of the analysis is gender determination. This is crucial for further analysis of unidentified human remains, because the techniques of assessment for age and stature differ for each gender. The pelvic bone provides the most definitive gender information, although the long bones of the arms and legs (humerus and femur) also provide important information for this determination.

Gender determination from preadult bones is very difficult. Some consideration can be given to the rate of tooth calcification and its relationship to maturation. It is thought this rate is the same in girls and boys; however, postcranial bones in males tend to mature more slowly than in females. This difference allows the determination of a growth-gender index between the two systems. For decades the skull was thought to be the most discriminating guide for gender determination, but this has now been superseded by the observed differences in the pelvic bones. Although gender information can be obtained from the skull, this information is generally used to complement information derived from the pelvis.

Estimating the age of death in preadult skeletons is also very difficult and relies on the examination of criteria such as the development of teeth, the appearance of ossification centers, the joining of the epiphyses, and the length of the long bones of the arms and legs. The calcification and eruption of teeth provide the most accurate age information in preadults.

All human bones develop from ossification centers. Early in utero there are about 800 centers of bone growth, which unite as the fetus grows. At birth there are about 450 centers, whereas the adult skeleton has only 206 centers.

The appearance of ossification centers and the process of epiphyseal union follow definite sequences and can be quite a reliable indicator for estimation of age. Ossification and epiphyseal union begins earlier in females than in males, leading to a shorter period of growth in females, and their smaller adult size. Ossification centers appear from birth to about fifteen years of age, and can be easily compared against appropriate standards. In skeletal remains, the ossification centers are extremely fragile and are apt to be damaged or destroyed. Age determination of immature bones requires examination of the developmental epiphyseal union that occurs from ten to twenty years of age. Again, results can be compared to available standards showing changes in epiphyseal union for most long bones.

The size of long bones can be used to estimate age in prenatal and postnatal skeletal remains. In prenatal remains, the length of the long bone shaft is measured to estimate length of the body, from which fetal age can be estimated. With postnatal skeletal remains, the length of the long bones can be used directly to estimate age. This method is used to estimate age of prenatal skeletal remains and children up to about six to seven years of age.

See also Anthropology

References

Byers, S. N. *Introduction to Forensic Anthropology: A Textbook.* Boston: Allyn and Bacon, 2001.

Maples, W. R., and M. Browning. *Dead Men Do Tell Tales: The Strange and Fascinating Cases of a Forensic Anthropologist.* New York: Doubleday, 1995.

Smokeless Powder

Smokeless powders are a class of explosive propellants that produce very little smoke on explosion and consist mostly of gelatinized cellulose nitrates. Powders in which this is the only explosive component are classified as single base. Double-base smokeless powder is an explosive consisting of nitrocellulose and nitroglycerin. The addition of nitroguanidine to double-base powder gives triple-base smokeless powder. The latter is not common and is normally only found in large-caliber munitions.

Smokeless powders are encountered in forensic science in the form of residues from gunshots or explosive devices. Explosives are classified according to the speed of the gas expansion. Low explosives are really extremely rapid-burning reactions, resulting in a pressure wave traveling at speeds up to 1,000 meters per second (3,281 feet per second). Low explosives are the basis of propellants, but can be highly destructive if contained. Commercial low explosives include smokeless powders, but almost any fuel and oxidizing agent combination can be made into an explosive.

Nitrocellulose is the polymer that gives body to the powder and allows it to be handled. In double-base powder, the nitroglycerin raises the energy content. In triple-base powder, the addition of nitroguanidine reduces flame temperature and so improves the energy-to-flame properties of the propellant. Diphenylamine and centralites are stabilizers commonly contained in smokeless powders to prevent the nitrocellulose and nitroglycerin from decomposing. Other ingredients of smokeless powders include plasticizers, flash suppressants, and opacifiers. Dyes may be added for identification purposes.

The shape and size of the powder particles have a profound effect on the burning rate and power generation of a smokeless powder. Most single-base powders consist of tube-shaped or cylindrical particles, while double-base powders are often composed of particles in the shape of discs or balls.

A single-base powder is made from nitrocellulose mixed with the selected additives, dissolved in organic solvent, and then extruded and cut into lengths. The resulting granules may be coated with graphite, then dried and screened to a consistent size.

Double-base powders can be made from organic solutions of nitroglycerin and nitrocellulose treated to form blocks that can be extruded and cut. Alternatively, an aqueous paste can be prepared and dried before extrusion and cutting. Triple-base powders are made by a similar process.

Approximately 10 million pounds of commercial smokeless powders are produced each year in the United States. Most of the powder is sold to the manufacturers to be used for manufacturing ammunition.

Smokeless powder is classified as a low explosive because it has a detonation rate of under 1,000 meters per second. It is, nonetheless, a powerful agent. It provides the explosive propulsive force for ammunition, and when confined in a small container it makes a very dangerous improvised explosive device. The pipe bomb is the most common of such devices. Typically, it will contain around one-half pound of smokeless powder within a pipe nine to twelve inches in length and capped at each end. Any type of material can be used for the pipe, although metal is the most common. The power of the device comes from the containment of the smokeless powder.

A National Research Council report published in 1998 found that during the five-year period from 1992 to 1996, there were an average of 653 incidents per year involving the use of black and smokeless powders. Bombs containing black or smokeless powders were responsible for an average yearly count of about 10 deaths, 83 injuries, and almost $1 million in property damage for each of the five years.

See also Black Powder; Explosions and Explosives
References
"Black and Smokeless Powders: Technologies for Finding Bombs and the Bomb Makers." National Academies Press. Washington, DC, 1998, 1.

De Forest, P., R. E. Gaensslen, and H. C. Lee. *Forensic Science: An Introduction to Criminalistics.* New York: McGraw-Hill, 1983.

Saferstein, R. *Criminalistics.* 8th ed. Upper Saddle River, NJ: Pearson Education, 2004.

White, P. *Crime Scene to Court: The Essentials of Forensic Science.* Cambridge: Royal Society of Chemistry, 1998.

Soil

Soil is one of the types of potential trace evidence materials encountered in forensic science. It can be found on the shoes or clothing of someone who was present at a crime scene, and may therefore be used to provide associative evidence.

Soil is a highly heterogeneous material, which reflects the organic and inorganic history of the area from which it originated. Weathered minerals and decomposed plants, the various fauna that pass over or live within the soil, and extraneous factors such as human contamination influence the nature of the complex mix termed *soil*. The gross composition of the soil material will reflect the history of the larger locale—loamy or sandy, for example—but the extraneous materials may introduce sufficient characteristics to a smaller locality to permit near individuality.

Visual comparison of color and texture, along with pH, can distinguish most broad classes of soil. Color is moisture dependent, and comparisons should utilize samples at the same degree of wetness.

The next level of discrimination is achieved through low-power microscopy. Larger particles originating from animal and vegetative matter and any artificial debris are identified this way. High-power microscopy, coupled with phase-contrast examination, will provide even more discrimination by identifying the many minerals and rocks that may be present. There are about forty common minerals that can be distinguished by physical attributes such as size, color, crystal form, refractive index, and birefringence properties. Rocks can be identified using their mineral content and grain size. Density-gradient and particle-size distribution measurements are also used in soil and mineral sample comparisons.

Soil is readily lost from items such as footwear or clothing, either through wear or by cleaning. Evidence should be collected as soon as possible and must be packaged securely. Samples from vehicles can be collected from the treads of the tires, from the wheel arches, and from fender recesses in older vehicles. Control samples from the scene need to be collected from a wide enough area to provide a representative cross-sample of the microenvironment that may exist. Sampling from the scene needs to be conducted carefully from another aspect, namely, to be sure that any impression evidence left by footwear or tires is not destroyed.

References

Saferstein, R. *Criminalistics.* 8th ed. Upper Saddle River, NJ: Pearson Education, 2004.

White, P. *Crime Scene to Court: The Essentials of Forensic Science.* Cambridge: Royal Society of Chemistry, 1998.

Species Identification

Species identification is an intermediate step in the sequence of tests applied to the comprehensive examination of body tissues. Traditional serology progressed from presumptive or screening tests, such as benzidine, through species identification to blood grouping by ABO typing or enzyme identification. Typing was conducted only if the species of origin was shown to be human. DNA has changed this because most of the markers used are primate, if not human, specific. Testing can therefore proceed from a positive screening result to typing, and the results reported as: "Material of human origin that gave a positive screening test for blood was typed and found to . . . etc."

Precipitin reactions form the basis of species identification testing in traditional serology. Plasma proteins of various animals are used as antigens, and the resulting antiserum is screened for specificity to the species of origin of the proteins. The plasma proteins and antibodies in the antiserum are large molecules but still soluble in water. However, because the antigen proteins include a wide range of sites that will bind to the antibodies, and because each antibody molecule has at least two binding sites, a network or lattice of antibody-antigens is formed when the bloodstain extract and the antiserum are mixed. The lattice is insoluble and precipitates out, hence the term "precipitin" test.

The earliest precipitin tests were conducted in tube tests in which the stain extract was layered on top of a small amount of antiserum. If the extract contained plasma proteins from the species identified by the antiserum, then a precipitin ring gradually formed at the interface between the two liquids. An increase in sensitivity and in the information content of precipitin testing was obtained by the radial diffusion technique named after Orjan Thomas Gunnarson Ouchterlony, a Swedish bacteriologist. Here, wells of 1 to 2 mm in diameter are punched in agar gel in the form of a ring of 5 or 6 wells surrounding a central well and about 1 cm away from it. Antiserum is added to the center well and stain extracts to the radial wells. The proteins diffuse radially from each well and where there is a concordance between species and antiserum, a precipitin arc is formed between wells. The test is reliable and sensitive, and can give information on species mixtures from the pattern of the precipitin arc. As with all biological identification tests in forensic science, controls from an unstained area near the suspect stain must be run along with the stain extracts.

A variant on the Ouchterlony test uses an electric current to force the proteins and antibodies together. The reaction depends on a property called electroosmotic force whereby at the pH of the test the direction of mobility of the gamma globulin molecules that include the antibodies is opposite to albumin and beta globulins from the stain extract. The test is fast and sensitive. In both forms of agar-based tests the precipitin line may sometimes be seen by the naked eye as an off-white band, but the plates are usually stained with a general purpose protein stain such as amido black.

The above addresses species identification from the perspective of testing for human origin. However, there are many other applications in forensic science. For example, the U.S. Fish and Wildlife Forensic Science Laboratory has an extensive range of programs for identifying biological species (plant and animal) and parts from them. Applications include identification of deer meat in poaching, identification of bear gall bladders in traditional medicines, and identification of ivory. The range of applications is matched by the range of tests, but as with human biological forensic science, modern testing is dominated by DNA. However, the value of taking a holistic view is well illustrated by the laboratory's work on ivory identification. Ivory was traditionally the name given to the teeth or tusks of elephants. All mammalian teeth have the same overall physical and chemical structure, and ivory came to mean any large tooth or tusk that could be made into an artifact by carving. Elephants and mammoths, walruses, certain whales, the hippopotamus, and the warthog all have been hunted for their ivory tusks. Assigning the source of an ivory artifact or fragment can be achieved by microscopic examination of the material. For example, elephant tusks contain characteristic markings called Schreger lines. The dentine in whale tusks may show prominent concentric rings, and the large lower teeth of the hippopotamus are triangular in cross section.

References

De Forest, P., R. E. Gaensslen, and H. C. Lee. *Forensic Science: An Introduction to Criminalistics.* New York: McGraw-Hill, 1983.

Ouchterlony, O. Antiger-antibody reactions in gels. *Acta Pathol Microbiol Scand* 26 (1949): 507.

Saferstein, R. *Criminalistics.* 8th ed. Upper Saddle River, NJ: Pearson Education, 2004.

U.S. Fish and Wildlife Service, National Fish and Wildlife Forensics Laboratory; http://www.lab.fws.gov/ (Referenced July 2005).

White, P. *Crime Scene to Court: The Essentials of Forensic Science.* Cambridge: Royal Society of Chemistry, 1998.

Spermatozoa

See *Semen Identification*

Sudden Infant Death Syndrome (SIDS)

Sudden infant death syndrome, crib death, cot death, and *sudden infant death syndrome unexplained* are all terms used to label the death of an infant less than one year of age when there is no plausible cause of death determined at autopsy. SIDS is a rare event but can be devastating for the family, who search for something on which to blame the death. There is, by definition, no known cause. The presence of recognized risk factors predisposes to the assumption that there is indeed a single cause rather than a range of unexplained factors. Posture (sleeping face down), and low birth weight are examples.

There are no warning symptoms. Typically the baby is healthy when put to bed and shows no sign of having suffered any distress. There may be a link with periods of apnea (pauses in breathing) but there is no compelling evidence that utilization of monitors has any influence in the incidence of SIDS deaths.

A painstaking and thorough investigation by the forensic pathologist is vital in a suspected SIDS case. This is one of the few situations in which the absence of information is used to determine a cause of death. It is probably also best that the pathologist is one with experience in child pathology, or at least is assisted by someone with that expertise. Infection, trauma, biochemical abnormalities, and poisoning are some of the alternatives that have to be ruled out.

Suicide

Death investigation is the most demanding of the many events investigated through forensic science. Whatever the cause of death, families have personal emotional and business details to deal with. The difference between an accident, suicide, and homicide is of the utmost significance to them. And, of course, the same

Marilyn Monroe was perhaps one of the most famous cases of suicide because of her success as a Hollywood icon. But controversy has continued since her death was ruled a suicide, with some contending that her connections with J. F. Kennedy and the mob put her in the middle of a very dangerous group. (The Illustrated London News Picture Library)

differences are of the utmost importance to public safety and the prosecution of justice.

Some suicides are easy to identify—for example, by a suicide letter and modus operandi such as drug overdose or self-inflicted gunshot wounds. However, some are more difficult, such as suicide by driving an automobile off the road over a cliff when it may be impossible to reconstruct the vehicle sufficiently well to show whether or not a component failure caused the fatal incident.

It is especially important to guard against being fooled by a homicide presented to appear like a suicide. For example, inhalation of carbon monoxide could be induced deliberately by a third party and then the scene doctored to make it look like suicide.

References

Byard, R., T. Corey, C. Henderson, and
J. Payne-James. *Encyclopedia of Forensic and Legal Medicine.* London: Elsevier Academic, 2005.

Knight, B. *Simpson's Forensic Medicine.* London and New York: Oxford University Press, 1997.

Payne-James, J., A. Busuttil, and W. Smock. *Forensic Medicine: Clinical and Pathological Aspects.* London: Greenwich Medical Media, 2003.

Sweat

Sweat is often present on areas of clothing that come into direct contact with the skin. Although the forensic identification of sweat is rarely necessary, it can be identified by its high salt content. Sweat may also contain ABO grouping substances. For this reason control tests should also be conducted on unstained areas close to the stain to determine the presence of ABO contaminants that might confuse other grouping results.

Sweat is also a component of most latent fingerprints. Secretions from the sweat pores in friction ridges under normal conditions contain 98.5 percent water and 1.5 percent dissolved solids. The dissolved solids are made up of organic and inorganic materials that in turn include lactic and fatty acids, riboflavin and pyridoxin, glucose and other sugars, ammonia and urea, creatinine and albumin, peptides, proteins and isoagglutinogens, and sodium, potassium, calcium, chloride, phosphate, carbonate, and sulfate ions.

References

De Forest, P., R. E. Gaensslen, and H. C. Lee. *Forensic Science: An Introduction to Criminalistics.* New York: McGraw-Hill, 1983.

Saferstein, R. *Criminalistics.* 8th ed. Upper Saddle River, NJ: Pearson Education, 2004.

White, P. *Crime Scene to Court: The Essentials of Forensic Science.* Cambridge: Royal Society of Chemistry, 1998.

T

Taggants

Taggants is the name given to a wide range of microscopic powders of characteristic composition that can be used to identify contact with objects tagged with them. The obvious example is tagging currency in a situation where there has been a history of theft. A typical taggant consists of lycopodium powder in which the spores have been dyed a range of colors including dyes that fluoresce. Screening can be carried out by shining a fluorescent light on the hands or possessions of a suspect.

More specific association with objects can be obtained by microscopic examination of possible contact surfaces and comparison of the range of particle types and colors with the taggant.

Two special cases of the use of taggants are in safes and inks. The application in safes consists of a dye reservoir constructed to spray the taggant dye in the event of the safe being forced open. The contents (currency notes) are indelibly marked and so of no use to the thief. To help in the dating of inks, most manufacturers now include fluorescent tags in their ink compositions. By changing the tag each year, it is possible to identify the year the ink was manufactured.

Teeth

All teeth have an exposed crown extending beyond a root that is embedded in the gum and jawbone. The crown surface is covered with a layer of hard, bonelike enamel. Under the enamel is a layer of dentine, similar to bone but not as hard as enamel. The dentine surrounds an inner pulp-filled cavity, which is fed with nerves and vasculature via the root canal. A layer of bony material called cementum surrounds the root. Most mammals first produce a set of small, weak teeth called milk teeth or deciduous teeth, which erupt soon after birth. The milk teeth consist of incisors, canines, and premolars; molars come in later as part of adult dentition. Replacement of milk teeth and eruption of molars takes place in a relatively fixed sequence.

There are four basic kinds of teeth: incisors, canines, premolars, and molars. Adult humans have two incisors, one canine, two premolars, and three molars in each quadrant of their jaws.

Knowledge of dentition is used in identifying the age of skeletal remains. The exact dentition is recorded in dental charts and can be expressed as a formula for each quadrant thus: I2/2 C1/1 P2/2 M3/3, which shows both incisors, the one canine, both premolars, and all three molars intact. The formula can be used to convey dentition status without needing photographs or x-rays. The charts can be used in screening skeletal remains or mass-disaster victims for identity.

Cases such as that of Ted Bundy (see **Bundy, Ted**) depend on the physical correspondence between teeth and the consequent bite marks. The assumed uniqueness depends on the combination of individual features, such as jaw size, tooth shape and damage, dentition, and jaw shape. There is little dispute about the individuality, but considerable concern about the precision with which the images of a bite mark can be compared to molds of actual teeth. Today, collection of saliva traces from the bite site and testing for DNA type is a much more reliable and discriminating way to deal with bites.

See also Bite Marks; Odontology

References

Bowers, C. M. *Forensic Dental Evidence: An Investigator's Handbook.* Amsterdam and Boston: Academic, 2004.

Bowers, C. M., and G. Bell. *Manual of Forensic Odontology.* 3rd ed. Saratoga Springs: American Society of Forensic Odontology. 1995.

Dorion, R. B. J. *Bitemark Evidence.* New York: Marcel Dekker, 2004.

Thin Layer Chromatography (TLC)

Thin layer chromatography (TLC) is a screening test that has many uses in the field of forensic science, including the analysis of drugs, fiber dyes, and inks.

The name *thin layer* comes from the nature of the technique. A thin layer (about 0.1 mm) of a solid stationary phase is coated onto an inert support medium. This so-called stationary phase is often silica gel and the support is a glass or plastic plate. Chromatographic separation is achieved by the action of a mobile phase consisting of a single solvent or a mixture of solvents.

To perform thin layer chromatography, a line is drawn at the bottom of the plate, and a small amount of the questioned sample (dissolved in a solvent) is transferred to the plate as a small spot on the line. A known standard of the suspect material is also spotted on the plate adjacent to the unknown.

The plate is placed vertically into a developing tank that contains a little of the mobile phase. The tank is covered so that the chromatography takes place in a saturated atmosphere. The mobile phase then moves up the plate by capillary action. The components of each of the samples will move up the plate depending on their solubility in the mobile phase and the degree to which they partition between mobile and stationary phases. After the mobile phase has traveled a certain distance up the plate, the plate is removed and the solvent front (the distance the mobile phase traveled) is marked. Any spots that are present are located with the use of ultraviolet light and visualizing reagents. If all components of both the known and the unknown sample travel the same distance, this indicates they may be the same substance.

Thin layer chromatography is often used because it is such a cheap and easy test to perform. It also allows simultaneous running of multiple samples. It has lower sensitivity and resolution than gas and high performance liquid chromatography.

Spot Location

Because many organic compounds are colorless, the separated sample components must be made visible. Many compounds can be located using a fluorescent indicator under short-wave ultraviolet light—absorbing spots appear as dark spots on a green background. Drugs that naturally fluoresce can be observed using long-wave ultraviolet light. Spots visualized by ultraviolet light need to be outlined in pencil.

Some components may not fluoresce and so require to be visualized by some other means. In the case of drugs, iodine vapor may be sprayed on the plate as a visualizing agent, or alternatively, presumptive test chemicals may be used because they will result in a color change when in contact with specific substances.

Rf Values

The Rf value is the basic chromatographic measurement of a substance in TLC. Rf is defined as the distance a substance travels from the origin divided by the distance the solvent front travels from the origin. The distance traveled by the substance is measured from the center of the spot if the spot is round. For spots with tails the distance traveled should be determined from the center of the densest area. The value varies from 0 to 1, but usually Rf × 100 is reported to avoid using decimals.

As with all forensic procedures, the test sample should always be run with appropriate controls or reference standards on the same plate. This is because the Rf value can be altered by a number of things, including laboratory temperature, stationary phase used, mobile phase used, and amount of sample loaded onto the plate.

References
De Forest, P., R. E. Gaensslen, and H. C. Lee. *Forensic Science: An Introduction to Criminalistics.* New York: McGraw-Hill, 1983.
Saferstein, R. *Criminalistics.* 8th ed. Upper Saddle River, NJ: Pearson Education, 2004.
White, P. *Crime Scene to Court: The Essentials of Forensic Science.* Cambridge: Royal Society of Chemistry, 1998.

Time of Death

The physical and biochemical changes that occur following death are described in the entry on **death**. Here we will describe the ways in which estimation of time of death is approached in a normal investigation and give some typical case examples. However, the reader should never lose sight of the dictum that an estimated time of death is an informed guess at best.

Examination of the body and pronouncements as to the cause of death and related matters are the responsibility of qualified medical practitioners with experience in forensic medicine. There will be an autopsy, but in many instances the first person to see and examine the body is not the pathologist performing the autopsy. As with any first responder at a scene, there will be a conflict between preservation and timely information.

For example, there is potentially valuable information recording rectal temperature from as early as possible and at least before the body is moved. This will give a measure of the cooling rate and assist in interpretation of the observed temperature. However, the death may have sexual overtones, including anal intercourse, and every insertion and withdrawal of the thermometer can result in loss of information from biochemical testing of the semen.

Recording of temperature, degree of rigor, body color, and lividity (see **Death** for explanations of these terms) must all be performed at the scene of a recent death. The postmortem examination can add information such as stomach content and potassium estimations. It is also wise to consider all available information because there may have been a substantial predeath interval. In obviously older deaths, putrefaction and infestations can provide additional information.

Illustrative Cases
Workmen found the partly mummified body of a middle-aged man in the crawl space below the floor of a house in a large town in Scotland. He was a known alcoholic and drug addict. Suspicion fell on his wife, who eventually made a statement to the police that she had gone to bed one night some months previously, leaving him asleep in a presumed alcoholic or drug-induced stupor. When she got up the following morning, he was dead. She had had enough problems with him and disposed of the body in the crawl space rather than inform the police and go through the trauma of an investigation. If that story was true, then she was guilty only of the minor offense of concealing a death. However, she could not prove to the police that she had not deposited the body of her unconscious husband and that his death was a consequence of her actions.

Toxicology showed the presence of two drugs, amylobarbitone (amobarbital) and quinalbarbitone (secobarbital), the ingredients of a prescription-only sleeping medication

A missing woman is found dead in Florida. Forensic investigators at the scene must determine the DNA of the victim for positive identification and establish a time of death. January 1999, Hamilton Country, Florida. (Karen Kasmauski / Corbis)

available in Britain as Tuinal. Amobarbital and secobarbital are removed from the blood at different rates, and the relative concentrations of each gives an indication of the time since the medicine was taken. In this case, the ratio indicated a time of less than twelve hours. This, together with the lack of any obvious signs of attempt to escape, such as injuries to the hands or marks on the underside of the flooring, led to the prosecution dropping the charge soon after the trial began.

The second case involved the homosexual assault and murder of a teenage boy, whose body was found in parkland. Evidence from eyewitnesses showed that the boy had been picked up by the accused, and blood grouping of semen confirmed this. However, the accused denied the homicide and said he had driven the boy back to where he had picked him up. The time of death was critical in regard to evaluation of the story of the accused. Temperature measurement, supported by examination of stomach content,

showed that death had occurred sufficiently close to the time that witnesses saw the boy being picked up to make it highly unlikely that there had been a subsequent pickup, drive to the park, and homicide.

See also Autopsy; Pathology
References
Byard, R., T. Corey, C. Henderson, and
    J. Payne-James. *Encyclopedia of Forensic and Legal Medicine.* London: Elsevier Academic, 2005.
Knight, B. *Simpson's Forensic Medicine.* London and
    New York: Oxford University Press, 1997.
Payne-James, J., A. Busuttil, and W. Smock. *Forensic Medicine: Clinical and Pathological Aspects.* London: Greenwich Medical Media, 2003.

Time since Intercourse

DNA has proven to be an invaluable tool for identification of the donor of semen found in samples from a rape investigation. Interpretation of all the evidence may, however, benefit from an estimation of the postcoital interval, or time since intercourse. Consider a simple illustration: A swab taken from the vagina of a murder victim tests positive for semen. Without further qualification, that finding would lead investigators to consider a scenario of rape-murder. Time of death is known to be recent. DNA typing matches a known associate of the deceased. What can we conclude from these observations? If further investigation shows that the semen was deposited more than two days earlier, then there is no time linkage between the events of death and intercourse. If the semen was deposited within a few hours of death, then there is. These circumstances have dramatically different implications for the investigation and for the associate of the deceased.

Reliable estimation of time since intercourse is therefore a potentially valuable tool. Techniques used mainly depend on the microscopy of the vaginal samples. After intercourse, ejaculate is lost from the vagina by physical drainage. The individual components of the ejaculate suffer degradation; specifically, the spermatozoa lose their motility, then their tails. It is therefore possible to derive information based on the number of spermatozoa in the sample and their physical features. Unfortunately, there is such wide variation in the rate of loss of material, and indeed in the nature of the contents of the original semen deposited, that only the most broad estimates can be relied on. A sample with high numbers of intact spermatozoa can be described as "recent"—perhaps up to six hours old. One with very few sperm heads can be described as "old"—perhaps more than thirty-six hours.

See also Date Rape; DNA in Forensic Science; Rape

References
Butler, J. *Forensic DNA Typing.* 2nd ed. Burlington, MA: Elsevier, 2005.
Hazelwood, R. R., and A. W. Burgess. Practical Aspects of Rape Investigation: A Multidisciplinary Approach. 3rd ed. Boca Raton, FL: CRC, 2001.
Seneski, P. C., N. Whelan, D. K. Faugno, L. Slaughter, and B. W. Girardin. *Color Atlas of Sexual Assault.* St Louis: Mosby, 1997.
Turvey, B. E., and J. Savino. *Rape Investigation Handbook.* Amsterdam and Boston: Elsevier Academic, 2004.

Tire Tracks

Tire-track comparison is a process similar to footprint or shoe-print comparison. The same techniques of evidence collection used in recording and preserving footprint or shoe-print evidence are applied to tire tracks and include lifts of prints, casts of impressions, and specialized lifting techniques for the recovery of tracks left in snow or even underwater.

When a tire track is compared with a known tire, the first class characteristics that are compared are tire width, arrangement of land (the raised part of the tread pattern) width, the number of lands, land and groove patterns, and land and groove locations. These characteristics can be used to place the tire into a specific class of tires. Accurate measurements of the tire tracks can also provide useful information on the make of tire, the year of production, and model of a suspect vehicle. Useful measurements include the wheelbase, which is the distance between the rear and front axles, and the front and rear track width. This distance relates to the length of an axle and is the distance between the middle of the two front or rear tires. These class characteristics can be used to provide investigative leads by identifying the possible make and model of vehicle that left the tracks.

After determining the class characteristics, individual characteristics are then compared. Individual characteristics develop according to the types of surfaces on which the vehicle is commonly driven, the physical and mechanical condition of the vehicle, tire inflation pressures, and the driver's habits. Marks made by wear processes are important in individualizing a suspect tire and subsequently the suspect vehicle.

Skidmark Analysis

The forceful application of brakes when making an emergency stop will lock the wheels unless the car is fitted with an antilock braking system (ABS) or driven by an expert driver able to control the braking just short of locking. The locking results in skid marks, which occur from the combination of heat and torn rubber left as dark marks with clear beginnings and ends on the road surface. These marks can often be used to determine the minimum speed at which the vehicle was traveling. Tire imprints and scuff marks tend to show tread marks and are not as affected by driver actions as skid marks. Skid marks are measured from the terminal point backwards. Marks from the front wheels must be distinguished from marks by the rear wheels and each wheel mark should be measured separately, if possible. Yaw marks occur when a tire tries to rotate in one direction while sliding in another, and these can also yield information about speed, direction of travel, and point of impact.

The full examination of a crash scene is a very specialized technical procedure. It should be carried out by experienced and trained personnel with good knowledge of the physics and mathematical manipulations required. There are several computer programs available to assist with the computations, but interpretation is best left to experienced investigators. However, an idea of the process can be obtained by illustrating the most simple situation: estimation of the speed of the skidding vehicle.

The first step is to understand the relationship between the coefficient of friction, force and mass of the surface and vehicle:

$$f = \frac{F}{W} = \frac{3000}{5000} = 60\%$$

f is the coefficient of friction (sometimes called drag factor);

W is the vehicle mass (say 5,000 pounds);

F is the force required to overcome resistance and make the vehicle slide on the surface.

Experience has shown that f, the coefficient of friction, is typically about 60 percent for most tire-road surface combinations in dry weather. It is less in wet weather or on greasy surfaces.

The value of f and the measured skid distance can now be factored into the following basic speed equation:

$$S = \sqrt{30Df}$$

where S is speed in miles per hour, D is the skid distance in feet, and f is the drag factor or coefficient of friction.

See also Individual Characteristics

References
De Forest, P., R. E. Gaensslen, and H. C. Lee. *Forensic Science: An Introduction to Criminalistics.* New York: McGraw-Hill, 1983.
James, S. H., and J. J. Nordby. *Forensic Science: An Introduction to Scientific and Investigative Techniques.* Boca Raton, FL: Taylor and Francis, 2005.
Lee, H. C., T. Palmbach, and M. T. Miller. *Henry Lee's Crime Scene Handbook.* London and San Diego, CA: Academic, 2001.
Saferstein, R. *Criminalistics.* 8th ed. Upper Saddle River, NJ: Pearson Education, 2004.
White, P. *Crime Scene to Court: The Essentials of Forensic Science.* Cambridge: Royal Society of Chemistry, 1998.

Tool Marks

Tool marks result from contact between a tool and some other surface. Both class and individual characteristics can be found in tool marks. For example, a tool mark made by a screwdriver in a windowsill may reveal the width and shape of the blade, as well as striations unique to that screwdriver.

There are two types of tool marks: impressions and striations.

Tool Impressions

A tool impression is made when a tool is compressed into the surface of an object, leaving a negative impression. The material in which the impression is made determines how well the irregularities of the tool surface are reproduced. Softer, smoother surfaces produce better marks than harder, brittle surfaces that can break or chip.

Trace evidence from the surface itself can also be transferred to the tool. Impression marks usually are not identifiable to a specific tool.

As with tool impressions, the surface determines how well the tool's imperfections will reproduce.

Striated Marks

Striated marks are made by sliding or scraping tool actions, such as occur with a crowbar, or by cutting, shearing, or pinching action. A striated mark is formed when the tool is moved across a surface with pressure. The edge of the tool has imperfections that create striations that can be compared with a comparison microscope. Test tool marks must be made that duplicate the way the questioned mark was made and are usually made in lead.

Laboratory examination of tool marks is best carried out on the actual object itself.

When this is not possible, a Mikrosil cast made from the tool mark can be submitted to the laboratory. In all cases, photographs of the scene showing where the tool mark was located in relation to other objects aids the examiner in determining how the tool was used.

See also   Casts; Individual Characteristics

References

De Forest, P., R. E. Gaensslen, and H. C. Lee. *Forensic Science: An Introduction to Criminalistics.* New York: McGraw-Hill, 1983.

James, S. H., and J. J. Nordby. *Forensic Science: An Introduction to Scientific and Investigative Techniques.* Boca Raton, FL: Taylor and Francis, 2005.

Lee, H. C., T. Palmbach, and M. T. Miller. *Henry Lee's Crime Scene Handbook.* London and San Diego, CA: Academic, 2001.

Saferstein, R. *Criminalistics.* 8th ed. Upper Saddle River, NJ: Pearson Education, 2004.

White, P. *Crime Scene to Court: The Essentials of Forensic Science.* Cambridge: Royal Society of Chemistry, 1998.

Toxicology

Forensic toxicology centers on the determination of toxic substances in human tissues, organs, and body fluids such as urine and blood, and the subsequent determination of the role any toxic agents may have had in contributing to or causing death. In general, forensic toxicology cases entail some form of drug or alcohol abuse. The ultimate goal of the forensic toxicologist is to provide answers to questions that may be asked during the investigation or at a subsequent legal hearing. Typically these questions include: Was a poison involved in the case? If so, what was the nature of the poison? How was it administered? And was it at a potentially lethal concentration?

Having identified the presence and amount of a drug and/or its metabolites in a person's body fluids, the toxicologist may be called upon to describe the effects that these levels may have elicited in the subject.

The specimen most commonly used for analysis and to draw any conclusion concerning behavioral effects of drugs is blood. Blood drug concentrations represent what is affecting the body at that particular time, whereas drugs in the urine, for example, could reflect what was exerting a pharmacological effect hours or even days previous.

See also   Poisoning; Poisons

References

Goldfrank, L., N. Flomenbaum, N. Lewin, M. A. Howland, R. Hoffman, and L. Nelson. *Goldfrank's Toxicologic Emergencies.* 7th ed. New York: McGraw-Hill Professional, 2002.

Klaassen, C. *Casarett and Doull's Toxicology: The Basic Science of Poisons.* 6th ed. New York: McGraw-Hill Professional, 2001.

Levine, B. *Principles of Forensic Toxicology.* Washington, DC: American Association for Clinical Chemistry, 1999.

Trace Evidence

Trace evidence deals with the physical and chemical characterization of materials transferred between objects. Usually the quantities transferred are small, hence the term *trace evidence*. The exchange process is sometimes described as the Locard exchange principle, which is widely quoted as "every contact leaves a trace." Examination of the transfers is used to associate people with other people or people with places. Several different methods can be employed in the collection of trace materials:

Lifting. Adhesive tape is repeatedly and firmly patted or rolled over an item, picking up loose trace evidence. The lifts are placed on a transparent sheet (clear plastic sheeting, glass slides, and clear plastic or glass petri dishes) to protect against contamination and allow the easy viewing of samples and for identification and comparison.

Scraping. A spatula or other tool is used to dislodge trace materials from an item onto clean paper. The collected debris is packaged in a way to avoid sample loss. This is usually conducted within the laboratory in a controlled environment that reduces the risk of contamination or loss of the trace evidence.

Vacuum Sweeping. A vacuum cleaner equipped with a filter trap is used to recover trace evidence from an item or area. The filter and contents are packaged immediately to avoid sample loss. All parts, filter, and trap must be changed and rigorously cleaned between vacuumings to avoid contamination.

Combing. A clean comb or brush is used to recover trace evidence from the hair of an individual. The comb and debris from the hair should be packaged together.

Clipping. Trace evidence can be recovered from fingernails by nail clipping and scraping. Usually fingernails or scrapings from the right and left hands are packaged separately.

References
De Forest, P., R. E. Gaensslen, and H. C. Lee. *Forensic Science: An Introduction to Criminalistics.* New York: McGraw-Hill, 1983.
James, S. H., and J. J. Nordby. *Forensic Science: An Introduction to Scientific and Investigative Techniques.* Boca Raton, FL: Taylor and Francis, 2005.
Lee, H. C., T. Palmbach, and M. T. Miller. *Henry Lee's Crime Scene Handbook.* London and San Diego, CA: Academic, 2001.
Saferstein, R. *Criminalistics.* 8th ed. Upper Saddle River, NJ: Pearson Education, 2004.
White, P. *Crime Scene to Court: The Essentials of Forensic Science.* Cambridge: Royal Society of Chemistry, 1998.

Trinitrotoluene (TNT)

Trinitrotoluene (TNT), also known as Triton, Trotyl, Trilite, Trinol, and Tritolo, is a constituent of many high explosives, including amatol, pentolite, tetrytol, torpex, tritonal, picratol, ednatol, and composition B. In its refined form, TNT is very stable and capable of being stored for long periods of time, because it can withstand most blows or friction. TNT is nonhygroscopic and doesn't form sensitive metal complexes, but reacts readily with alkalis forming unstable compounds very sensitive to heat and impact. TNT may exude a flammable, oily brown liquid, which oozes out around the threads at the nose of the shell, forming pools. TNT is used as a booster or as a bursting charge for high-explosive shells and bombs.

See also Explosions and Explosives
References
Saferstein, R. *Criminalistics.* 8th ed. Upper Saddle River, NJ: Pearson Education, 2004.
White, P. *Crime Scene to Court: The Essentials of Forensic Science.* Cambridge: Royal Society of Chemistry, 1998.

Typewriting and Printing

Mechanical impressions are impressions made on paper by machines or tools. Many mechanical devices are used to produce all different sorts of documents. The most common types of mechanical impressions are made by typewriters, copying machines, printing presses, and stamping machines. Desktop printers and fax machines may also make marks and impressions.

Typewriters
Four types of typewriting examination are generally requested:

- Determination of the type, make, and model of the machine that produced the questioned document
- Determination of whether a particular typewriter produced a specific document
- Determination of when the questioned document was produced
- Determination of whether the typed material was produced continuously

In typewriter examinations the examiner must have access to the complete reference collection of typefaces used by all typewriter manufacturers. These typefaces determine the structure of the letters (serif, sans serif) and provide the class characteristics for comparison purposes.

Each typewriter has its own individualizing characters—physical defects present on the typeface—that enable differentiation of the characters typed on it from characters typed on other machines. These imperfections come from manufacturing processes and wear and tear due to usage.

Even straight from the manufacturer, each new typewriter will have its own identifying characteristics. Common defects include misalignment of the type, irregular impressions of the type bars, cavities and accidental defects on the characters, and mechanical imperfections.

The same techniques used in typewriter comparisons can be used to compare machine printing or stamped impressions on documents.

Document examiners may also be called to examine documents produced by a variety of business machines; adding machines, stamping machines, check-writing machines, and printing presses are some examples. Again, they may be asked to determine:

- The type of machine that produced the document
- The make of a particular machine used to produce a document
- Whether a particular machine was in fact used to produce a particular document

The same general principles using class and individual characteristics apply to the examination of these types of machines.

See also Document Examination

References

Hilton, O. *Scientific Examination of Questioned Documents*. Boca Raton, FL: CRC, 1993.

Saferstein, R. *Criminalistics*. 8th ed. Upper Saddle River, NJ: Pearson Education, 2004.

Vastrick, T. W. *Forensic Document Examination Techniques*. Institute of Internal Auditors Research Foundation, Alpharetta, GA, 2004.

White, P. *Crime Scene to Court: The Essentials of Forensic Science*. Cambridge: Royal Society of Chemistry, 1998.

U

Ultraviolet (UV) and Visible Spectrometry

Ultraviolet (UV) and visible absorption spectrometry are used in pharmaceutical and biomedical analysis for the quantitation and characterization of drugs, diluents, metabolites, and other related substances. This technique works on the principle that substances absorb or emit electromagnetic radiation at different wavelengths. The energy absorption or molecular absorption of a substance can be recorded to provide an absorbance spectrum, and used to determine the concentration of a substance (or drug) in a solution. Ultraviolet-visible spectrometry is little used in its own right today in forensic science. However, many detectors used in high performance liquid chromatography (HPLC) operate on the same principles. It is important that users understand the basis of the detection's system, and the material presented below focuses on how ultraviolet and visible light are employed in chemical analyses.

The technique depends on the properties of radiation in a very small region of the electromagnetic spectrum. Electromagnetic radiation is characterized by two physical quantities: a wavelength λ and a frequency f that are related to the velocity of light (c) by the equation $f = c/\lambda$. The velocity of light in a vacuum is constant at 2.99792×10^8 meters per second.

Electromagnetic radiation can be thought of as a stream of energy particles or packets of energy known as photons. The energy of the photon or energy packet is proportional to the frequency of the radiation. Electromagnetic radiation travels as a wave and when it impinges on matter, interactions occur that affect the wave and the material. Some materials absorb the energy, causing their molecules to oscillate, inducing electrical and magnetic fields that in turn also generate an electromagnetic wave or light, which is emitted by the material. Changes in the electromagnetic radiation can be detected and used to generate absorption and emission spectra for a substance.

Various wavelengths and their interactions with atoms and molecules make up the electromagnetic spectrum. Ultraviolet light lies in the wavelength range of 200 to 400 nanometers (nm), and visible light lies in the range of 400 to 800 nm.

The molecular group in a compound that is responsible for energy absorption is called a chromophore. Chromophores are usually conjugated systems with a delocalization of electrons. Any saturated chemical group that has little or no intrinsic absorption, but modifies the absorption spectrum when attached or conjugated directly to a chromophore via its polarizable lone pair of electrons, is called an auxochrome.

Different chromophores absorb at different wavelengths, and so ultraviolet-visible spectrometry can provide some information as to the structure of chemicals. However, its main application is in measuring the amount of substance present in a solution. This can be done because it is known that absorbance is directly proportional to concentration. Stated another way, the greater the absorbance measured for a particular substance, the greater the concentration of that substance. Of course, there are some conditions in which this is not the case, such as high concentrations of analyte, scattering of radiation due to particles in the solution, reemission of absorbed light as fluorescence, and use of nonmonochromatic light.

For ultraviolet and visible spectrometry a continuum light source is used. The power of this source does not change over a wide range of wavelengths. For an ultraviolet range of 160 to 375 nm, sources are usually hydrogen or deuterium lamps. The light emitted from the source is then passed through a monochromator to select and produce light of a specific wavelength. Some alternative light sources use simple filters to produce light in a specific narrow wavelength band so that they can be tailored to the application. For example, an ultraviolet light at 254 nm will aid detection of semen, as the fluid fluoresces when irradiated with ultraviolet light. For any type of spectrophotometry, the cuvettes or sample cells, as well as the sample solvent, must be transparent to incident radiation. For sample observation in the ultraviolet range of less than 350 nm, cuvettes are made of quartz or fused silica, both being transparent to ultraviolet and visible-range wavelengths. Silicate glass and some plastics can be used in the visible range above 350 nm. The most common cell length for sample measurement is 1 cm. Matching cuvettes should be used for the measurement of reference and test samples, taking care not to contaminate the outer surface of the cuvettes with finger grease or oils, as this will affect the quality of the absorbance data. Data obtained from an ultraviolet spectrometer is in the form of a graphical representation showing wavelength along the x-axis and absorbance on the y-axis.

See also High Performance Liquid Chromatography (HPLC)

References

De Forest, P., R. E. Gaensslen, and H. C. Lee. *Forensic Science: An Introduction to Criminalistics.* New York: McGraw-Hill, 1983.

Saferstein, R. *Criminalistics.* 8th ed. Upper Saddle River, NJ: Pearson Education, 2004.

White, P. *Crime Scene to Court: The Essentials of Forensic Science.* Cambridge: Royal Society of Chemistry, 1998.

Urine

Urine is encountered in forensic science in three different settings. The most important is when it is used in toxicology examinations to screen for the presence of drugs (and sometimes to estimate the amount taken). Urine itself is sometimes the direct object of testing, for example, in a burglary where the perpetrators relieve themselves on the furnishings of the house as an act of vandalism. Finally, traces of DNA or ABO blood types can be detected in urine and so identify the source of an evidence item, such as a pair of jeans with urinal splashback or bladder leakage onto the crotch area.

Urine has mixed value as a fluid for toxicology testing. Nearly all drugs are eliminated in urine either as the unchanged drug, or as by-products formed during metabolism in the liver, or (mostly) a mixture of these. Urine can be obtained without performing technically invasive procedures such as venepuncture. Relatively large volumes of fluid are available and the concentration of drug can be quite high.

Disadvantages include the need for caution in interpretation of results. For example, using urine in preemployment drug screening or drug screening for driving while intoxicated can produce a positive for apparent marijuana use that resulted from ingestion many days previously. This is because the active component

Drug-testing employees is becoming more and more popular among both British and American companies. (iStockphoto.com)

(THC) is removed very rapidly from the bloodstream, but the metabolite persists for several weeks. This may be acceptable in an employment-screening context where there is an absolute prohibition on drug abuse, but it will invalidate testing that seeks to demonstrate current intoxication.

Urine has been specifically identified as an acceptable body fluid for alcohol analysis in some jurisdictions (primarily the United Kingdom). This is about the only situation where quantitative drug testing in urine has any validity. The reason is that the way in which the kidneys remove drugs from the bloodstream does not normally result in the concentrations in urine being proportional to those in blood. The defining relationship is that the rate of elimination (that is, the amount excreted into the urine) is proportional to the concentration in blood. The proportionality constant is called the *clearance* of the drug. However, for alcohol, the clearance value is not fixed but varies with the rate of formation of the urine itself, resulting in an approximate proportionality between concentration in blood and that in urine. The early United Kingdom drink-driving legislation set an equivalence of 80 milligrams of alcohol per 100 milliliters of blood, being the same as 107 milligrams of alcohol per 100 milliliters of urine.

Sometimes the laboratory has to identify a fluid or stain as being urine. Urine contains a large amount of urea, a chemical by-product of normal metabolic processes in the body. Identification of high levels of urea can therefore serve as a screening test for urine in fluids or stains. The presence of creatinine is also used for identification purposes. Creatinine forms a red compound with picric acid (the Jaffe test). Urine also has a characteristic odor, which can help in locating its presence. Gentle heating of urine-stained materials gives rise to the distinctive odor.

Urine may also contain traces of DNA in the minute number of cells usually present as they are lost from the walls of the bladder and urethra. Soluble ABO grouping components may also be present. The ability to type these materials in urine is a mixed blessing. Although it provides a tool to identify the origin of a sample or stain—for example to prove the authenticity of a urine sample presented for drug screening—it may also produce background contamination in the testing of a blood or semen stain on clothing with urine traces.

Water and oil do not mix. Neither do gasoline and urine. There are many tales of police officers taking the opportunity to relieve themselves into the fuel tank of a superior's automobile. Many are probably apocryphal, and it is hard to find a modern vehicle that does not have a locking mechanism on the cap to the fuel filler. However, it does happen, and the writer has had direct experience of examining fuel-tank contents for urine and subsequently attempting to group the urine and identify the culprit. For the record, the identification of the urine was successful but not the grouping, so in a way everyone was happy.

The USS Iowa *during a firing exercise in 1984, five years before an accident killed 47 men during a similar training episode. Near Puerto Rico. (U.S. Navy Photograph/Department of Defense Still Media Collection)*

References

Levine, B. *Principles of Forensic Toxicology.* Washington, DC: American Association for Clinical Chemistry, 1999.

Saferstein, R. *Criminalistics.* 8th ed. Upper Saddle River, NJ: Pearson Education, 2004.

White, P. *Crime Scene to Court: The Essentials of Forensic Science.* Cambridge: Royal Society of Chemistry, 1998.

USS *Iowa*

On April 19, 1989, during a firing drill, the number-two turret on the U.S. Navy battleship the USS *Iowa* exploded, killing forty-seven of its crew. An investigation held immediately afterwards was unable to identify any cause of the explosion and concluded that the explosion had been a case of sabotage. For the three guns on each turret to fire shells weighing 2,700 pounds to distances of up to 24 miles, bags of nitrocellulose explosive must be added to the shells. The shells are loaded into the guns and then, using a rammer, bags of explosive are pushed close to (but not touching) them. The investigation found that on the day of the explosion, the left gun was loaded in 44 seconds, while the other was loaded in 61 seconds. It was also found that the operator of the rammer had reported a problem a minute or so before the explosion occurred. The investigation revealed the presence of steel wool, brake fluid, and calcium hypochlorite in the gun that exploded, and concluded that these had been placed there deliberately so that the explosion occurred when pressure from the rammer was applied.

It was then decided by the Senate Armed Services Committee to request that the Sandia National Laboratory carry out a full forensic investigation into the incident. This investigation found that similar traces of steel fibers, calcium, and chlorine were found in the other gun turrets, as well as on other battleships, including USS *Wisconsin* and USS *New Jersey.* During the investigation, it was found that the rammer pushed the explosives two feet further up the gun than it should have done, and also that the operator had been inexperienced. It was also found that extra sticks of explosives had been placed loosely on top of the bag of explosives in order to make up weight, which had been the cause of the explosion. The report concluded that the explosion had been an accident and not sabotage. The loading procedure for these guns was subsequently changed by the navy in order to ensure that such an accident does not happen again.

Reference

Owen, D. *Hidden Evidence: Forty True Crimes and How Forensic Science Helped Solve Them.* Willowdale, Ontario: Firefly, 2000.

V

Validation

Validation, or the process of establishing the reliability of a test procedure, is an essential part of the operations of a forensic science laboratory. The topic is addressed in the "Guidelines for Forensic Testing" produced by the International Laboratory Accreditation Cooperation (ILAC).

Validation studies can be conducted by the scientific community (as in the case of standard or published methods) or by the forensic science laboratory itself (as in the case of methods developed in-house or where significant modifications are made to previously validated methods).

A specific implementation of validation in one area—drug analysis—can be developed using other resources, such as the FDA's "Guidelines for Good Laboratory Practice." Once the forensic scientist has decided upon an analytical method to verify the identity and purity of a drug substance, the analytical method must be validated. Many approaches may be taken to validate an analytical procedure. The main objective of method validation is to demonstrate that the procedure is adequate for its intended use, and some general guidelines can be described for this purpose. These guidelines can be easily applied to the validation of methods for drug analysis. An analytical procedure can generally be considered to include the following components:

- Extraction of the test analyte from a matrix
- Construction of a standard curve
- Quantitation of the test analyte
- Quality control of the analytical method

Parameters to be considered when validating such a process include the purity and identity of the neat test sample, homogeneity of solutions, extraction recovery, range of standard curve and control samples, linearity and precision of standard curves, sensitivity, specificity, accuracy and precision, stability, and quality control.

References

International Laboratory Accreditation Cooperation, Guidelines for Forensic Science Laboratories; http://www.ilac.org/downloads/Ilac-g19.pdf (Referenced July 2005).

U.S. Food and Drug Administration, Center for Drug Evaluation and Research; Guidance for Industry, Bioanalytical Method Validation; http://www.fda.gov/cder/guidance/4252fnl.htm (Referenced July 2005).

Vollman, John

Of significance in this case was the admissibility of hair neutron activation analysis and the importance of the evidence in the case.

The body of Gaetane Bouchard was found in an area outside of Edmunston, New Brunswick, in 1958 by her father. He had previously looked for her at the home of John Vollman, who according to Gaetane's friends was her new boyfriend. However, Vollman told the father that he was no longer involved with her. Investigators, however, found a single chip of green paint at the murder site that they found to match exactly a missing chip of paint from John Vollman's car. Also, in Vollman's car police found a half-eaten bar of chocolate with lipstick on it—the same kind that Gaetane had been seen buying earlier in the day.

A second autopsy of the body revealed a single hair around Gaetane's fingers that it was thought she may have pulled from her attacker in a struggle. Neutron activation analysis was performed on the hair they found around her finger, as well as Gaetane's hair and John Vollman's hair. The proportion of sulfur to phosphorus radiation was measured and found to be the same in the hair found around Gaetane's finger and John Vollman's hair. The proportion found in Gaetane's own hair was very different. After much debate, the evidence was determined to be admissible in court, following which Vollman changed his plea to guilty of manslaughter. He was found guilty of murder.

See also Admissibility of Scientific Evidence; Hair

References

Evans, C. *The Casebook of Forensic Detection: How Science Solved 100 of the World's Most Baffling Crimes.* New York: Wiley, 1998.

Owen, D. *Hidden Evidence: Forty True Crimes and How Forensic Science Helped Solve Them.* Willowdale, Ontario: Firefly, 2000.

W

World Trade Center Bombing

Although today, inevitably, the name "World Trade Center" conjures images of the terrorist attacks of September 11, 2001, there was an earlier attack on the building in which carefully applied forensic science produced compelling results.

On February 26, 1993, at 12:18 p.m., a van containing an improvised explosive device exploded in the underground garage of the World Trade Center. Six people were killed and more than a thousand were injured.

Bomb technician special agents from the FBI were first to respond; they concluded that the explosion was due to an improvised device. This was confirmed by special agents from the FBI Laboratory Explosives Unit, who were subsequently dispatched to examine the crime scene. The crater formed by the explosion was determined to be 150 feet in diameter and more than 5 stories deep. This, together with the damage done to the surrounding building and automobiles, suggested to investigators that the explosive had a detonation velocity of 14,000 to 15,500 feet per second, and that the explosive was probably 1,200 to 1,500 pounds. Due to the estimated size of the explosive and the height restrictions on the parking garage, it was suspected that the explosive had been transported in a pickup truck or van.

Two days after the attack, explosive residues were collected and taken for analysis. At this time, the explosives experts examining the crime scene found a large fragment from a vehicle from which they were able to obtain a vehicle identification number. This identified the vehicle as being a rental van that had been reported stolen the previous day. This vehicle was soon identified as having been that which contained the explosive. While FBI agents were interviewing the Ryder Rental Agency manager from the location the vehicle had been rented, a man named Mohammed Salameh telephoned to request that his security deposit for a truck he had rented be returned. Arrangements were made for him to do this, and when he did so, he was placed under arrest by FBI agents. A search of Salameh's belongings led agents to a chemist named Nidel Ayyad. He and Salameh were found to have a joint bank account, and were also connected via telephone records. Through scrutiny of Ayyad's records and receipts, an apartment was found in Jersey City in which acids and other chemicals had been used in the manufacture of explosives. Agents determined that there were traces of nitroglycerine and urea nitrate present on the carpet of the apartment. A nearby storage shed was searched and items found there included 300 pounds of urea, 250 pounds of sulfuric acid, empty and full gallon containers of nitric acid, and sodium cyanide, fuse materials, and a trash can in which were found traces of urea nitrate. A third man, Mahmud Abouhalima, was identified during investigations in the neighborhood of this bomb factory. He fled to Egypt, but was soon arrested and extradited to the United States.

On March 3, 1993, a letter was received by the *New York Times* from someone claiming responsibility for the attack in the name of Allah. DNA tests of saliva revealed that the envelope of this communication had been licked by Ayyad.

Analysis of the debris from the World Trade Center revealed the presence of three AGL Welding Company gas cylinders. A fragment of red paint from one of these was analyzed and found to be the same as the red paint used by that company on their hydrogen cylinders. Also, tiny pieces of blue plastic found on the vehicle fragment were examined and found to be the same as blue plastic containers found at a storage location rented by Salameh.

The two men who were thought to be the "masterminds" behind the operation, Ramizi Ahmed Yousef and Eyad Izmoil, fled the country immediately after the attack. A $2 million reward was posted. Yousef was arrested two years later in Pakistan, and Izmoil was arrested in Jordan. These men were extradited to the United States, where they stood trial with the other four men involved in the attack. Each of them was sentenced to life imprisonment.

It was believed that the men had actually intended to collapse one of the towers on top of the other, and in doing so they had hoped to release a deadly cloud of cyanide gas. Luckily they did not achieve their aim, and the cyanide gas was consumed by the fire rather than being vaporized. Had they achieved their aim, the results might have been devastating.

See also Explosions and Explosives
References
Owen, D. *Hidden Evidence: Forty True Crimes and How Forensic Science Helped Solve Them.* Willowdale, Ontario: Firefly, 2000.

Reeve, S. *The New Jackals: Ramzi Yousef, Osama Bin Laden, and the Future of Terrorism.* Boston: Northeastern University Press, 2002.

Wounds

Crimes against the person involving physical violence cause wounds. Earlier entries emphasized the importance of understanding injury cause and effect in order to differentiate correctly (where possible) between accidental and deliberate causation, and between those that are self-inflicted and those inflicted by another person. This section will give an overview of wounds, providing a background for their interpretation. Note that technically a wound is an event that results in breaking of the continuity of the skin. However, injuries such as bruising and bone fractures will also be considered here.

Abrasions are wounds in which the surface of the skin is damaged by contact with a rough object. Abrasions can result from falling, dragging, scratching, or violent contact with any rough surface such as a wall. They are relatively easy to identify and interpret. The location can be of considerable significance—around the genitalia, for example, in a sexual assault or the neck in a strangulation. Abrasions inflicted postmortem can be differentiated from those inflicted before death by whether or not there is any gross or microscopic evidence of bleeding. Redness and pinpoint hemorrhages will only be found in abrasions inflicted during life.

Bruises arise as a result of bleeding beneath intact skin. In a forensic context, the cause is almost always blunt force. The color of a bruise develops from invisible just after it is inflicted, through the typical blue coloration, and finally fading through various shades of yellow-brown. Deep bruises may take some hours to become visible, so no conclusion should be drawn about the absence of obvious bruising in an examination conducted soon after someone reports severe beating to the abdomen, for example. A blow after death will not result in a bruise because there is no blood flowing to escape through the capillaries. A bruise may show the pattern of the blunt instrument that caused it, especially marks arising from ligatures.

Incisions are clean cuts produced by sharp objects. An incision need not be caused by a knife—broken glass can cause major injury in the form of incisions. Defense wounds on the hands or arms in response to a knife attack are incisions in which the nature and location of the wound can produce helpful information. Self-inflicted injuries with sharp objects do not typically produce the deep and clean incisions of an attack.

Lacerations are wounds that split the skin, but are not caused by sharp objects and do not result in the clean and deep wound of an incision. They typically involve some degree of force, and there will be bruising associated with the wound site. Like bruises, it can be difficult to distinguish lacerations caused by accidental injury from those caused deliberately.

Stab wounds are, in a way, a special example of incisions. They can be made by sharp objects (including glass) but are deeper than they are long. Stab wounds can also be made by more blunt objects, provided they are applied with sufficient force. Great care must be taken in making any inferences about the nature or dimensions of the weapon used on the basis of the dimensions and characteristics of the wound or covering clothing. The width and depth of the stab wound is influenced by the force, direction, and action of the stabbing incident. For example, a downward stab with vertical movement of the knife will result in a wide incision. A stab into a body region where the skin was under tension at the moment of stabbing will also produce a wider entry than the dimensions of the blade. Conversely, as blood dries on the clothing, the actual size of the incision will decrease.

In the early 1990s in South Australia, three youths were charged with causing the death of a fourth by beating him with a pipe and then dragging him along behind their truck. The defense argued that the deceased had been traveling in the back of the truck and, drunk, had fallen out. They claimed that his injuries were caused by impact with the ground and subsequent abrasions from the movement of his body over the road surface, resulting from the momentum of the fall. However, the pathologist and crime-scene investigators found bruises among the many abrasions that revealed the image of a pipe with a threaded end. The pattern of the bruising was sufficiently close to that of metal piping found in the truck to have been caused by the deceased having been beaten with it. The jury decided that this evidence disproved the defense explanation and the three were convicted of murder.

Near Greenock, Scotland, in the early 1980s, a young woman accepted a ride home from work offered by a youth that she recognized but did not know well. Instead of taking her home, he took her to a deserted county lane, made her take off her clothes, burned her with a cigarette, and beat her with his belt. He did not have sex with her but drove off after the assault. She identified him to the police but the case was essentially her word against his. He had a version of what he had done that came close to an alibi. As there was no sex, there was no biological evidence. However, the bruises on her body included two with clear imprints of a belt buckle that was an exact match to the trouser belt the accused was wearing when interviewed by the police. Along with other laboratory evidence, including fiber transfers, a sufficiently strong case was put together to result in the defense pleading to the charge and saving the victim the trauma of a trial.

See also Autopsy; Pathology

References
Byard, R., T. Corey, C. Henderson, and J. Payne-James. *Encyclopedia of Forensic and Legal Medicine.* London: Elsevier Academic, 2005.
Knight, B. *Simpson's Forensic Medicine.* London and New York: Oxford University Press, 1997.
Payne-James, J., A. Busuttil, and W. Smock. *Forensic Medicine: Clinical and Pathological Aspects.* London: Greenwich Medical Media, 2003.

X-rays

X-rays feature in forensic medicine when used to identify and illustrate skeletal injuries. They are also used in mass disaster victim identification to record jaw and dental impressions. One little-used but important aspect of x-rays in forensic medicine is using postmortem imaging to search for bullets and bullet fragments in shootings.

In the laboratory, x-rays are the basis of a technique known as *x-ray fluorescence* that can be used to identify the nature and quantity of traces of metals in evidence such as glass fragments.

References
Payne-James, J., A. Busuttil, and W. Smock. *Forensic Medicine: Clinical and Pathological Aspects.* London: Greenwich Medical Media, 2003.
Saferstein, R. *Criminalistics.* 8th ed. Upper Saddle River, NJ: Pearson Education, 2004.

BIBLIOGRAPHY

Almirall, J., and K. Furton. *Analysis and Interpretation of Fire Scene Evidence.* Boca Raton, FL: CRC, 2004.

American Board of Criminalistics; http://www.criminalistics.com/ (Referenced July 2005).

American Polygraph Association; http://www.polygraph.org/journals.htm (Referenced July 2005).

American Society of Crime Laboratory Directors; www.ascld.org (Referenced July 2005).

American Society of Crime Laboratory Directors/Laboratory Accreditation Board; www.ascld-lab.org (Referenced July 2005).

Andreasson, R., and A. Jones. "Historical anecdote related to chemical tests for intoxication." *Journal of Analytical Toxicology* 20 (1996): 207–208.

Antibody Resource Page, The; http://www.antibodyresource.com/ (Referenced July 2005).

Antigen Presentation; http://users.rcn.com/jkimball.ma.ultranet/BiologyPages/A/AntigenPresentation.html (Referenced July 2005).

ATF Online, Bureau of Alcohol, Tobacco, Firearms and Explosives, U.S. Department of Justice; http://www.atf.treas.gov/ (Referenced July 2005).

ATF Online, Bureau of Alcohol, Tobacco, Firearms and Explosives, U.S. Department of Justice, Operation Ceasefire program; http://www.atf.treas.gov/pub/fire-explo_pub/ceasefire/index.htm (Referenced July 2005).

BBC Website, Crime Case Closed, Infamous Criminals; http://www.bbc.co.uk/crime/caseclosed/tedbundy1.shtml (Referenced July 2005).

Bevel, T., and R. M. Gardner. *Bloodstain Pattern Analysis: With an Introduction to Crime Scene Reconstruction.* 2nd ed. Boca Raton, FL: CRC, 2001.

BioForum Applicable Knowledge Center Fourier Transform Infrared; http://www.forumsci.co.il/HPLC/FTIR_page.html (Referenced July 2005).

Blitzer, H. L., and J. Jacobia. *Forensic Digital Imaging and Photography.* San Diego, CA: Academic, 2001.

Bodziak, W. J. *Footwear Impression Evidence.* New York: Elsevier, 1991.

Borkenstein, R. F., R. F. Crowther, R. P. Shumate, W. B. Ziel, and R. Zylman. "Report on the Grand Rapids Survey." Department of Police Adminstration, Indiana University, 1964.

Bowers, C. M. *Forensic Dental Evidence: An Investigator's Handbook.* Amsterdam and Boston: Academic, 2004.

Bowers, C. M., and G. Bell. *Manual of Forensic Odontology.* 3rd ed. Saratoga Springs: American Society of Forensic Odontology, 1995.

Bureau of Alcohol, Tobacco, Firearms and Explosives, National Integrated Ballistic Information Network; http://www.nibin.gov/ (Referenced July 2005).

Butler, J. *Forensic DNA Typing.* 2nd ed. Burlington, MA: Elsevier, 2005.

Byard, R., T. Corey, C. Henderson, and J. Payne-James. *Encyclopedia of Forensic and Legal Medicine.* London: Elsevier Academic, 2005.

Byers, S. N. *Introduction to Forensic Anthropology: A Textbook.* Boston: Allyn and Bacon, 2001.

Byrd, J. H., and J. L. Castner. *Forensic Entomology: The Utility of Arthropods in Legal Investigations.* Boca Raton, FL: CRC, 2000.

Caddy, B. *Forensic Examination of Glass & Paint: Analysis & Interpretation.* London and New York: Taylor and Francis, 2001.

California Association of Criminalistics; http://www.cacnews.org/ (Referenced July 2005).

Casey, E. *Digital Evidence and Computer Crime.* 2nd ed. London and San Diego, CA: Academic, 2004.

Christian, D. R. *Forensic Investigation of Clandestine Laboratories.* Boca Raton, FL: CRC, 2004.

Conan Doyle, A. *The New Annotated Sherlock Holmes: The Complete Short Stories.* Edited by J. Lecarre and L. Klinger. New York: W. W. Norton, 2004.

Court TV's Crime Library, Criminal Minds and Methods, Bundy; http://www.crimelibrary.com/bundy/attack.htm (Referenced July 2005).

Court TV's Crime Library, Criminal Minds and Methods, Crime Scene; http://www.crimelibrary.com/criminal_mind/forensics/crimescene/6.html?sect=21 (Referenced July 2005).

Court TV's Crime Library, Criminal Minds and Methods, Gacy; http://www.crimelibrary.com/serial/gacy/gacymain.htm (Referenced July 2005).

Court TV's Crime Library, Criminal Minds and Methods, Haigh; http://www.crimelibrary.com/serial/haigh/ (Referenced July, 2005).

Court TV's Crime Library, Criminal Minds and Methods, Mengele; http://www.crimelibrary.com/serial_killers/history/mengele/index_1.html (Referenced July 2005).

Court TV's Crime Library, Criminal Minds and Methods, Ramirez; http://www.crimelibrary.com/ramirez/ (Referenced July 2005).

Court TV's Crime Library, DNA in Court; http://www.crimelibrary.com/criminal_mind/forensics/dna/6.html?sect=21 (Referenced July 2005).

Curran, J. M., T. N. Hicks, and J. S. Buckleton. *Forensic Interpretation of Glass Evidence.* Boca Raton, FL: CRC, 2000.

Daubert on the Web; http://www.daubertontheweb.com/ (Referenced July 2005).

De Forest, P., R. E. Gaensslen, and H. C. Lee. *Forensic Science: An Introduction to Criminalistics.* New York: McGraw-Hill, 1983.

DiMaio, V. J. M., and D. DiMaio. *Forensic Pathology.* 2nd ed. Boca Raton, FL: CRC, 2001.

Dolinak, D., E. Matshes, and E. O. Lew. *Forensic Pathology: Principles and Practice.* Amsterdam and Boston: Elsevier Academic, 2005.

Dorion, R. B. J. *Bitemark Evidence.* New York: Marcel Dekker, 2004.

Dr. George Johnson's Backgrounders, DNA Fingerprinting; http://www.txtwriter.com/Backgrounders/Genetech/GEpage14.html (Referenced July 2005).

Drummer, O. H. *The Forensic Pharmacology of Drugs of Abuse.* New York: Oxford University Press, 2001.

Electronic Crime Scene Investigation: A Guide for First Responders. Washington, DC: U.S. National Institute of Justice, 2001.

Emerson, S., and B. Duffy. *The Fall of Pan Am 103: Inside the Lockerbie Investigation.* New York: Putnam, 1990.

Evans, C. *The Casebook of Forensic Detection: How Science Solved 100 of the World's Most Baffling Crimes.* New York: Wiley, 1998.

Exonerated by Science: Case Studies in the Use of DNA Evidence to Establish Innocence after Trial. Washington, DC: U.S. National Institute of Justice, 1996.

Fay, S., L. Chester, and M. Linklater. *Hoax: The Inside Story of the Howard Hughes–Clifford Irving Affair.* New York: Viking, 1972.

Federal Bureau of Investigation, Combined DNA Index System; http://www.fbi.gov/hq/lab/codis/index1.htm (Referenced July 2005).

Federal Bureau of Investigation, Integrated Automated Fingerprint Identification System; http://www.fbi.gov/hq/cjisd/iafis.htm (Referenced July 2005).

Federal Bureau of Investigation, National Crime Information Center; http://www.fas.org/irp/agency/doj/fbi/is/ncic.htm (Referenced July 2005).

Fisher, B. A. J. *Techniques of Crime Scene Investigation.* Boca Raton, FL: CRC, 2000.

Florida State University, Molecular Expressions, Exploring the World of Optics and Microscopy; http://micro.magnet.fsu.edu/ (Referenced July 2005).

Forensic Examination of Digital Evidence: A Guide for Law Enforcement. Washington, DC: U.S. National Institute of Justice, 2004.

Forensic Quality Services; www.forquality.org (Referenced July, 2005).

Forensic Technology, IBIS/Ballistic Identification; http://www.fti-ibis.com/en/s_4_1.asp (Referenced July 2005).

Fulton, C. C. *Modern Microcrystal Tests for Drugs.* New York: Wiley, 1969.

Golan, T. *Laws of Men and Laws of Nature: The History of Scientific Expert Testimony in England and America.* Cambridge, MA: Harvard University Press, 2004.

Goldfrank, L., N. Flomenbaum, N. Lewin, M. A. Howland, R. Hoffman, and L. Nelson. *Goldfrank's Toxicologic Emergencies.* 7th ed. New York: McGraw-Hill Professional, 2002.

Greenberg, B., and J. C. Kunich. *Entomology and the Law: Flies as Forensic Indicators.* Cambridge and New York: Cambridge University Press, 2002.

Gross, H. *Criminal Investigation.* 4th ed. Edited by Ronald Martin Howe. London: Sweet and Maxwell, 1949.

Gutheil, T. G., and P. S. Appelbaum. *Clinical Handbook of Psychiatry and the Law.* 3rd ed. Philadelphia, PA: Lippincott Williams and Wilkins, 2000.

Hamby, J. E., and J. W. Thorpe. "The History of Firearm and Toolmark Identification." *Association of*

Firearm and Tool Mark Examiners Journal, 30th Anniversary Issue. Vol. 31, no. 3 (1999).

Hazelwood, R. R., and A. W. Burgess. *Practical Aspects of Rape Investigation: A Multidisciplinary Approach.* 3rd ed. Boca Raton, FL: CRC, 2001.

Heard, B. J. *Handbook of Firearms and Ballistics: Examining and Interpreting Forensic Evidence.* New York: Wiley, 1997.

Helmer, W. J., and A. J. Bilek. *The St. Valentine's Day Massacre: The Untold Story of the Gangland Bloodbath That Brought Down Al Capone.* Nashville, TN: Cumberland House, 2004.

Hilton, O. *Scientific Examination of Questioned Documents.* Boca Raton, FL: CRC, 1992.

Holocaust, The; Crimes, Heroes and Villains, Victims of Mengele; http:// auschwitz.dk/Mengele/id17.htm (Referenced July 2005).

Houck, M. M. *Trace Evidence Analysis: More Cases in Mute Witness.* Burlington, MA: Elsevier, 2004.

International Laboratory Accreditation Cooperation, Guidelines for Forensic Science Laboratories; http://www.ilac.org/downloads/Ilac-g19.pdf (Referenced July 2005).

James, S. H., and W. G. Eckert. *Interpretation of Bloodstain Evidence at Crime Scenes.* 2nd ed. Boca Raton, FL: CRC, 1998.

James, S. H., and J. J. Nordby. *Forensic Science: An Introduction to Scientific and Investigative Techniques.* Boca Raton, FL: Taylor and Francis, 2005.

Jasanoff, S. *Science at the Bar: Law, Science, and Technology in America.* Cambridge, MA: Harvard University Press, 1997.

Kakis, F. J. *Drugs: Facts and Fictions.* New York: Franklin Watts, 1982.

Kind, S. S. *The Scientific Investigation of Crime.* Harrogate, England: Forensic Science Services, 1987.

Klaassen, C. *Casarett & Doull's Toxicology: The Basic Science of Poisons.* 6th ed. New York: McGraw-Hill Professional, 2001.

Kleiner, M. *Handbook of Polygraph Testing.* San Diego, CA: Academic, 2001.

Knight, B. *Simpson's Forensic Medicine.* London and New York: Oxford University Press, 1997.

Knight, B., and P. Saukko. *Forensic Pathology.* London: Hodder Arnold, 2004.

Komarinski, P. *Automated Fingerprint Identification Systems (AFIS).* Amsterdam and Boston: Elsevier, 2005.

Kuhn, T. S. *The Structure of Scientific Revolutions.* Chicago: Chicago University Press, 1996.

Lee, H., and J. Labriola., Jerry, *Famous Crimes Revisited.* Southbury, CT: Strong Books, 2001.

Lee, H. C., and R. E. Gaensslen. *Advances in Fingerprint Technology.* 2nd ed. Boca Raton, FL: CRC, 2001.

Lee, H. C., T. Palmbach, and M. T. Miller. *Henry Lee's Crime Scene Handbook.* London and San Diego, CA: Academic, 2001.

Levine, B. *Principles of Forensic Toxicology.* Washington, DC: American Association for Clinical Chemistry, 1999.

Macaulay, David. *Motel of the Mysteries.* Boston: Houghton Mifflin, 1979.

Maples, W. R., and M. Browning. *Dead Men Do Tell Tales: The Strange and Fascinating Cases of a Forensic Anthropologist.* New York: Doubleday, 1995.

Microscopy U, Michael-Levy Birefringence Chart; http://www.microscopyu.com/articles/polarized/michel-levy.html (Referenced July 2005).

Microscopy U, Polarized Light Microscopy; http://www.microscopyu.com/articles/polarized/polarizedintro.html (Referenced July 2005).

Murray, R. C. *Evidence from the Earth: Forensic Geology and Criminal Investigation.* Missoula, MT: Mountain, 2004.

National Research Council. *The Evaluation of Forensic DNA Evidence.* Washington, DC: Academy, 1996.

Nic Daeid, N. *Fire Investigation.* Boca Raton, FL: CRC, 2004.

Office of National Drug Control Policy; http://www.whitehousedrugpolicy.gov/ (Referenced July 2005).

Office of National Drug Control Policy, Drug Facts, Cocaine; http://www.whitehousedrugpolicy.gov/drugfact/cocaine/index.html (Referenced July 2005).

Office of National Drug Control Policy, Drug Facts, Marijuana; http://www.whitehousedrugpolicy.gov/drugfact/marijuana/index.html (Referenced July 2005).

Office of National Drug Control Policy, Drug Facts, Methamphetamine; http://www.whitehousedrugpolicy.gov/drugfact/methamphetamine/index.html (Referenced July 2005).

Office of National Drug Control Policy, Drug Policy Information Clearing House; http://www.whitehousedrugpolicy.gov/publications/factsht/mdma/index.html (Referenced July 2005).

Owen, D. *Hidden Evidence: Forty True Crimes and How Forensic Science Helped Solve Them.* Willowdale, Ontario: Firefly, 2000.

Pan Am 103; http://plane-truth.com/pan_am_103.htm (Referenced July 2005).

Payne-James, J., A. Busuttil, and W. Smock. *Forensic Medicine: Clinical and Pathological Aspects.* London: Greenwich Medical Media, 2003.

Popper, K. *The Logic of Scientific Discovery.* London: Routledge, 2002.

Reeve, S. *The New Jackals: Ramzi Yousef, Osama Bin Laden, and the Future of Terrorism.* Boston: Northeastern University Press, 2002.

Reference Manual on Scientific Evidence. Washington, DC: Federal Judicial Center, 2000.

Robertson, J. *Forensic Examination of Human Hair.* London: Taylor and Francis, 1999.

Robertson, J., and M. Grieve. *Forensic Examination of Fibers.* 2nd ed. London and Philadelphia: Taylor and Francis, 1999.

Rosner, R. *Principles and Practice of Forensic Psychiatry.* 2nd ed. New York: Chapman and Hall, 2003.

Royal Commission of Inquiry into Chamberlain Convictions. "Report of the Commissioner the Hon. Mr. Justice T. R. Morling." Government Printer of the Northern Territory, 1987.

Rudin, N., and K. Inman. *An Introduction to Forensic DNA Analysis.* 2nd ed. Boca Raton, FL: CRC, 2003.

Russ, J. C. *Forensic Uses of Digital Imaging.* Boca Raton, FL: CRC, 2001.

Saferstein, R. *Criminalistics.* 8th ed. Upper Saddle River, NJ: Pearson Education, 2004.

Saks, M. J. *The Daubert/Kumho Implications of Observer Effects in Forensic Science: Hidden Problems of Expectation and Suggestion with Risinger, Rosenthal & Thompson.* 90 CAL. L. REV. 1 (2002).

Saukko, P., and B. Knight. *Knight's Forensic Pathology.* 3rd ed. London: Arnold, 2004.

Schwoeble, A. J., and D. L. Exline. *Current Methods in Forensic Gunshot Residue Analysis.* Boca Raton, FL: CRC, 2000.

Seneski, P. C., N. Whelan, D. K. Faugno, L. Slaughter, and B. W. Girardin. *Color Atlas of Sexual Assault.* St Louis: Mosby, 1997.

Smith, S. *Mostly Murder.* New York: Dorset, 1989.

Tebbett, I. *Gas Chromatography in Forensic Science.* New York: Horwood, 1992.

Turvey, B. E., and J. Savino. *Rape Investigation Handbook.* Amsterdam and Boston: Elsevier Academic, 2004.

Twain, Mark. *Mississippi Writings: Tom Sawyer, Life on the Mississippi, Huckleberry Finn, Pudd'nhead Wilson.* New York: Library of America, 1982.

U.S. Department of State, International Information Programs; http://usinfo.org/usia/usinfo.state.gov/topical/pol/arms/stories/01061323.htm (Referenced July 2005).

U.S. Department of the Treasury; http://www.ustreas.gov/topics/currency/ (Referenced July 2005).

U.S. Drug Enforcement Administration; http://www.usdoj.gov/dea/ (Referenced July 2005).

U.S. Drug Enforcement Administration, Drug Descriptions, Barbiturates; http://www.usdoj.gov/dea/concern/barbiturates.html (Referenced July 2005).

U.S. Drug Enforcement Administration, Drug Descriptions, Cocaine; http://www.usdoj.gov/dea/concern/cocaine.html (Referenced July 2005).

U.S. Drug Enforcement Administration, Drug Descriptions, Heroin; http://www.dea.gov/concern/heroin.html (Referenced July 2005).

U.S. Drug Enforcement Administration, Drug Descriptions, Marijuana; http://www.dea.gov/concern/marijuana.html (Referenced July 2005).

U.S. Drug Enforcement Administration, Drug Descriptions, MDMA; http://www.usdoj.gov/dea/concern/mdma/mdma.html (Referenced July 2005).

U.S. Drug Enforcement Administration, Drug Descriptions, Methadone; http://www.whitehousedrugpolicy.gov/publications/factsht/methadone/ (Referenced July 2005).

U.S. Drug Enforcement Administration, Drug Descriptions, Methamphetamine/Amphetamines; http://www.usdoj.gov/dea/concern/amphetamines.html (Referenced July 2005).

U.S. Drug Enforcement Administration, Drug Intelligence Brief, Khat; http://www.usdoj.gov/dea/pubs/intel/02032/02032.html (Referenced July 2005).

U.S. Fish and Wildlife Service, National Fish and Wildlife Forensics Laboratory; http://www.lab.fws.gov/ (Referenced July 2005).

U.S. Food and Drug Administration, Center for Drug Evaluation and Research; Guidance for Industry, Bioanalytical Method Validation; http://www.fda.gov/cder/guidance/4252fnl.htm (Referenced July 2005).

University of Glasgow, Forensic Medicine Archives Project, Case Against Dr. Buck Ruxton, Lancaster, Moffet and Manchester (1935); http://www.fmap.archives.gla.ac.uk/Case%20Files/Ruxton/Case_File9.htm (Referenced July 2005).

Vastrick, T. W. *Forensic Document Examination Techniques.* Altamonte Springs, FL: Institute of Internal Auditors Research Foundation, 2004.

Wambaugh, J. *The Blooding.* New York: William Morrow, 1989.

Warlow, T. *Firearms, the Law, and Forensic Ballistics.* 2nd ed. London and Bristol, PA: Taylor and Francis, 1996.

Wecht, C., M. Curriden, and B. Wecht. *Cause of Death.* New York: E. P. Dutton, 1993.

White, P. *Crime Scene to Court: The Essentials of Forensic Science.* Cambridge: Royal Society of Chemistry, 1998.

Wilson, Colin. *Written in Blood: The Trail and the Hunt.* New York: Warner Books, 1989.

Winger, G., F. G. Hofmann, and J. H. Woods. *A Handbook on Drug and Alcohol Abuse: The Biomedical Aspects.* 3rd ed. New York: Oxford University Press, 1992.

Wonder, A. Y. *Blood Dynamics.* San Diego, CA: Academic, 2001.

Yinon, J. *Forensic Applications of Mass Spectrometry.* Boca Raton, FL: CRC, 1994.

————. *Advances in Forensic Applications of Mass Spectrometry.* Boca Raton, FL: CRC, 2004.

INDEX

Note: Boldface page numbers indicate main encyclopedia entries; italic page number indicates picture; t. indicates table.